PoliTO Springer Series

Springer, in cooperation with Politecnico di Torino, publishes the PoliTO Springer Series. This co-branded series of publications includes works by authors and volume editors mainly affiliated with Politecnico di Torino and covers academic and professional topics in the following areas: Mathematics and Statistics, Chemistry and Physical Sciences, Computer Science, All fields of Engineering. Interdisciplinary contributions combining the above areas are also welcome. The series will consist of lecture notes, research monographs, and briefs. Lectures notes are meant to provide quick information on research advances and may be based e.g. on summer schools or intensive courses on topics of current research, while SpringerBriefs are intended as concise summaries of cutting-edge research and its practical applications. The PoliTO Springer Series will promote international authorship, and addresses a global readership of scholars, students, researchers, professionals and policymakers.

THE SERIES IS INDEXED IN SCOPUS

Andrea Borroni · Guido Perboli ·
Gian Maria Piccinelli
Editors

Metaverses: Reshaping Law, Economy, and Society

A Comprehensive Analysis of Virtual Worlds and Their Real-World Implications

POLITECNICO DI TORINO

Springer

Editors
Andrea Borroni
Dipartimento di Scienze Politiche
Università della Campania L. Vanvitelli
Caserta, Italy

Guido Perboli
DIGEP
Politecnico di Torino
Turin, Italy

Gian Maria Piccinelli
Dipartimento di Scienze Politiche
Università della Campania L. Vanvitelli
Caserta, Italy

ISSN 2509-6796 ISSN 2509-7024 (electronic)
PoliTO Springer Series
ISBN 978-3-032-08663-1 ISBN 978-3-032-08664-8 (eBook)
https://doi.org/10.1007/978-3-032-08664-8

This Springer imprint is published by the registered company Springer Nature Switzerland AG
The registered company address is: Gewerbestrasse 11, 6330 Cham, Switzerland

Contents

Immersive Futures: Charting the Legal, Ethical, and Technological Frontiers

Andrea Borroni, Guido Perboli, and Gian Maria Piccinelli

Abstract The Metaverse marks a new stage in digital evolution, merging virtual, augmented, and mixed realities with artificial intelligence, blockchain, and cloud computing. This volume explores its technological, legal, and ethical dimensions, examining how this immersive ecosystem reshapes social interaction, work, education, and commerce. The first section focuses on theoretical and methodological issues, addressing regulation, privacy, taxation, and labor transformations, while introducing frameworks such as the LETO methodology for developing responsible Metaverse applications. The second section highlights practical applications in fields like finance, maritime law, cultural heritage, and tourism. The contributions reveal both the opportunities and the challenges of this emerging digital frontier, calling for new legal and ethical paradigms to ensure innovation aligns with human rights and social inclusion.

After a long phase in which cloud computing, virtual reality, and artificial intelligence have been the focus of media and institutional attention, interest has now shifted to a new approach that unites these technologies, creating a co-presence that leads to the exploration of innovative ways to carry out everyday activities. The transition is evident, particularly from the moment Mark Zuckerberg announced in 2021 that his company's name would change from Facebook to Meta, symbolizing a paradigm shift toward the Metaverse—a concept that has its roots in the cyberpunk culture of the 1980s and 1990s, with Neal Stephenson's novel "Snow Crash" anticipating it as early as 1992.

A. Borroni · G. M. Piccinelli
Università degli Studi della Campania "Luigi Vanvitelli", Caserta, Italy
e-mail: andrea.borroni@unicampania.it

G. M. Piccinelli
e-mail: gianmaria.piccinelli@unicampania.it

G. Perboli (✉)
Politecnico di Torino, Turin, Italy
e-mail: guido.perboli@polito.it

A. Borroni et al. (eds.), *Metaverses: Reshaping Law, Economy, and Society*, PoliTO Springer Series, https://doi.org/10.1007/978-3-032-08664-8_1

The Metaverse is not just a digital immersive environment, but a vision of an ecosystem in which virtual, augmented, and blockchain reality come together to enable highly interactive and realistic virtual interactions. On a theoretical level, an ideal Metaverse should have features such as persistence, the ability to move freely between environments, a robust digital economy, and the integration of social interactions in a space that fosters cooperation, learning, and play.

A crucial aspect of this development is that the Metaverse ties in closely with technological evolutions already underway, such as 5G and immersive technologies that, for example, can make remote collaborations more fluid and natural, leading to remote work experiences that seem almost tangible.

Synthetically, the Metaverse concept builds on previous experiences that have paved the way for this evolution. Among these, we recall, if only for a little mnemonic note:

(i) MMORPGs (Massively Multiplayer Online Role-Playing Games): Online role-playing games such as "World of Warcraft" have demonstrated the potential of persistent virtual worlds where thousands of players interact simultaneously.
(ii) Second Life: Launched in 2003, this virtual world anticipated many aspects of the Metaverse, allowing users to create avatars, build environments, and even manage a virtual economy.
(iii) Virtual and Augmented Reality: The evolution of these technologies, with devices such as Oculus Rift (now owned by Meta), has facilitated the development of even more immersive digital experiences.

In its ideal form, each and every Metaverse should possess some minimal and unavoidable basic characteristics:

a. Persistence: The virtual world continues to exist and evolve even when users are not connected.
b. Interoperability: Users can move freely between different platforms and environments, carrying their own avatars and digital assets.
c. Functioning economy: An economic system that enables the creation, ownership, and exchange of digital goods, often based on cryptocurrencies and NFTs (Non-Fungible Tokens).
d. Sociality: Virtual spaces that foster social interaction and community formation.
e. Immersiveness: Enhanced sensory experiences that make staying in the virtual world more and more like reality.

Given these premises, the Metaverse represents an extraordinary horizon of possibilities, with applications that extend far beyond our current understanding and remain largely unexplored.

In the field of education and training, the potential of the Metaverse is extraordinary. Students can, for example, participate in realistic simulations that immerse them in practical scenarios and help them develop skills in a more engaging and challenging way than the traditional classroom lecture. In such an environment, international collaboration and digital content creation become increasingly integrated, creating fertile ground for research and innovation.

However, as we move into these new frontiers, inevitable questions emerge. One of the most important concerns is information management: in the Metaverse, data can be manipulated and shared with such ease that it is difficult to trace back to the original source or to protect users' privacy. The issue of privacy, which already plagues current digital reality, takes on an even more complex dimension in the context of immersive virtual environments, where the level of interaction and data collection could lead to unprecedented control.

The entertainment sector represents another area of transformation, with events such as virtual concerts and art exhibitions that could open up new modes of enjoyment, transcending the limits of the physical world. Similarly, digital preservation of cultural assets within the Metaverse becomes an opportunity to preserve artistic and historical heritages, turning digitization into a powerful safeguarding tool.

In commerce, the Metaverse has the potential to radically reshape the way we shop, taking the online experience to an unprecedented level of engagement. Consumers will no longer just be viewers of images and descriptions but will be able to explore and interact with products in three-dimensional virtual environments. This kind of approach could give e-commerce a boost, making it more interactive and personalized.

Areas such as tourism, which is heavily impacted by the pandemic, may also find an ally in the Metaverse. BuzzAR, a Singaporean startup, is bringing with Augmented Reality and AI new hospitality solutions that incorporate gamification mechanisms and immersive 3D environments, allowing avatars to explore tourist destinations.

Real estate is also one of the fields where the Metaverse and AI could have a new use. It was just a few years ago that a group of virtual real estate in the online world of Decentraland—a Metaverse where customers can buy land, visit buildings, walk around, and meet people as avatars—was sold for $2.4 million in cryptocurrency.

In healthcare, the ability to combine augmented and virtual reality for diagnosis and treatment is another important step with physicians operating in virtual environments rich in real-time information, or patients receiving more precise remote therapies through the ability to interact with advanced tools and simulations. While the potential is obvious, it is critical to consider the risks associated with increasing access to sensitive data.

Admittedly, the reality of the metaverse inherits some problems that had already arisen with the semantic web and poses others, among which emerges the issue of identifying the subject producing sources as they are manipulated continuously. In the semantic web, content is often merely information decontextualized from the subject issuing the source and the historical and temporal context in which the source was produced. Another legacy is the right to privacy, the right to be forgotten.

In essence, concerns about the proper control and use of information already existed. However, with the advent of metaverses, there is a strong acceleration.

Thus, the discourse shifts to sources and their preservation and transmission in the digital shift.

Despite the excitement surrounding the Metaverse, this new technological frontier brings with it a number of challenges that cannot be ignored. User privacy and security top the list of anxieties, as the immersive nature of these virtual environments could result in an unprecedented collection of personal data, raising new and complex questions about privacy protection. There is also growing concern that the Metaverse could exacerbate mental health issues and the problem of digital addiction, pushing people to immerse themselves even more deeply in virtual experiences than they already do with existing social media platforms.

It is also essential to consider the possibility that the Metaverse will intensify digital inequalities, with a portion of the population likely to be excluded from these new opportunities due to unequal access to technology, internet infrastructures, and the necessary knowledge and skills.

Thus, regulation becomes a crucial priority, as does control over data, to avoid irresponsible use of potentially invasive technologies.

In the legal context, the Metaverse brings with it significant challenges related to privacy, intellectual property and civil liability. These issues must be addressed urgently, as the evolution of technology often exceeds the ability of legal systems to adapt and protect fundamental rights.

Privacy, already a problematic matter in the context of social media, could become an even more critical topic in the Metaverse. The constant and immersive interaction between users and digital environments involves an unprecedented collection of data: not only personal data but also sensory, behavioral, and psychological information that could reveal intimate aspects of an individual. Current regulations, such as the GDPR in Europe, may not be sufficient to protect users in such dynamic and complex environments. For example, who owns the data collected in the Metaverse? How is it protected? The implications of behavior tracking and profiling are equally concerning, as the risk of manipulating virtual experiences for commercial purposes could undermine the right to individual freedom and privacy.

Another relevant issue concerns intellectual property rights, which could be difficult to manage in the Metaverse. Who owns the rights to an avatar, a digital artwork, or an object created within these environments? The possibility of "ownership" of digital goods through NFTs (Non-Fungible Tokens) has raised questions about the legal validity of such assets, as well as their protection against digital piracy. Moreover, the production and sharing of content by users raise issues related to copyright and attribution of works. In an environment where anyone can create, exchange, and sell digital content, the law must respond to a crucial question: how can creators' rights be ensured while respecting copyright law?

Civil liability represents another complex field. If a user suffers damage in a virtual environment—for example, if an avatar is hacked or if offensive content damages a person's reputation—who is responsible? The platforms hosting the Metaverse could be held liable for damages caused by their users, but laws regarding liability in virtual environments are still unclear. In many cases, current civil liability regulations were not designed to address situations occurring in digital worlds, and the creation of specific case law is still under development.

The legal uncertainty surrounding the Metaverse makes legislative intervention necessary to establish clear rules for protecting users' rights, managing data, and resolving disputes. The European Union, the United States, and other jurisdictions are beginning to consider specific regulations to address these new scenarios. For example, in the United States, the administration has proposed creating a bill of rights to protect citizens from abuses stemming from the use of advanced technologies, including those in the Metaverse. Similarly, the European Union is working on regulations to protect digital rights, but the speed at which these technologies evolve could make it difficult for institutions to keep up.

China and Russia are also moving in this direction, and the advent of the Metaverse adds a sense of urgency in establishing a new set of rules to protect people and companies.

Charters of rights and declarations of intent on this front are certainly important steps in setting priorities, but it is also vital that these instruments find concrete regulatory application, giving those who develop or will develop AI solutions even in the Metaverse a framework of laws to comply with. The United States has begun to regulate the use of AI, particularly by blocking the use of algorithms that do not function clearly in decisions that affect consumers.

Legal solutions could include new forms of digital rights regulation, the creation of a code of ethics for Artificial Intelligence and the Metaverse, and the development of legislation that more effectively protects intellectual property and privacy rights. At the same time, the platforms themselves must be actively involved in creating self-regulatory systems that can ensure a safe and respectful use of technologies while maintaining control over users and sensitive data.

Currently, the Metaverse remains more of a futuristic vision than a tangible reality, but numerous tech companies are working hard to lay the foundations of this new digital world. At the forefront is Meta, which is investing billions of dollars in the development of virtual and augmented reality technologies, with the goal of creating immersive social platforms such as Horizon Worlds. At the same time, Microsoft is integrating mixed reality elements into its suite of work and collaboration products, aiming to redefine the concept of the virtual office. Meanwhile, Epic Games is expanding its popular game Fortnite into broader social experiences, including events such as virtual concerts that attract millions of participants. On a more experimental front, platforms like Decentraland and The Sandbox are exploring the potential of blockchain-based virtual worlds, offering users the opportunity to own and monetize virtual land, thereby opening new frontiers in the digital economy.

These various approaches and experiments are gradually shaping what could become the Metaverse of the future, each contributing with its own vision and technology to this ambitious collective project. As a result, the Metaverse is not only a technological matter, but a phenomenon that raises profound philosophical questions about the nature of reality, identity, and human relationships in the digital age. As we venture into this new territory, it will be crucial to maintain an open and critical dialogue about how we want this technology to shape our future.

From a technological perspective, there are several tools today contributing to the creation of metaverses, including virtual reality, mixed reality, artificial intelligence, blockchain, and the internet infrastructure and processors for devices. In particular, artificial intelligence can create realistic-looking users who can interact with others in the Metaverse. In addition to the virtual version in the Metaverse, new AI-managed figures could animate new worlds and interactions.

The company Meta Human Creator already promises to create digital people in a simple and realistic way. Furthermore, AI can create entire environments within the Metaverse—landscapes, buildings, infrastructures—where one can operate in the Metaverse. Already today, some AI technologies are used in gaming, and games like Minecraft and Fortnite rely on AI solutions to create parallel worlds, in this case for educational and entertainment purposes. For the Metaverse, as in other fields, artificial intelligence helps automate software development processes, reducing the time required to create new applications and solutions within the Metaverse. The link between AI and blockchain technologies will also enable the creation, control, and protection of transactions within the Metaverse.

The production of content and the related implications for copyright in sharing and using these contents will be the subject of extensive discussions. There is great potential regarding productivity or entertainment offered by Metaverse scenarios, but there are also many ethical questions arising from these developments. Experts, in fact, are already raising questions about what the new rules will be for these digital worlds and individuals.

The Metaverse is a vision made up of many 3D worlds where everything happens in real time. Worlds and people can interact with each other, and data, identities, payments, and digital resources can be transferred between them, creating a "parallel" space. The Metaverse can also interact with the real world through, for example, digital glasses or sensors that could create a bridge between the two worlds.

While the Metaverse offers countless opportunities, from immersive experiences to technological innovation, it also raises complex legal challenges that cannot be overlooked. Legal solutions must be thought forward to ensure a balance between technological development and the protection of fundamental rights. It is crucial that law evolves along with technology, to prevent innovation from undermining individual rights and freedoms.

Issues related to privacy, intellectual property, and civil liability require special attention from legislators, lawyers, and technologists. An integrated approach, considering technical as well as ethical and legal aspects, will be fundamental to ensure that the Metaverse can become a tool for progress without compromising the rights and security of its users.

This transition is not solely a technological issue but raises deep reflections on the nature of reality and the evolution of human identity in the digital age. The Metaverse (along with the related emerging technologies more in general) challenges us to think about how we will define authenticity, relationships and social interaction in a world where the physical and the virtual are intertwined. The challenge is not only technological, but also ethical: we must establish a balance between innovation and the protection of our privacy, rights, and well-being.

Finally, although the Metaverse has the potential to revolutionize many areas and aspects of our daily life, we must continue to ask ourselves: are we prepared to navigate a world where reality is fluid and constantly changing? And, most importantly, can we ensure that this new technological frontier serves to improve our society, making it more inclusive and sustainable, rather than exacerbating existing inequalities? These questions remain unanswered as we venture into a future that could be as fascinating as it is uncertain.

Authors' Contributions

The authors contributed to the book along with two main lines of development. Section I is addressed to answer to the Methodological development issues.

A. BORRONI (*The Metaverse as an Autonomous Legal System: a Study of Comparative Law*) opens the volume exploring the profound transformations brought about by the Anthropocene, focusing in particular on the transition from the "Great Acceleration" to the rise of the intangible domain. The discussion centers on the Metaverse as a key example of this shift—a complex and expanding network of virtual environments that challenge traditional notions of reality and expose significant gaps in current legal systems. By analyzing the defining features of the Metaverse, the author highlights the urgent need for a "smart legal system" capable of addressing issues such as jurisdiction, data protection, virtual assets, platform liability, and cybercrime. Ultimately, the paper frames the Metaverse as a testing ground for rethinking legal paradigms in order to strike a balance between innovation and regulation.

G. PERBOLI, A. BORRONI, S. PRATALI, C. VANDONI (*The Metaverse and Web3.0: a pathway towards value generation*) explore the shift from Web 2.0 to Web 3.0, focusing especially on the Metaverse and related technologies like cryptocurrencies and blockchain. As these terms gain widespread adoption in the digital world, the article aims to clarify the key differences between Web 2.0 and Web 3.0, emphasizing how Web 3.0 is centered around decentralization, privacy, and empowering users. Through a detailed technological analysis, it looks at the business models and strategies for value creation within Web 3.0. Additionally, the article highlights the need for a comprehensive legal framework—an often overlooked aspect in current discussions. It provides practical examples and managerial perspectives, while also considering future developments and ongoing challenges within Web 3.0 and the Metaverse.

Then, G. PERBOLI, A. BORRONI, M. E. BRUNI (*Developing Metaverse use cases: the LETO Methodology*) introduce a new methodological approach. The Metaverse, the transformative concept of a continuous, shared virtual space, is quickly attracting attention from various industries. However, creating practical use cases for this technology demands a well-rounded approach that takes into account multiple factors, while existing research on Metaverse applications typically concentrates mainly on the technological side. This paper introduces the LETO Methodology, a comprehensive framework for developing Metaverse use cases. LETO offers a holistic approach to designing and validating Metaverse solutions, ensuring alignment with business strategies. The methodology incorporates four key aspects represented by the acronym LETO: Legal, Environmental & Social, Technological, and Organizational. While much of current research tends to prioritize the technological components, LETO provides a more balanced approach, guiding developers in creating use cases that are not only technologically feasible but also legally sound, socially responsible, and organizationally viable. The paper outlines the structure of the LETO Methodology, illustrates its application through a case study, and highlights its potential advantages for Metaverse implementation. It concludes by suggesting directions for future research in the rapidly evolving field of Metaverse applications.

Then, A. BORRONI (*Banks in the Metaverse: the Legal Framework between Evolution and Revolution*) steps into the financial world. The ongoing advancement of technology and communication systems continues to reshape the defining characteristics of the global economy. Factors such as the rise of multinational corporations, the new international division of labor, and the emergence of tax and financial havens have played a key role in creating an economy with transnational features. In this context, national borders have evolved from being mere structural elements to actual obstacles. As a result, the need for a more integrated global economy, relying on universally accepted exchange mechanisms, has become evident. This has led to the growing acceptance of cryptocurrencies and decentralized currencies, built on blockchain technology. In this context, discussions around decentralized finance have gained traction, with concerns that it could undermine the stability of traditional financial institutions. These institutions, dependent on fees and intermediaries, may no longer be the most efficient way for individuals to conduct transactions. Consequently, a transformation of the banking, insurance, and financial sectors is necessary. Investing in the Metaverse, digital intelligence, Web3, blockchain, and related products has become essential for staying competitive in the market. However, several issues related to this phenomenon are still pending legislative action. Therefore, this research aims to contribute to understanding the relationship between these technologies and financial institutions, by examining the current regulatory framework and exploring potential developments in this area.

P. VALENTE (*Taxless Metaverse. Conceptual Paradigms for the Reappearing Digital Taxpayer*) takes care of the tax domain. The author writes that Geotaxation focuses on international tax relations and their development under the influence of geographical factors. This includes considerations such as collaborations among states and the international impact of national or local tax policies. Today, geotaxation must expand to new spaces where human activity is evolving, such as cyberspace. Unlike traditional geo-tax subjects, cyber-reality challenges established social structures, norms, and international relations, including those in the tax domain. This article explores the implications of cyberspace and modern geotaxation for the state and the potential development of the international tax landscape.

P. ROSSI and G. ROASIO investigate the emerging role of the Metaverse in the context of outer space operations, a field still largely unregulated and governed by outdated international treaties. Highlighting the transformative potential of this technological paradigm, the author addresses key challenges, such as the lack of data transparency and sharing frameworks, which currently hinder the development of a metaverse suited for space-related applications. The chapter explores how overcoming these barriers could facilitate advancements in areas like Space Traffic Management, collision avoidance protocols, and the regulation of insurance, liability, and compensation mechanisms. Drawing on progress made in commercial metaverse applications, the author presents the Metaverse as a promising tool for bridging technological innovation and regulatory evolution in outer space governance.

E. SIGNORINI (*Implications of Artificial Intelligence for the Labour Market*) exploring the labour law market observes that technological innovation has undoubtedly transformed the way labour relations are managed. Assuming that generative artificial intelligence (AI) is more likely to enhance existing jobs rather than replace them, automation is having a significant impact on work performance, altering its intensity and methods due to new AI-driven organizational models. Furthermore, AI is reshaping traditional employer powers in areas such as organization, management, and control, introducing new risks for workers. These risks have major implications for data processing, monitoring, and evaluation techniques, potentially threatening worker privacy and job security. Given these considerations, the article aims to explore the changing landscape of labor relations in the age of AI and underscores the need for policies that strike a balance between fostering innovation and protecting workers' rights.

R. A. QIBLAWI, L. SLIMAN, Z. OSMAN, A. M. HAIDAR (*MirageXR: A Framework for Metaverse-Enabled Cross-Platform Integration of Physical Hardware and XR Environments*) present the MirageXR Framework, a novel modular solution aimed at overcoming key challenges in integrating physical hardware with virtual environments within Extended Reality (XR) systems. The author addresses issues such as fragmented tools, platform-specific constraints, and the lack of standardized communication protocols by proposing a framework that enables real-time synchronization between physical devices like sensors and actuators and their virtual counterparts in AR, VR, and mixed reality applications. The framework features two main components: an Arduino-compatible library for ESP32 microcontrollers that abstracts hardware into virtual pins, and a Unity package that links these pins to virtual objects using TCP/IP and JSON. Designed to support a variety of applications—from tactile VR gaming to IoT prototyping and immersive training—the framework simplifies development through familiar Arduino-like syntax while ensuring broad compatibility. Experimental results demonstrate low latency and effective synchronization, positioning MirageXR as a scalable and platform-independent tool that bridges physical and virtual experiences, thereby enhancing accessibility to immersive technologies.

P. TORTORANO, in his paper (*From Virtual Realms to Tangible Reality: Privacy at the Nexus of Artificial Intelligence and the Metaverse*) explores the complex and rapidly evolving field of personal data protection, focusing particularly on the significant challenges posed by the Metaverse and artificial intelligence (AI). It begins by outlining the core principles of privacy and data governance, providing a foundational framework for the reader. The discussion then shifts to examine the mechanisms designed to protect privacy within the Metaverse, a new digital space filled with immersive and interconnected virtual experiences. The article further analyzes the transformative impact of AI, highlighting its extraordinary ability to process and analyze data, which raises unprecedented privacy concerns. In this context, the article critically assesses data minimization strategies, emphasizing their crucial role in maintaining ethical and regulatory standards in the face of these technological advancements. In conclusion, the paper calls for adaptive and forward-thinking regulatory approaches to address the complexities of data protection in an era of rapid digital progress.

M. EL BIZRI, A. M. HAIDAR, A. M. EL-HAJJ, and L. SLIMAN (*Quantum-Powered Metaverse Opportunities, Challenges, and Future Directions in Virtual Worlds*) explore the convergence of quantum computing and the metaverse, highlighting how emerging quantum technologies could address the growing computational, security, and scalability challenges faced by increasingly complex virtual environments. As classical computing proves insufficient for the demands of these digital ecosystems, the author presents quantum computing—through mechanisms like superposition and entanglement—as a transformative solution. The chapter examines potential applications including quantum-enhanced artificial intelligence, post-quantum cryptography, and ultra-secure, high-speed quantum networking. At the same time, it reflects on critical challenges such as hardware limitations, integration with classical systems, and the ethical implications of quantum-driven digital economies. Offering a comprehensive overview of current developments, research gaps, and future directions, the chapter aims to guide researchers, developers, and policymakers in designing a secure, efficient, and quantum-enabled metaverse.

Section II is devoted to the Application-oriented developments.

Here, E. M. PAONE (*Evaluating the Potential of the Metaverse: The Chinese Perspective*) concentrating on commerce with the Far East observes that the Metaverse, envisioned as an interoperable network of virtual worlds, is characterized by its 3D, real-time, persistent, and synchronous nature, allowing for an unlimited number of users to interact within it. In this virtual space, users can maintain the continuity of essential data such as identity, rights, communications, payments, and assets. This system holds the potential to revolutionize the global technological landscape. Recognizing this potential, China has focused on markets related to key technologies that are integral to the Metaverse, such as 5G, augmented reality (AR), and virtual reality (VR). However, the Metaverse does not merely represent a new model for social interaction; more crucially, it serves as the foundation for a new economic paradigm, with cryptocurrencies and blockchain underpinning most transactions within this space. In response, China is not only making significant technological investments but also ramping up legislative efforts to better regulate and control this rapidly evolving domain. This paper aims to analyze the development of the Metaverse from China's perspective, examining how the unique sociopolitical context of the country shapes its approach to this emerging phenomenon and the barriers it faces in fully realizing its vision.

F. DE ROBERTIS (*Navigating the law of the sea and ports—The impact of Technologies in the maritime sector and navigation law*) shifts the attention to safety in transportation, particularly in the harbor. The development of new technologies, along with the growing integration of virtual or augmented reality applications that increasingly blend with real-world experiences, is now extending to the maritime sector. Maritime law is distinguished from common law by its emphasis on public interest, particularly regarding the safety of navigation, and the unique responsibilities of entities overseeing the placement of seafarers. The adoption of emerging technologies holds the potential to gradually reduce human error in navigation and port operations. While this may not entirely eliminate the unpredictability of unforeseen events, it represents an investment towards improving the quality of services provided, combining technological innovation, transparency, and training. Additionally, artificial intelligence (AI) presents significant potential for enhancing vessel traffic management, optimizing work sequences, and effectively addressing unforeseen emergencies, with each situation being handled according to prioritized objectives. This paper aims to provide an initial analysis of how "sea and port law" has responded to and incorporated these new technological developments in key areas.

Then, F. ZAMBARDINO (*Cultural heritage in the metaverse and the impact of AI: amplifying creativity and legal protection in the virtual world*) sustains that the integration of cultural heritage into the Metaverse, enhanced by artificial intelligence (AI), marks the beginning of a transformative era for both creative expression and legal protection within virtual spaces. This development allows for the preservation and celebration of diverse cultural heritages through immersive and dynamic experiences. AI-driven tools provide unprecedented opportunities for artists and creators, enabling them to push new boundaries within these digital environments. However, these technological advancements necessitate strong legal frameworks to prevent the exploitation of cultural assets and ensure their rightful ownership and authenticity. Striking a balance between innovation and the preservation of cultural integrity is crucial to respecting heritage while encouraging technological progress. With this in mind, the article aims to explore the complex relationship between cultural heritage, AI, and the Metaverse, and to propose strategies for safeguarding cultural assets while encouraging limitless creative exploration.

A. RUSSO examines how the rise of the Metaverse is reshaping traditional notions of contract law, as digital interactions increasingly involve economic exchanges conducted through avatars and virtual interfaces. The chapter focuses on how classical contract principles are being adapted to virtual environments where users negotiate and transact digital goods and services. Key issues such as consent, identity verification, and the enforceability of agreements in immersive settings are critically analyzed. By referencing real-world disputes stemming from virtual transactions, the author illustrates the limitations of existing legal frameworks and the need for their evolution. Ultimately, the chapter calls for the development of straightforward and flexible contractual models that reflect the dynamic nature of the Metaverse, aiming to protect users' rights without stifling innovation.

M. CAPELLI–G. ROASIO (*Metaverse and Tourism: current applications and future perspectives*) underline that tourism is an economic sector that is closely connected to the physical territory, not centered on producing tangible goods but on offering services that provide travelers with unique and memorable experiences at specific destinations. As such, the concept of tourism within the Metaverse, a virtual realm powered by advanced technologies like augmented reality (AR), virtual reality (VR), and mixed reality (MR), might initially seem implausible. However, it's crucial to note that the pandemic has introduced new forms of "travel" that have gained significant popularity, enabling tourists to "visit" destinations without leaving the comfort of their homes. This development has sparked the creation of new services and innovative forms of cyber-tourism, indicating that competition in tourism is no longer confined to physical locations. Consequently, operators and destinations must adjust to this evolving landscape. This paper examines the intersection between tourism and the Metaverse, shedding light on both the opportunities it presents and the challenges it poses.

F. PETRUCCIANO (*The value and exchange of wealth in the Metaverse. The juridical frontier of a law we do not own*) clearly states that the Metaverse presents a unique and challenging environment, offering users and observers a fresh perspective on the concept of reality itself. As a virtual yet permanent space where wealth is exchanged and transactions have tangible consequences in terms of assets, defining the legal nature of the value produced or lost in this environment becomes crucial. In the Metaverse—a space "beyond" the physical world—traditional principles like the monopoly of state-issued currency or the territoriality of law may need to be reconsidered. The "otherness" of this virtual realm, which exists "after" the normal physical ecosystem, is central to this discourse. This contribution examines the conceptual nature of value within the Metaverse and discusses the current legal framework, describing it as a "frontier" that serves as both the last regulatory boundary of the physical world and potentially the first glimpse of a new legal domain for private law. Could this lead to the emergence of a "metalaw"?

L. CHIARIGLIONE (*Seamless experiences across metaverse instances*) remarks that similar to how digital media and the web have become integral parts of the lives of billions, the Metaverse is poised to play a significant role in shaping our future. However, its development is still in the early stages, with many variables—such as technological advancements, business models, and standards—yet to be fully determined. This paper focuses on the issue of Metaverse standards, using the MPAI Metaverse Model (MPAI-MMM) recently published by Moving Picture, Audio, and Data Coding (MPAI) as a case study. MPAI has approached the standardization of the Metaverse by initially addressing a subset of functional requirements applicable across several domains, with plans to expand as more applications emerge. The paper outlines the key steps in the development of the MPAI-MMM, starting with the analysis of potential application domains and the specification methodology. It then delves into the core components of the MMM, including the processes that perform actions on items, the specification of profiles, and the verification of the standard's usability.

The Metaverse as an Autonomous Legal System: A Study of Comparative Law

Andrea Borroni

Abstract The Anthropocene marks an era of profound transformation, characterized by large-scale human impact on the planet, encapsulated by the notion of the "Great Acceleration". However, as technological innovation continues to advance at an unprecedented rate, this idea is starting to give way to a new archetype: that of the intangible domain. The Metaverse—an intricate network of virtual worlds—is now challenging the foundations of our understanding of reality while posing novel challenges for existing legal frameworks, which were not designed to govern dematerialised, borderless environments. In light of increasing investment and growing integration, the absence of a uniform regulation to face matters such as jurisdictional ambiguity, protection of personal data and virtual assets, platform liability, and cybercrimes is alarming and particularly striking. This paper examines the definitory profiles and basic characteristics of the Metaverse to assess the viability of creating a "smart legal system", framing this new reality as a test site for a new legal approach, one that needs reconfiguring in order to preserve a careful balance between innovation and regulation.

Keywords Metaverse · Metaverse legal system · Neuroscience · Artificial intelligence

1 From the Myth to the Network: The Cultural and Philosophical Genesis of the Metaverse

That of the Anthropocene [1], the advent of which is, according to some scholars, the result of the human processes of the so called 'Great Acceleration' [2], already seems to be a notion in the process of being surpassed in the light of the whirlwind of changes that the man-technology hendyad is bringing through hyperloops. In fact, the transition towards a new era (straddling the virtual and the digital) increasingly

---

A. Borroni (✉)
University of Campania, "Luigi Vanvitelli", Caserta, Italy
e-mail: andrea.borroni@unicampania.it

A. Borroni et al. (eds.), *Metaverses: Reshaping Law, Economy, and Society*, PoliTO Springer Series, https://doi.org/10.1007/978-3-032-08664-8_2

characterised by the domain of the intangible, in which the 'physical' dimension is progressively losing importance, is evident [3]: icastically, in the digital fabric that constitutes the immanent vesture of our times, the metaverse emerges as an intricate connectome of virtual worlds and interactions, endowed with the Copernican power to critically question the traditional boundaries of the human being's understanding of 'reality'.

Certainly, for some time now, man has, at least unconsciously, been fond of the idea at the basis of the metaverse: indeed, although this term indicates a novelty on the linguistic landscape even before the technological one, the perception of a 'parallel' world, an illusory representation of reality, has very ancient roots. The greatest thinkers, starting with Plato's myth of the cave [4] and arriving at Arthur Schopenhauer's world that exists only as a representation [5], questioned the mind's effective capacity to elaborate and internalise simulated, hypothetical, and surreal worlds [6] from what is concrete, tangible, and perceivable through one's senses [7]

In more recent times, neuroscience itself has advanced the hypothesis of the impossibility of understanding the world, except to a partial and arbitrary extent, our nervous system would create, through its own filtering and processing of data, a universe that does not coincide with the real world, excluding some of the existing frequencies [8]; in this respect, the metaverse is therefore a 'meta-reality', but not in the sense of an al-beyond matter, but rather al-beyond the way our mind, connected to the brain, understands matter [9]. The first to give these visions a name and a plausibly tangible form were two works of fiction from the turn of the last century: William Gibson's Neuromancer (1984) and Neal Stephenson's Snow Crash (1992). The former contains a preliminary description of the concept of cyberspace, understood as a 3D manifestation that is projected straight into the brain and allows users to 'navigate' the current of data and information coming from computer networks; the author of the second, on the other hand, is known for having coined in his work the forerunner definition of the 'metaverse', taken as a digitally constructed environment accessible through special visors, that enables users to engage in social interaction, professional activities, and the construction of an alternative mode of existence that transcends the physical limitation of the material world, via a self-representation known as an avatar [10].

Neuromancer and Snow Crash ended up stimulating the creativity and ferment of ideas of those working in the technology sector who, in an attempt to translate these visions into concrete solutions, gave impetus to a fervent and constant process of innovation and development: for instance, during the design and construction of the first Xbox console in the 1990s, members of the design team were required to read Snow Crash in order to gain a comprehensive understanding of Microsoft's vision of online multiplayer gaming [11].

Then, in 1999, the plot of 'The Matrix' tells of a dystopian future, anticipating the meta-verse: the protagonist is unable to have consciousness of his sleeping or waking state, he was generated and lived inside the Matrix, a virtual world created by a computer programme to keep human beings under control, through interactive neuro-simulation [12].

However, the real keystone in the establishment and spread of the meta-verse concept was Mark Zuckerberg's decision to rename Facebook to Meta in 2021, promising the creation of an immersive platform through which you can engage in activities strongly reminiscent of the logic behind the Snow Crash metaverse: this platform is based on the idea of creating an internet where is possible not just watching but also experiencing first-hand, being able to do almost anything: meeting friends and family, working, learning, playing, buying. [13].

A development, this, that—while attracting much scepticism—has contributed to increased investment in the sector by tech giants around the world, who have grasped its potential beyond gaming. And so, thanks to the concretive support given to the 'techno-social' visions sprung from science fiction, the meta-verse, from a simple idea, gradually became a reality, representing a valid expression of the concept of 'self-fulfilling prophecy' or, as it has been said, 'hyperstition': "the future is the narrative of the future", and hyperstitions, in fact, are "all those hypotheses of the future that become captivating enough to operate by transforming the present, thus influencing the outcome of history, in an attempt to come true" [14].

2 The Architecture of Cyberspace

Due to its nature as a constantly evolving phenomenon [15], however, a uniformly accepted definition of the metaverse is still in the process of being established. In an attempt to explain what it represents today, the metaphor of the rhizome comes in handy: in botany, in its original meaning, this term denotes a portion of the stem of herbaceous plants that develops horizontally below ground level and possesses its own buds capable of proliferating independently. Due to the peculiarity of its ramifications, the rhizome can be associated with the common thread that connects the virtual experiences, interactions and digital identities of those who access the metaverse, giving rise in the new dimension to a patchwork of interconnections.

More technically, the metaverse as an emerging technology consists of a digital infrastructure that, from an operational point of view, is supported by advanced virtualisation systems—in detail, AR, VR, sensors and joypads—and the Web [16]; in the metaverse, the Internet, virtual worlds, and augmented reality converge and grant users from all over the world the possibility to interact and interoperate not only between virtual but also real activities they conduct, up to the point of interconnection between the various platforms [17]: "what is clear […] is that there are multiple metaverses stretched across different platforms, rather than a singular, all encompassing (capital-M) Metaverse" [18].

As a matter of fact, the characteristics most commonly attributed by doctrine [19] to this reality at its earliest stages are the following: (i) spatio-temporal persistence, understood as the capacity to exist indefinitely; (ii) the synchronicity of connections, which occur in real time; (iii) the possibility of being usable by an unlimited number of users at the same time, each one endowed with its own individuality; (iv) being endowed with an independent and functioning economy that allows users to purchase goods and services, invest, earn, and exchange; (v) interoperability with the physical world, with particular reference to the transposition of data and information; (vi) the portability of data, digital assets, content across the various experiences and platforms operating in the metaverse; (vii) the creation and management of content and experiences by a wide range of collaborators [20].

Apparently, these peculiarities seem to delineate a model of the virtual world containing a promise and perspective of 'eternal duration' [21]; however, upon closer inspection, 'meta-experiences' conceal a character of innate fragility [22]: in fact, and contrary to what has just been said, they are ephemeral, destined to last as long as those who manage the servers on which they take place allow it [23]; and equally ephemeral are the digital assets that can be acquired there, which might not be compatible with possible future updates to the platform employed by its developers [24].

In addition to this, not all the particularities of the metaverse are attractive to companies: for instance, the use of avatars whose style is reminiscent of the world of video games—and which, in reality, finds its own logic in the need to minimise the constraints linked to graphic rendering and, consequently, to bandwidth—is often associated with an image that lacks professionalism [25]; similarly, the conduct of online users, who often underestimate the consequences of verbal attacks, insults and, harassment solely because of their occurrence in a virtual context [26], represents a problem of no small importance: "[t]he very thing that makes the metaverse interesting—the very real sense of sharing a virtual space—is the thing that brings out the worst in all too many men" [27].

Further critical issues derive, then, from the impact that augmented reality devices have on the individual's mental and physical health: in addition to potentially causing addiction [28], these tools condition man's sensory perception and may, therefore, entail effects such as nausea or dizziness and foster visual vices, with the risk of slipping into an (intriguing as much as disturbing) short-circuit between real and virtual [29].

Ultimately, as far as the more strictly legal profiles are concerned, it has become indispensable to outline a regulatory framework for a controlled development of the 'metaverse' ecosystem: in the absence of adequate regulation, in the long run, it could easily transmute into a dystopian reality where the protection of civil rights and personal data is compromised and transactions escape any kind of monitoring [30].

3 Regulatory Crisis and Legal Horizons: Smart Law or Arbitrary Power? Rethinking Norms in the Era of the Metaverse and AI

The inherent complexity in the elaboration of such a framework can be traced to the dynamics of regulators constantly adapting existing regulations in response to the rapid and often unpredictable technological developments that feature this new virtual environment [31]; it is further accentuated by the fact that the experience in the metaverse bears more marked similarities to real life than the two-dimensional web, whose guidelines are insufficient for the specificities of the metaverse [32]; moreover, on a case-by-case basis, therefore, it will be important to determine whether to implement co-regulation, hetero-regulation, or self-regulation (self-regulation of platforms). Is the regulation uniform or sector-specific? Is it useful or structural? What ought to be subject to platform regulation and what ought to fall under state regulation? [33].

From a macroscopic point of view, in assessing the impact of the technological frontier on the offline world, comparative law has questioned the nature of virtual space which, despite its evident aterritoriality, is not to be understood, economically, politically, or even less legally speaking, as totally separate from material space, but rather as an extension of it [34]; however, "the Internet frontier has its own law, but we cannot say prima facie whether, as an economic, political, or social territory, it is ruled by the common law, the civil law, or any other recognisable family of legal systems, nor whether it is governed by any pattern of law at all" [35]. The implications that this consideration entails are relevant in systemological terms: for example, the most attentive doctrine on the subject has recently revisited its position, which identified three factors of social control—specifically, law, politics, and tradition—that portray all legal systems, albeit with different intensity and with case-by-case variations concerning the relationship of force and attraction between them [36].

Analysing the interactions between these, therefore, it was possible to arrive at a new typology of classification of legal systems that, in overcoming the shortcomings of the existing ones (among all, Eurocentrism and staticity), placed them in one of three possible areas: the area of professional hegemony of law (rule of professional law), the area of hegemony of politics (rule of political law) and the area of hegemony of religious or philosophical tradition (rule of traditional law). In the light of the technological advances of the new era, therefore, the question arises as to whether an area characterised by the 'rule of smart law' [37] can be identified today: the existence of a real smart legal system or a smart legal family does not seem to be such a distant hypothesis.

And, in fact, the idea of the metaverse as a smart legal system, in which the most advanced technologies can constitute a valid support to the legal framework, would fit well with its parallel conception as a (atechnically) mixed legal system: a system capable of reflecting the relationships and complexities of all the virtual (but also material) communities coexisting within it thanks to its flexibility and adaptability, which make it easy to combine such heterogeneous elements and institutions.

However, on the other hand, the ultimate degeneration of this idea, which sees the total replacement of metaverse law with algorithms and artificial intelligence systems, is not a viable path at present: the ways in which such tools make connections between the data contained in their immense databases still remain, at least partially, obscure [38]; moreover, this would create a 'pattern of law' unhinged with respect to what historically, factually, has happened and based on predictions that, however accurate, would represent nothing more than a pathology of the system in a logic of "arbitrary and brute power" [39].

Shifting the focus to the microscopic profile, however, among the various aspects to be investigated, the protection of privacy in the metaverse with reference to the collection and processing of personal data is one of the most discussed. Such platforms can, in fact, access a large amount of information, monitoring consumption habits, trends, opinions, emotional states, as well as 'biometric data' such as, for instance, those linked to body gestures, eye movements, facial micro-expressions, or tone of voice [40]; such activity could imply a process of 'bending', manipulating reality and adapting the virtual experience to the perceptions and opinions of users in order to profile, 'target' and ultimately persuade them to make certain choices—which is especially, but not exclusively, the case in the commercial sphere [41]. Similarly, copyright, the protection of cultural heritage, sustainability, and economic transactions raise important legal issues; finally, the metaverse is the ideal substratum for the emergence of new types of cybercrime (think, for instance, of deep fakes or financial crimes linked to the use of cryptocurrencies [42]), which need to be traced back to the avatar's real identity in order to prevent them.

4 The Future of the Metaverse Between Utopia and Hyperstitions: Charting a Reliable Blueprint for the Ungraspable World

As can be deduced from the above picture, the applications and evolutionary prospects of the metaverse show an exceptionally vast potential for development, and the implications of such advances elude complete and accurate foresight for the time being. It is also plausible that the metaverse in its current conception may undergo slowdowns, redefinitions, and metamorphoses; however, the intention to completely curb the emergence of parallel experiential realities [43] seems questionable: what we are witnessing, on the other hand, is merely the prelude to what has so far only been imagined. The implementation of a fully interoperable and all-encompassing system of all existing virtual worlds is still in its infancy: time, huge investments, and even more high-performance experiential technologies are necessary elements for it to function as actually envisaged; what is certain is that, despite the progress already made, the temporal hiatus between the being and the becoming of the metaverse is still a deep fault, which technological developments still fail to bridge, making the tech pillar slippery and inaccessible to most [44].

Although we only stand at the dawn of the metaverse, the convergence of technologies and the growing maturity of the solutions underpinning these emerging environments suggest the prospect of rapid developments in the near future. Developments that will hopefully contribute to making the physical world a more inclusive and sustainable place, creating more cohesion between individuals. Yet, one must ask: is this an achievable objective within our tangible reality, or merely an idealized projection, a scenario confined to virtual realms?

All the issues that are still 'open' and in the process of being defined contribute to making the phenomenon particularly interesting for the scientific community, and law is waiting but, surprisingly, perhaps the most mature branch for ready utilisation: the positive effects that consist in the natural fallout of technological evolution, cultural development, and regulation can and must converge in a future that, in addition to revolutionising socio-economic dynamics, is truly aligned with those fundamental values that represent the cornerstone of contemporary society [45]. Therefore, it is incumbent upon legal practitioners to take charge of the critical issues outlined above, in seeking to put in place regulatory milestones that are capable of orienting, in a proleptic sense, the future development of this reality in continuous and frenetic transformation [46].

Among the various questions raised, the plausibility of a metaverse first is debated. Technological progress and transformation have shifted the angle of observation from the desktop to the web, with appreciable effects on hardware as well. And the mutation has certainly not stopped. In the words of Mark Zuckerberg "the metaverse will be the successor of the mobile internet". But, even if we are here waiting for it with great expectations, at the moment, the promised future has not yet manifested itself.

References

1. Specifically, the idea of the 'Anthropocene', developed and proposed by the biologist E. F. Stoermer in the 1980s as a plausible continuation phase of the Holocene, indicates how human actions have become a characterising factor of geological epochs. It is, therefore, a new worldview, a Weltanschauung that places subjective experience and, specifically, the many and still expanding effects of human activity on the planet and its atmosphere at its very core. Crutzen PJ, Stoermer EF (2000) The 'Anthropocene'. IGBP Newsletter 41, p. 17. On the evolution of the conception of Weltansschaunng see, also, Ponsetto A (1994) Genesi dell'idea di Weltanschauung e recupero dell'idea originaria di filosofia. Idee: rivista di filosofia 26–27:35–68
2. Among the most relevant are population growth, the exploitation and consumption of energy resources, technological development, the process of industrialisation and, of course, globalisation. Borselli D, Ravaioli G (2021) Tracce dal futuro. Leggibilità e virtualità in The Anthropocene Project. piano b. Arti e culture visive 6(1), p. 87. Available at https://doi.org/10.6092/issn.2531-9876/14302, last accessed 2025/10/04
3. According to McLuhan's reconstruction, machines (or, more generally, mediums) came into being as a projection and extension of the human body reproduced in a fabric other than that of which human beings are made. McLuhan M (2008) Capire i media. Gli strumenti del comunicare, p. 58. Milano, Il Saggiatore. On the other hand, the later technologies, the

electronic media, rather than an extension of the physical organs, represent an extension of the central nervous system, a means to improve the quality of life but also a tool for further technological developments. Id. (2008) Rimorso di incoscienza. Lettera internazionale: rivista trimestrale europea 98. The question, therefore, arises as to whether the internet, artificial intelligence, and the digital replica of people and material objects in the metaverse can actually embody, within their inner boundaries, the next stage comparable to the extension of the human mind, consciousness, and spirit

4. By means of the allegory of individuals chained inside a cave, whose knowledge of the outside world is limited to the shadows cast on the sides by a source of light, Plato illustrates how man's perception of reality is limited to his own experience and conditioned by his imagination; with reference to the metaverse, in a critical sense, it has been argued that the more sophisticated and inventive 'technological twin' of empirical reality provides a cozy grotto in which to stay engrossed, but we fail to notice who, how, and why it seeks to imprison us in a setting whose creators and owners are "elsewhere." It is difficult to say if the technological cave is an open or closed society since it turns into a social space where a portion of human existence and activity finds a place. Pessina A (2023) L'essere altrove. L'esperienza umana nell'epoca dell'intelligenza artificiale. Mimesis, Milano-Udine. On the topic of the allegory of the cave as applied to legal research, see Yntema HE (1956) Comparative Legal Research—Some Remarks on 'Looking out of the Cave'. Michigan Law Review 54 (7)
5. Similarly, in asserting that life is nothing but a representation, Schopenhauer intends to highlight how subjective perceptions influence and shape our understanding of the world Schopenhauer A (1928) Il mondo come volontà e rappresentazione, translated by Savj-Lopez P, De Lorenzo G. Laterza, Bari. A concept that Schopenhauer actually derives from Immanuel Kant, who described with the term "phenomenon" the mere representation of reality as it appears to us, while referring with the term "noumenon" to the "thing-in-itself," existing beyond the constraints of the human mind: reason's knowledge pertains only to appearances, while the thing-in-itself certainly exists as something real on its own, but remains unknown to us. Kant I (1976) Critica della ragione pura, translated by Colli G. Adelphi, Milan. However, the references to Schopenhauer's philosophy regarding the Veil of Maya are evident: the barren real-world hides behind a veil of illusion that anachronistically replaces the truth. The seeds planted by philosophy in the cultural soil that gave rise to the idea of the metaverse are, however, numerous enough to make a detailed reconstruction impossible in this context. It suffices here to briefly mention Descartes, who questions the ontological nature of reality and whether one exists beyond one's own thought and self, and Kierkegaard, with the existentialist theme of choice and the connection between possibility and freedom, in light of which one may wonder whether, as Leibniz asserted, we indeed live in the best of all possible worlds. Lastly, Sturzo L (1978) La vera vita. Sociologia del soprannaturale, pp. 5–6. Bologna, Zanichelli, conceptualizes supranatural life as what he calls "true life": not merely as an addition to the natural dimension, but a transformative synthesis of human existence, which integrates various layers of being—our innermost aspirations, our capabilities, our activities. Although the metaverse does not involve a spiritual transcendence in the theological sense, it nonetheless introduces a meta-level of social interactions, which may be interpreted—by analogy—as a new synthetic dimension within legal systems. In this sense, the emergence of the metaverse as a new experiential space contributes to the further development of the theory of knowledge, inviting humanity to reflect on the plausibility of a partial or total transposition of one's beliefs into the virtual world
6. Again, according to Buddhism, we live in an illusionary cosmos made by a magical power known as Maya; according to Pedro Calderón de la Barca, life is a dream; and according to the Upanishads, the whole cosmic process is the dream of a God. Cerrina Feroni G (2023) Il metaverso tra problemi epistemologici, etici e giuridici. Media Laws 1, p. 21. Available at https://www.medialaws.eu/rivista/il-metaverso-tra-problemi-epistemologici-etici-e-giuridici/, last accessed on 2025/10/04
7. Parallel to advancements in computer science, the insights of various scientific figures of the 20th century expanded the horizons of this concept, ultimately identifying a symmetry between physical and computerized reality. In particular, in a famous speech, the inventor of the first

programmable computer, Konrad Zuse, compared the behaviour of digital particles (bits) to that of subatomic particles, and his arguments proved so persuasive that everyone was much more certain by the end of the meeting that the adage "the Cosmos is a Great Computer" was a legitimate heuristic tool rather than just a metaphor. Longo GO, Vaccaro A (2014) La nascita della filosofia digitale. Mondo Digitale 52, p. 4

8. As was already mentioned, the use of cognitive biases and heuristics—that is, mental constructs based on a biased understanding of reality—may confirm this viewpoint because they enable us to make decisions and draw conclusions quickly, even though they run the risk of making us mistaken. Cerrina Feroni G, Il metaverso tra problemi epistemologici, etici e giuridici… cit., p. 22
9. Ibidem. The idea of the metaverse extends into the field of metaphysics, sparking multiple avenues of reflection. If metaphysics, as a discipline, questions "being" in its concreteness and integrity—going beyond, as the etymology of the term suggests, the contingent elements of sensory experience—its meaning is instead influenced by shifts in perspective imposed by interactions with other fields of knowledge. In other words, "one needs to leave this world, transcend it, and "become aware that this world is not everything" in order to get at the fundamental causes of the things that we observe (Wittgenstein). Therefore, anytime the realm of experience is completely transcended, metaphysics emerges: when the leap beyond is accomplished" (author's translation). Mondin B (2007) Manuale di filosofia sistematica, vol. 3, pp. 8–9. ESD, Bologna. When the boundaries of physical reality are transcended, one may question the actual validity of identity and, more generally, of reality within the virtual context, to the point of wondering whether it can be undermined solely by the digital nature that defines it. For a broader discussion on these topics, see Cacciari M (2023) Metafisica Concreta. Adelphi, Milan
10. The Zeitgeist is also common in science fiction literature and film representations. See, among others, the works of Philippe K. Dick and his Pseudomondes, in which experiences are narrated only as a seamless mixture of the real and the unreal, in which worlds are projected in a distorted manner (on all, Dick PK (1957) Eye in the Sky. Ace Books, New York). There is a continuous perceptual schizophrenia in which reality and illusion intertwine, unravelling the intricate question of what is actually real
11. As noted, Stephenson himself was hired as 'Chief Futurist' for the AR start-up known as Magic Leap. Jones P (2023) Geographies of the metaverse. In: Osborne T, Jones P (eds.) A Research Agenda for Digital Geographies, p. 177. Edward Elgar Publishing, Cheltenham-Northampton
12. The influences exerted and the parallels evident are also found when looking at the metaverse through the prism of 'Le avventure di Alice nel Paese delle Meraviglie' by Carroll: Alice falls down the rabbit hole and ends up in a universe of curious and strange events. Carroll L (1865) Alice's Adventures in Wonderland. Macmillan, London
13. Meta (2021) The Metaverse and How We'll Build It Together, Youtube. In particular, the great attraction of the metaverse is the possibility to enjoy easy and ubiquitous access to information (not deferred in time), which also allows for a simultaneous reduction in transaction costs. This is evident in specific areas, among others access to public services and interaction with legislators and government—one of the best-known examples is certainly that of the South Korean capital Seoul, which has announced a five-year plan (known as Metaverse Seoul Basic Plan) for the creation of a virtual town hall, square, and civil service centre. In addition to this, we are witnessing the gradual emergence of virtual or hybrid courts and courtrooms that are revolutionising the justice sector. In most cases they come to life on videoconferencing platforms such as Zoom; however, in 2022, the court in Paraiba, Brazil, held the first national court hearing in the metaverse in relation to proceedings initiated in 2018 and concluded on this very occasion by the parties represented by customised 3D avatars. Similar experiments were conducted in Colombia, where AI and the metaverse were declared admissible in judicial proceedings through Ley 2213 of 2022. This is undoubtedly an efficient modality, capable of bringing together parties even physically far apart, but which raises questions in relation to aspects such as the impossibility of capturing nuances related to facial expressions and

body language. European Parliament's Policy Department for Citizens' Rights and Constitutional Affairs (2023) Metaverse. Available at https://www.europarl.europa.eu/RegData/etudes/STUD/2023/751222/IPOL_STU(2023)751222_EN.pdf, last accessed 2025/10/04
14. Goggi Z (2022) Eternità del metaverso. Domino 9, p 54 (author's translation)
15. It is estimated, in this regard, that the gradual establishment of the Metaverse will not lead to concrete results for another 10–15 years at the earliest, as full 'telepresence' cannot be guaranteed at present; in fact, "[e]xtended reality tools are not fully pervasive, and latency remains an important factor on internet networks". Council of the European Union General Secretariat (2022) Metaverse—Virtual World, Real Challenges, p. 7. Available at https://www.consilium.europa.eu/media/54987/metaverse-paper-9-march-2022.pdf, last accessed 2025/10/04. Nevertheless, technological advancement is in itself a process of a non-linear but exponential nature: "[…] we won't experience 100 years of progress in the 21st Century—it will be more like 20,000 years of progress (at today's rate). […] Within a few decades, machine intelligence will surpass human intelligence, leading to The Singularity—technological change so rapid and profound it represents a rupture in the fabric of human history". Kurzweil R (2004) The Law of Accelerating Returns. In: Teuscher C (ed.) Alan Turing: Life and Legacy of a Great Thinker, pp. 381–416. Springer, Lausanne
16. More precisely, as explained in Kuai L, Lacity M, Mullins JK (2023) Web 2 vs. Web 3 Paths to the Metaverse: Who Is Leading? Who Should Lead?. The JBBA 6 (2), the current 'metaverse' seems to develop along two different lines: that of Web2.0 and that of Web3.0. In the case of the companies that follow the first path, their main aim is, from a purely capitalist perspective, to become providers of services operating on centralised platforms, through which they could monitor every gesture, every form of communication (whether verbal or paraverbal), every movement and interaction with other avatars or assets. The authors place Meta's experiment in this scenario—after examining the privacy statement accompanying the Quest visor—along with other leading companies in the sector such as Citigroup, Aries I Acquisition, Atari, and Vision Sensing Acquisition. Web3.0, on the other hand, is a notion that, building on the generally recognised notion of the World Wide Web, incorporates concepts such as decentralisation, blockchain technologies and token-based economy, an independent virtual economy made possible by digital currencies and non-fungible tokens (NFT). Rossi A (2023) Il buono, il brutto e il cattivo: il "triello" del metaverso. Documenti Geografici 2, p. 673. The path that sees the development of metaverse in this view is the one generally followed by communities and non-profit foundations that envision for the final product a decentralised architecture controlled by both developers and users. Moreover, thanks to the use of blockchain, these structures are more resilient in the field of cybersecurity and more powerful the more nodes are added to the network. Among the most important Web 3 communities are ApeCoin, The Sandbox, Render Token and Decentraland. Similar conclusions are reached by Dolata M, Schwabe G (2023) What is the metaverse and who seeks to define it? Mapping the site of social construction. Journal of Information technology 38 (3), pp. 258–259, according to which the conception and definition of the metaverse changes according to the interests pursued by different actors. In fact, it is a system in which those who developed it are actually testing the limits of their influence, having an interest in exploring: "the emergence of new business and investment models […]. Mixing definitions and proposing various metaphors at the same time, some actors (un)consciously imply that they may have no interest in stabilizing the meaning [of metaverse], such that they can keep exploring. This attitude further increases the interpretative and design flexibility". Thus, a metaverse dominated by the pursuit of purely economic goals and by the representations of large tech companies or major videogame producers could coincide with a centralised and closed platform; a metaverse as a space controlled entirely by the state, on the other hand, would configure a faithful virtual replica of the real world
17. Giovanardi E (2022) Il Metaverso, le nuove frontiere del mercato disintermediate. In Deotto D, De Simone D, Giovanardi E, Rampone F, NFT e Metaverso. Cosa sono, come funzionano, caratteristiche e aspetti tecnici. Guida operativa alle regole giuridiche e fiscali, p. 41. Il Sole 24 Ore Professional
18. Jones P, Geographies of the metaverse… cit., p. 179

19. Among the vast literature on the subject, see in particular Ball M (2022) The Metaverse: and How it will Revolutionize Everything. Liveright Publishing Corporation, New York; Id. (2020) The Metaverse: What It Is, Where to Find it, and Who Will Build It, available on the website https://www.matthewball.vc/all/themetaverse, last accessed 2025/10/04; Epicoco IO, Pierro M, Sarzana Di Sant'ippolito F (2022) Il diritto del metaverso. NFT, DeFi, GameFi e privacy, p. 7 ff. Giappichelli, Torino; Natale G (2022) Metaverso: necessità di un diritto reale per un mondo virtuale. Gli aspetti giuridici rilevanti del Metaverso. Rassegna Avvocatura dello Stato 1, pp. 2–3
20. Although theoretically, attempts to define the metaverse are wasted, even to the point of detail, it is not yet possible to establish what its future and definitive connotations might be; more precisely, above all in terms of its structure or governance model (centralised or polycentric, unitary or multipolar), so that scholarly writing started thinking of a potential pluriverse. Stanzione P (2023) Il Metaverso tra utopie e distopie. In: Atti del Convegno 30 gennaio 2023 "Il Metaverso tra utopie e distopie. Orizzonti e sfide della protezione dei dati". Available at https://www.garanteprivacy.it/web/guest/home/docweb/-/docweb-display/docweb/9939113, last accessed 2025/10/04. And, in fact, there persists an intricate conglomeration of terminology that includes, in addition to the metaverse, the more nuanced concepts of multiverse—understood as the simultaneous existence of independent virtual universes—and omniverse—which finds its most recent practical application in the NVIDIA platform of the same name—until it culminates in the idioverse, which represents, finally, the ultimate degeneration of these ideas, embodying the legitimisation of the narration and vision of the world that people literally constructs to their own taste and spirit, with all the problems that this entails in terms of univocity and, therefore, intolerance of what is outside one's own schemes. On the concept of the idioverse, see Cera A (2022) L'idioverso ovvero della post-realtà. Genesi di un'ontologia a-sociale. Available at https://www.leparoleelecose.it/?p=43176, last accessed 2025/10/04. More specifically, among the best-known platforms are The Sandbox e Decentraland; However, many of the features listed by Ball can be attributed to various video game platforms: one thinks of Second Life, where one can buy virtual objects created and sold by other players on the appropriate marketplace to customise one's experience, or Fortnite, which has incorporated its own payment system based on a currency created for this purpose (the so-called v-Bucks). On the first experiences of alternative digital currency, see Borroni A (2015) Bitcoins, a New Frontier of Money?. Ianus, pp. 69–97
21. Start-ups are trying to translate this concept into preserving an everlasting memory (with a 'Live forever' feature) that allows users to crystallise their selves in a hyper-realistic avatar, capable of reproducing their appearance, movements, voice and facial expressions even after their disappearance
22. In fact, developers of metaverse platforms attempt to project a sense of stability using "such words as world, land, objects, reality, place do live"; however, what is actually sold to the end user is nothing more than a "digital simulation, an information technology service depending on providers' IP address, servers and ICT services". Hence the need to exercise particular caution when investing in these types of projects, which may not be particularly sustainable in the long run. European Parliament's Policy Department for Citizens' Rights and Constitutional Affairs, Metaverse… cit., p. 13
23. In this regard, consider, icastically, the way in which the terms of use concerning the so-called continuity of service are formulated. In the case of Decentraland, for example, it is provided that "[t]he DAO [decentralized autonomous organization] may decide to terminate or suspend all or part of the Site and the Tools and your access to the Site and the Tools immediately, without prior notice or liability. You will not receive any refunds if you cancel your Account, or if these Terms are otherwise terminated. You agree that the DAO, in its sole discretion and for any or no reason, may terminate these Terms and suspend your Account(s) for the Tools. You agree that any suspension of your access to the Site or the Tools may be without prior notice, and that the DAO and/or the Foundation (and its officers and employees) will not be liable to you or to any third party for any such suspension". Ivi, p. 114

24. And this is related to what happens in the world of gaming in general: even, trivially, in offline video games, so-called modders must carefully monitor any updates released by developers so that their virtual products are compatible with them and users can continue to use them; another problem linked to these dynamics arises when a modder decides to stop updating his content. The risk of technological and economic obsolescence (due to unsustainable costs) are cases that are likely, and all the more so, to affect a constantly developing reality such as the metaverse
25. Lee C, Joo H, Jun S (2021) Social VR as the new normal? Understanding user interactions for the business arena. In: Extended abstracts of the 2021 CHI conference on human factors in computing systems, CHI EA 2021 (Conference on Human Factors in Computing Systems—Proceedings). Association for Computing Machinery. https://doi.org/10.1145/3411763.3451740. Accessed 04 Oct 2025
26. Such attitudes also have repercussions on the training of experimental chatbots: think of the case of Microsoft Tay, ChatGPT's precursor, which was contaminated by racist and misogynist materials inserted into the system by so-called Internet trolls. A problem that was solved by substantially limiting the creativity of future models: ChatGPT itself, when faced with more ethically complex questions, now admits that it does not possess moral beliefs or the capacity to make moral judgements such as to be able to offer adequate answers. On this subject, see https://www.nytimes.com/2023/03/08/opinion/noam-chomsky-chatgpt-ai.html, last accessed 2025/10/04
27. Jones P, Geographies of the metaverse… cit., p. 181
28. In the literature devoted to the study of technological development, the loss of control over time spent in the digital environment is equated with more traditionally debated forms of addiction: in other words one could look at the concept of 'addiction,' a term that describes dependency on the digital world, along with the idea of 'pruning,' which refers to the process of neural pruning in the brains of younger generations, who have more brain plasticity. There's also the 'dopamine effect,' meaning that the digital world can create an addiction similar to that of drugs, and the 'Flynn effect,' which points to the decline in cognitive abilities among younger generations due to excessive use of digital tools. Natale G, Metaverso: necessità di un diritto reale per un mondo virtuale… cit., p. 4
29. More generally, these symptoms have been described as cybersickness, a particularly delicate phenomenon especially where minors whose nervous system is not yet fully developed are involved. For this reason, optical viewers should not be used under the age of 13 and, in any case, caution should be exercised, and the time of use should be limited. It is worth noting, however, how in the metaverse, and more generally on the Internet, the age verification process remains highly unreliable and is often found unable to provide accurate authentication. On this point, see European Parliament's Policy Department for Citizens' Rights and Constitutional Affairs, Metaverse… cit., p. 44. See also Stanney K, Lawson BD, Rokers B, Dennison M, Fidopiastis C, Stoffregen T, Weech S, Fulvio JM (2020) Identifying Causes of and Solutions for Cybersickness in Immersive Technology: Reformulation of a Research and Development Agenda. International Journal of Human-Computer Interaction 36(19): 1783–1803
30. In this regard, as a globally accessible environment requiring a coordinated regulatory framework, the metaverse is potentially able to radically transform the roles held by individual users, corporations and States. Specifically, the Council of the European Union has identified three possible regulatory approaches: i) an approach supported by States, which would inevitably experience differences in the way they view the regulation of "issues such as personal freedom, surveillance, and protection"; ii) an approach supported by tech activists, whose goal is the creation of a "free, decentralised, and open web for everybody" in which "digital money and property rights will be embedded in automated protocols, whereas the existing web remains controlled by corporate giants and under surveillance by states"; iii) a "business-oriented and profit-driven model backed by the tech sector" in which ownership of the Metaverse is sought by controlling its individual components. More generally, moreover, the absence of borders and of a comprehensive legal framework in the Metaverse is challenging the concept of territorial sovereignty: in fact, it is reported that President Erdogan has requested studies on this

phenomenon as a consequence of the purchase on Metaverse platforms "of thousands of virtual colonies", most of them located in Istanbul. Council of the European Union General Secretariat, Metaverse-Virtual World, Real Challenges… cit., p. 9

31. In general, legal regulation 'travels' much slower than technological innovation, and case law plays a crucial role in its integrating role to the black letter rule of the legislation. Among the many, see Ciacci G (2023) Le fonti del diritto dell'informatica. In: Bocchini E (ed.) Manuale di Diritto Privato dell'Informatica, p. 32. Edizioni Scientifiche Italiane, Napoli
32. Celotto A (2022) Tema del "metaverso" e delle sue implicazioni per l'ordinamento giuridico. GiustAmm.it 7
33. Natale G, Metaverso: necessità di un diritto reale per un mondo virtuale... cit., p. 2. On this point, see also Macchia M (2023) Quale regolazione per il metaverso? I caratteri dell'approccio europeo. In Stanzione P, Violante L (eds.) Il Metaverso. Diritti-Libertà-Antropologia, Edizioni Scientifiche Italiane, Napoli
34. In considering the metaverse—and, more generally, the virtual—as an extension of the real, a semantic, heuristic, and epistemological operation typical of law is in fact performed; so typical that it was first transposed into the Italian Civil Code and Navigation Code and subsequently the subject of analysis by case law. In particular, the reference is to the combined provisions of Articles 832 and 840 of the Civil Code: according to the former, the owner has the right to fully enjoy and dispose of property, within the limits and in accordance with the obligations established by the legal system. The second provision, on the other hand, established that ownership of land extends to the subsoil and all that it contains, and the owner may carry out any excavation or construction work that does not cause harm to neighbouring property, with the exception of the cases specified by the law. Furthermore, the landowner may not oppose third-party activities carried out at such a depth below ground or such a height above it that he has no legitimate interest in excluding. It follows that the legislator intended to project the right to property, in theory, vertically, creating a sort of 'cylinder' of unlimited extension in a logic that is reflected in the phrase "usque ad sidera usque ad inferos", albeit with the condition of the existence of a specific interest. The metaverse, on the other hand, is a derived world, but derived from the physical and material; the regulation of its effects and products can easily take place by means of a form of adaptation—or, to borrow a terminology familiar from private law, of 'subsumption'—of existing law, which already possesses the necessary foundations without the need for excessive upheaval
35. Mattei U (2024) The Legal Metaverse and Comparative Taxonomy: A Reappraisal. The American Journal of Comparative Law XX, p. 18. The Internet is a border that, paradoxically, is "far from being a universal frontier open to all" because of limitations such as, first and foremost, linguistic ones: on closer inspection, it represents a space in which the cultural (and political) peculiarities of the communities with the greatest virtual (and material, considering the enormous expenses in economic and energy terms that derive from it) presence are reflected, and which, from a geopolitical point of view, aim to maintain their influence within this space as well. It is not inconceivable, therefore, that in the future real 'regional' metaverse will develop—as is already happening in the case of China—as comprising those systems that, sharing the same language, have access to the same 'portion' of the Internet. Ivi, p. 29 ff. In one of the most speculative passages of his work, Sturzo L, La vera vita. Sociologia del soprannaturale… cit., pp. 235–236, reflects on the existence of several systems (or groups of systems) inside the cosmos and, consequently, on the presence of 'intelligent and conscious beings' scattered across a myriad of those habitable worlds but participating together in the unfolding of a universal ascension to God. Leaving aside, once again, the teleological dimension, this vision resonates—metaphorically—with how, at present, the metaverse consists of multiple parallel social (and, maybe, legal) domains, each potentially endowed with its own internal rules and identities In this sense, Sturzo's ontological pluralism may be read as a precursor to contemporary reflections on the emergence of smart legal systems populated by digital entities and their need, on the other side, for a potentially decentralized digital jurisdiction
36. Mattei U, Monateri PG (1997) Introduzione breve al diritto comparato, Padova, CEDAM, 1997. The graphical transposition elaborated by Werner F. Menski in 2006 also has traits in common

with this reconstruction, using the shape of the triangle; in this case, the illustration places at the apexes of the geometric figure, respectively, the normative influences stemming from State, Society and Religion/Ethics/Morality. These factors concur, to varying degrees depending on the case, to determine the institutional peculiarities of each society, and for the author, an advocate of a pluralist perspective of analysis, each of these represents in itself a plural entity; hence the futility of the search for a uniform law for the world order, since the only element actually in common between all legal systems is what he calls 'plurality-consciousness'. In 2010, Menski reworked his theory by adding a fourth vertex, international law, proposing a new figure—a kite or tetrahedron. On this topic, see Menski MF (2006) Comparative Law in a Global Context: The Legal Systems of Asia and Africa, p. 188. Cambridge University Press, Cambridge; Id. (2011) Flying kites in a global sky: New models of jurisprudence. SocioLegal Review 7(1): 1–22
37. Mattei U, The Legal Metaverse and Comparative Taxonomy... cit., p. 5
38. On this point, N. Chomsky, according to whom comparing tools such as ChatGPT to the human mind is incorrect, since such algorithms represent nothing more than "a lumbering statistical engine for pattern matching, gorging on hundreds of terabytes of data and extrapolating the most likely conversational response or most probable answer to a scientific question"; differently, man is able to reason from smaller amounts of data and his mind does not operate on the basis of 'brute correlations' between them, but rather tries to offer meaningful explanations. According to the scholar, man is also endowed with an "operating system" which, from an early age, enables it to develop and internalise grammar and language on the basis of a series of principles and parameters that are, more or less automatically and unconsciously, absorbed. Conversely, AI systems are as if 'stuck' in a "prehuman or nonhuman phase of cognitive evolution", thus failing to develop the critical capacity that characterises the very concept of intelligence: "to say not only what is the case and what will be the case—that's description and prediction—but also what is not the case and what could and could not be the case. Those are the ingredients of explanation, the mark of true intelligence. […] True intelligence is demonstrated in the ability to think and express improbable but insightful things". In other words, the author views ChatGPT and similar programs as "constitutionally unable" of achieving a proper balance between creativity and constraint and concludes by stating severely that, "given the amorality, faux science and linguistic incompetence of these systems, we can only laugh or cry at their popularity". See https://www.nytimes.com/2023/03/08/opinion/noam-chomsky-chatgpt-ai.html, last accessed 2025/10/04. Then, according to V. ZenoZencovich, the conception of AI is "tainted" by two important prejudices. Firstly, the term 'intelligence' is used ambiguously because it is to be understood differently depending on whether we are talking about human beings or AI: in the first case, it represents the complex of mental faculties and the human body's four senses and internal organs; in the second case, it is the so-called 'computational ability', the faculty to process billions of data and make decisions on the basis of the inferential conclusions to which this process has led. The second issue lies on the other hand in the fear of what is still partly unknown coupled with "the conceited idea to master it". What the author intends to emphasise here is that AI, at its core, remains the product of man and, as such, the quality of its outputs depends on the quality of human input, with all the strengths and limitations this entails. As pointed out, "behind each and every phase of an AI tool are human beings with their abilities and their limits. To put things bluntly, stupid people create artificial stupidity. Stupidity is not a feature that can—stupidly—be measured and curbed through some IQ test, but a much more nuanced notion that must take into account the context, the circumstances, the actors, the purpose". Therefore, one must not lose sight of the notion that since error is intimately part of man, it can be present, all the more so, in the products and tools he develops: new technologies "must be neither glorified nor damned, but simply properly used". Zeno-Zencovich V (2024) Artificial Intelligence, natural stupidity and other legal idiocies. Media Laws', available at https://www.medialaws.eu/rivista_category/1-2024/, last accessed 2025/10/04
39. This consideration is even more valid if one considers that, at present, the vast majority of data is controlled by private individuals—often, those companies (or multinationals) who have

devised the algorithm and who, therefore, can decide its mechanics and duration of operation. It follows that the affirmation of a rule of smart law can only take place where such processes are totally emancipated from the arbitrariness of those in control of the technologies; to this end, it is desirable to devise effective institutional mechanisms for determining responsibility and capable of guaranteeing the transparency and predictability of decisions taken by means of algorithms or AI

40. Epicoco IO, Pierro M, Sarzana Di Sant'ippolito F, Il diritto del metaverso... cit., p. 28
41. Ibidem. Exercising a high degree of control over individuals by means of smart devices connected to the Internet (including wearables or implantable devices, so much so that from the 'Internet of Things' we have moved on to speak of the 'Internet of bodies') is now extremely easy thanks to the tracking of consumer habits: the constant monitoring of data has made possible "the transition of smart control from the Internet frontier to the real world" already in countries like China and South Korea, where QR codes-based transactions have started to replace cash payments and where shehui xi-nyong tixi (social credit systems) are being implemented to monitor citizens' social behaviour and economic activity. Mattei U, The Legal Metaverse and Comparative Taxonomy... cit., p. 23. See also: Mac síthigh D, Siems M (2019) The Chinese Social Credit System: A Model for Other Countries? The Modern Law Review 82 (6). In relation to this topic, the growing use of synthetic data is marking a significant development in the evolution of the digital economy, providing a practical alternative to real data and its limitations (regulatory burdens, scarcity, and prohibitive costs). Created through advanced AI generative models, synthetic data retain the statistical properties of real datasets without being tied to any event that happened in real life—thus opening up new possibilities for training artificial intelligence models in different domains. At the heart of this development lies a deeper evolution: that of "data monetization" i.e., a new business model based on the transformation of data from passive byproduct to strategic economic asset, generated, sold, and used to fuel service innovation, enhance decision-making processes, and more in general, create new market opportunities. Yet the use of synthetic data is contingent on the careful and effective integration of simulation technologies, advanced computational capabilities, and data governance strategies so that it is scalable, reliable, and compliant with current regulations
42. At the heart of this development lies a deeper evolution: that of "data monetization" i.e., a new business model based on the transformation of data from passive byproduct to strategic economic asset, generated, sold, and used to fuel service innovation, enhance decision-making processes, and more in general, create new market opportunities. Yet the use of synthetic data is contingent on the careful and effective integration of simulation technologies, advanced computational capabilities, and data governance strategies so that it is scalable, reliable, and compliant with current regulations
43. In particular, in the context of the metaverse, it is interesting to analyse the relationship between these types of cyber-financial crimes and Islamic finance in light of the latter's growing importance as a "facilitator for fairer financial relationships and its connection to real value creation". Cryptocurrencies are generally considered to be a permissible medium of exchange despite their volatility and consequent speculative nature; however, a particularly stringent form of supervision would need to be incorporated to ensure that they are not used for activities considered illicit and clearly forbidden by Islamic principles (e.g. gambling and all activities that may relate to it). Katternauer K, Syed H, CleenewerckL (2022) Financial cybercrime in the Islamic Finance Metaverse. Journal of Metaverse 2 (2), p. 59
44. The transposition of property rights into the metaverse, for instance, entails a reinforcement of its ultimate meaning and purpose: it is no coincidence that the first successful virtual worlds—think of the aforementioned Second Life—are equipped with a real internalised economic system. Another phenomenon of growing relevance are the so-called gacha games, those video games that—as can be deduced from their name, which winks at gachapons, the Japanese term for those gadgets contained in capsules and obtained by means of special machines—allow one to convert real currency into game currency in order to obtain, in a totally random manner, virtual contents for one's account. Among the most popular gacha of the moment is Genshin Impact by the Chinese miHoYo, which offers, in addition to the mechanics that have just

described, the possibility for players to start cooperative (co-op) matches in one of their own virtual worlds. Considering how this game periodically generates substantial revenue streams, it is possible to state that interest in this type of assets and realities (in their various forms and typologies) is far from waning. Very recently, the. Consumer Protection Cooperation Network (CPC Network) of the European Union has published its Key Principles on In-game Virtual Currencies (available at https://commission.europa.eu/document/8af13e88-6540-436c-b137-9853e7fe866a_en, last accessed on 2025/04/10) as a tool aimed at enhancing the protection of EU consumers engaging in in-game transactions. The document poses a strong focus on the protection of vulnerable consumers, who are frequently mislead by practices that «make the implications of the purchasing decisions less clear for them»: indeed, Principle 1 advocates for a clear and transparent price indication, meaning that«when in-game virtual currency or in-game digital content or services are offered for sale, their price in real-world money should be clearly and prominently displayed. When in-game virtual currency is offered in exchange for another in-game virtual currency which the users can buy with real-world money, its price should be indicated also in real-world money. When in-game digital content or services are offered in exchange for in-game virtual currency that can be bought (directly or indirectly via another in-game virtual currency), their price should also be indicated in real-world money». Similarly, Principle 3 discourages the practice of offering bundles of in-game virtual currencies that do not match the actual value of the purchasable item The videogame sector, despite being the one mainly associated with the idea of the metaverse, is not the only one concerned: the so-called digital twin, virtual reproductions and faithful replicas of material assets, the development of simulations and remote assistance techniques by means of AR technologies, and the provision of training courses in VR (which offer training without the use of physical prototypes) are all promising examples of applications of metaverse technologies, being able to reduce the costs and risks associated with their performance in the real world. Some of these, in particular, can also be used in non-industrial contexts, to help reconfigure spaces: think of the so-called digital twin smart city models, which consist of an extremely accurate virtual facsimile of the physical territory of real cities and thus make it possible to verify the implications of any transformative processes in the same, as well as to know in real time and manage the status of subsystems such as traffic lights, for example

45. The small size of screens and user interfaces, the lack of large-scale interoperability as well as the—sometimes—limited catchment areas of applications using VR and AR technologies represent important barriers to the full enjoyment of truly immersive experiences
46. The belief that these values should form the foundation of the development and conceptualization of new technological devices dates back to the 1990s and is part of the concept known as value sensitive design, which originated "in the field of ethics of technology and has a strong pragmatic focus on making ethics work in a world of technology. It aims to help designers to put social and moral values at the heart of the design of new technologies". European Parliament's Policy Department for Citizens' Rights and Constitutional Affairs, Metaverse… cit., p. 68

The Metaverse and Web 3.0: A Pathway Towards Value Generation

Guido Perboli, Andrea Borroni, Serena Pratali, and Chiara Vandoni

Abstract This chapter delves into the evolution from Web 2.0 to Web 3.0, with a particular emphasis on the Metaverse and associated technologies such as cryptocurrencies and blockchain. In a context marked by the burgeoning adoption of these terms within the digital landscape, the article endeavors to elucidate the distinctions between Web 2.0 and Web 3.0, underscoring how the latter embodies a paradigm predicated upon decentralization, privacy, and user empowerment. Through a meticulous technological analysis, it examines the business models and value-creation strategies inherent to Web 3.0. Furthermore, the article accentuates the necessity of a holistic approach that incorporates a robust legal framework, an aspect frequently overlooked in extant literature. Concrete use cases and managerial insights are presented, along with a contemplation of prospective future developments and the unresolved challenges that persist within the realms of Web 3.0 and the Metaverse. Moreover, a factual methodology for managing Metaverse use cases is presented, namely, the GUEST methodology. It is a structured framework designed to guide organizations in identifying use cases, articulating value propositions, and implementing Web 3.0 and Metaverse initiatives. By integrating business model innovation with technological feasibility, GUEST provides a roadmap for navigating the complexities of decentralized ecosystems, ensuring alignment with stakeholder objectives and regulatory requirements.

G. Perboli (✉)
CIRRELT, Montreal, Canada
e-mail: guido.perboli@polito.it

A. Borroni
Università Degli Studi Della Campania Luigi Vanvitelli, Caserta, Italy
e-mail: andrea.borroni@unicampania.it

G. Perboli · S. Pratali · C. Vandoni
Politecnico Di Torino, Turin, Italy
e-mail: s.pratali@studiolegalepratali.it

C. Vandoni
e-mail: chiara.vandoni@polito.it

A. Borroni et al. (eds.), *Metaverses: Reshaping Law, Economy, and Society*, PoliTO Springer Series, https://doi.org/10.1007/978-3-032-08664-8_3

Keywords Metaverse · Business models · Web 3.0 · Blockchain · NFTs · Guest methodology

1 Introduction

Over the past two decades, the internet has undergone a profound metamorphosis. From the read-only static pages of Web 1.0 to the participatory, user-centric platforms of Web 2.0, online experiences have evolved dramatically [1]. This shift is exemplified by social media and content-sharing networks, which enabled individuals to create and disseminate their own content, forging new digital communities and unlocking innovative monetization channels. Nonetheless, as these platforms matured, concerns about data ownership, privacy, and the monopolistic power of large tech companies became increasingly apparent [2].

In response, a new paradigm, Web 3.0, is emerging. Underpinned by decentralization, user empowerment, and robust data-protection mechanisms, Web 3.0 aspires to return control and ownership of digital identities and assets to end users. By leveraging blockchain technologies, tokenized economies, and trustless protocols, this paradigm aims to foster more equitable digital ecosystems, reducing reliance on traditional intermediaries [3]. Alongside this shift, the concept of the Metaverse—a network of immersive, persistent virtual worlds—has gained prominence. The Metaverse merges elements of virtual and augmented reality with distributed digital ownership, offering new opportunities for social interaction, collaboration, and commerce [4].

Together, Web 3.0 and the Metaverse promise a reimagined internet that transcends the boundaries of physical and virtual spheres. They raise critical questions for businesses regarding how to capture and create value in these emerging environments, while also highlighting the importance of comprehensive legal and regulatory frameworks [5]. Although some organizations have begun to incorporate Web 3.0 technologies, such as cryptocurrencies, blockchain-based supply chain solutions, and NFT-driven marketplaces, relatively few have systematically mapped out clear use cases, outlined robust business models, or addressed the management challenges that decentralization entails [2, 3]. Moreover, the scholarly discourse is frequently fragmented, focusing either on isolated technological features or on high-level theory without detailing how these innovations can be practically implemented in complex business ecosystems [4]. Moreover, a factual methodology for managing Metaverse use cases is presented, namely, the GUEST methodology. It is a structured framework designed to guide organizations in identifying use cases, articulating value propositions, and implementing Web 3.0 and Metaverse initiatives. By integrating business model innovation with technological feasibility, GUEST provides a roadmap for navigating the complexities of decentralized ecosystems, ensuring alignment with stakeholder objectives and regulatory requirements.

The present article delves into the evolution from Web 2.0 to Web 3.0, emphasizing how the latter unlocks novel avenues for user-centric value generation. It showcases how businesses can harness decentralized technologies, including blockchain and NFTs, to reimagine their offerings and engage in emerging virtual environments. Furthermore, it underscores the necessity of incorporating legal and regulatory considerations into strategic planning, an aspect frequently overlooked in contemporary literature. Through use cases illustrating real-world applications, the article demonstrates actionable pathways for implementation and highlights significant unresolved challenges, such as scalability and interoperability.

The remainder of this paper is organized as follows. Section 2 provides a detailed technological analysis, focusing on decentralization, blockchain, and other foundational elements of Web 3.0 ecosystems. Section 3 examines the main characteristics of Web 3.0 and the Metaverse. Section 4 presents use cases that illustrate how businesses can generate value within these decentralized platforms and how these business models and use cases can be efficiently introduced by a structured methodology as the GUEST methodology. Finally, Sect. 5 offers managerial insights and explores possible future developments.

2 Technological Analysis

This section provides a comprehensive analysis of the technological building blocks of Web 3.0, their relevance to the Metaverse, and the challenges they face in delivering large-scale, secure, and sustainable solutions.

2.1 Distributed Ledgers and Blockchain Infrastructure

Regardless of Web 3.0 lies the concept of a distributed ledger, where data is replicated across multiple nodes rather than stored and managed by a single central authority [6]. Distributed ledgers can take different forms, such as a permissioned private ledger maintained by a consortium of stakeholders or a fully public blockchain accessible to anyone with an internet connection [7]. Regardless of the specific ledger type, the overarching principle is that data records, including transactions, asset ownership, or smart contract states are securely validated through consensus mechanisms rather than by a single controlling entity [4].

Compared to conventional database architectures, these distributed protocols significantly heighten security. A perpetrator seeking to tamper with data must compromise the majority of nodes or alter all copies of the ledger concurrently; a far more formidable task than attacking a single database [8]. Moreover, by distributing trust among many independent participants, distributed ledgers can reduce or eliminate dependence on intermediaries such as clearinghouses or escrow services [9].

This approach not only reconfigures the dynamics of user ownership but also challenges established business models that rely on proprietary data control for profit [10].

Public versus Private Blockchains

Blockchains fall into two categories: public (permissionless) and private (permissioned) [3]. Public blockchains, such as Ethereum or Bitcoin, permit any participant to read and write to the ledger, with consensus typically achieved by proof-of-work (PoW) or proof-of-stake (PoS) algorithms [1, 10]. These infrastructures operate on public domain software, allowing anyone who downloads and executes the software to participate. In some cases, even the code is developed further in the public domain [11]. Their core advantage is trustlessness: no pre-existing relationships between participants are needed, as the consensus mechanism and cryptographic proofs verify the accuracy of data.

In contrast, private blockchains refer «to a type of ledger that has several limitations in its procedures, especially in relation to control» [11], restricting access to known participants [6]. Although private blockchains sacrifice some of the openness that distinguishes Web 3.0, they can facilitate more efficient consensus (given fewer nodes) and offer a higher degree of control over transaction throughput, user permissions, and regulatory compliance [12]. Hence, many enterprises experimenting with blockchain for supply chain management or record-keeping adopt private or consortium-ledger models, citing confidentiality and performance needs [4]. Nonetheless, a hybrid strategy, where data resides on a private ledger but is periodically anchored on a public chain, can achieve a balance between efficient operations and verifiable external trust [13].

Scalability and Environmental Impact

While blockchain technologies have attracted significant enthusiasm, they also contend with notable constraints. Chief among these is scalability: many public blockchains support far fewer transactions per second than established payment systems [4]. Although recent upgrades, such as Ethereum's transition toward proof-of-stake consensus, promise improvements, achieving the throughput required for mass adoption remains an open question [14]. Furthermore, environmental concerns stemming from energy-intensive proof-of-work algorithms persist, prompting both industry and academia to explore more eco-friendly consensus models [15]. The need to reconcile decentralization, security, and energy efficiency underlines the ongoing R&D that fuels the evolution of Web 3.0 infrastructure.

2.2 Tokens, Cryptocurrencies, and Non-Fungible Tokens (NFTs)

Tokens lie at the heart of Web 3.0 value flows. Cryptocurrencies such as Bitcoin or Ether exemplify the earliest and most prominent digital tokens, functioning as mediums of exchange and stores of value in decentralized ecosystems [1]. Beyond pure cryptocurrency use [16], so-called "utility tokens" grant holders access to DApp features or preferential services. A governance token, by contrast, enables voting rights in a protocol's future development or a DAO's treasury allocations. These innovations have spurred entire micro-economies wherein user engagement, development contributions, and loyalty can be incentivized through token rewards [8, 17].

Nonetheless, volatility in token prices raises concerns about risk exposure, especially for users entering the space primarily for practical use rather than speculative trading. Additionally, some regulators view certain token sales as akin to unregistered securities offerings, triggering legal scrutiny and compliance obligations. As token-based systems evolve, balancing decentralized innovation against responsible risk management remains a core challenge for startups and established enterprises alike.

Non-Fungible Tokens (NFTs) represent the ownership of a unique asset, distinguishing them from fungible cryptocurrencies. An NFT's metadata can point to a digital file (e.g., an artwork) or a physical object, while the token on the blockchain verifies its provenance and authenticity. Over the past several years, NFTs have garnered mainstream attention for digital art sales, collectible trading cards, and metaverse real estate [8].

Their applications, however, extend well beyond art and collectibles. In the supply chain, NFTs could be used to tie a physical good (for example, a luxury item or a scarce wine bottle) to a digital certificate of authenticity. This facilitates seamless resale on secondary markets while protecting consumers from counterfeits [4]. In sports or events, NFTs can serve as tokens for ticketing, guaranteeing that holders receive exclusive merchandise or meet-and-greet opportunities with organizers.

Yet these novel use cases also trigger debates about content hosting, gas fees, and the possibility of "rug pulls" in an unregulated NFT market. Although an NFT itself is recorded on a blockchain, the underlying digital file may be stored on off-chain protocols or centralized services, introducing reliability and longevity questions [6]. Projects aiming to deepen user trust often integrate decentralized storage solutions, IPFS (InterPlanetary File System), or cross-chain bridging to mitigate the single-point-of-failure vulnerabilities of external servers [18].

2.3 Smart Contracts and Automated Governance

Smart contracts are executable programs stored on a blockchain that automatically enforce the rules and terms set forth by their code [6].

In the original idea, the creator of smart contracts (Nick Szabo) argued that smart contracts should be rooted in real life, translating into legal terms (*rectius*, clauses of contracts) ordinary actions whose execution «would be allowed based on the occurrence of certain conditions: if the input conditions were met, the software would allow the smart contract to be executed» [11].

Unlike traditional digital agreements—which often rely on external enforcement—a smart contract operates autonomously once it is deployed and validated on the network. This autonomous nature can significantly reduce transaction costs by eliminating the need for human intermediaries or paper-based procedures. For instance, if a decentralized application (DApp) stipulates that a user must deposit a certain token balance to access a premium feature, the smart contract enforces this rule, verifying the user's balance before executing the feature [4].

Because smart contracts are stored on the blockchain, they inherit transparency, immutability, and traceability from the ledger itself [19]. Each function call, along with its associated data, is chronologically recorded as a transaction. Users, regulators, and stakeholders can thus audit the contract's entire operational history, enhancing trust in the system's fairness [7].

The utility of smart contracts extends well beyond simple financial interactions. In the realm of supply chain management, for example, sensors and internet-of-things (IoT) devices can feed data to smart contracts that release payment upon the verification of product delivery [4]. Decentralized exchanges (DEXs) allow users to swap tokens without depending on centralized custodians, relying instead on automated market makers coded into a set of smart contracts [3]. Non-Fungible Tokens (NFTs) also leverage smart contract logic to manage ownership transfers, royalty payments, and provenance tracking [11].

However, designing robust and secure smart contracts requires expertise in both software engineering and cryptographic methods [6]. Even a minor bug can open the door to large-scale exploits, such as re-entrancy attacks or price manipulation in decentralized finance (DeFi). This risk underscores the importance of code audits, formal verification, and standardization efforts. Moreover, legal scholars and policymakers grapple with the status of smart contracts under existing laws, particularly regarding liability and enforceability should a contract's code conflict with broader statutory obligations [20].

Building on the foundation of smart contracts, many Web 3.0 communities have formed Decentralized Autonomous Organizations (DAOs) [19]. These are collectives governed by token holders who vote on proposals, resource allocation, and community rules. By distributing decision-making power among stakeholders, DAOs aim to minimize top-down control and foster more democratic ecosystems. In practice, DAOs have been used for open-source development, venture capital pooling, and community-driven projects, though questions persist about their real-world legal standing and capacity for effective governance [11]. Not all DAO decisions align with regulatory frameworks, and controversies have arisen when tokens equate to near-corporate equity but lack standard investor protections [21].

2.4 Interaction with the Metaverse

The synergy between Web 3.0 and the Metaverse arises from their shared emphasis on decentralization and user empowerment. In a fully realized metaverse, individuals require robust mechanisms for personal identity, property rights, and governance to facilitate trust-minimized interactions in immersive virtual worlds. Web 3.0 technologies fulfill this role by providing NFT-based proofs of ownership for land parcels, avatars, or digital clothing, while enabling direct peer-to-peer transactions using cryptocurrencies or stablecoins. Additionally, DAOs can manage virtual communities, overseeing development resources, shared in-world assets, and event planning [12].

A critical element of a thriving metaverse involves true interoperability—where assets or identity credentials can move seamlessly across different virtual environments. This requires standardized protocols that define how NFTs or tokens function and how user profiles and transaction histories are managed [6]. The complexities of bridging blockchains, storing user data in a decentralized yet privacy-preserving manner, and harmonizing dozens of competing platforms underscore that the metaverse remains in its infancy [11].

Yet, even partial interoperability fosters new economic opportunities. Virtual land or digital objects purchased in one game could be ported to another, or a user's metaverse avatar could maintain a consistent identity across multiple digital realms. As these systems expand, the role of cross-chain solutions, such as sidechains, layer-2 protocols, and bridging smart contracts, will become increasingly central to bridging various virtual ecosystems [22].

2.5 Legal and Regulatory Considerations

While the technological innovation around tokens and smart contracts has advanced rapidly, legal clarity has often lagged. Multiple jurisdictions treat certain tokens as securities if they confer profit rights, effectively subjecting them to capital market regulations. Others classify tokens as commodities, currencies, or intangible assets, depending on the underlying function. The lack of uniform standards leads to regulatory fragmentation that can hamper cross-border DeFi applications and hamper global adoption [23].

Efforts to bring clarity are evident in some legislative initiatives that set guidelines for stablecoins, licensing frameworks for exchanges, and consumer protection mandates in crypto markets [4]. Still, projects spanning multiple jurisdictions must navigate complex and sometimes contradictory rules regarding anti-money-laundering (AML) and "Know Your Customer" (KYC) protocols, taxation, and consumer rights [6]. The challenge for Web 3.0 and Metaverse developers involves structuring their operations to remain compliant without undermining the decentralized ethos that underpins their value proposition [12].

Another crucial legal dimension concerns digital identity and data privacy, particularly in immersive metaverse environments that collect vast amounts of user-generated content. It would be, in some cases, an oxymoron to support the concept of privacy in meta-reality. In fact, in the metaverse, it will be possible to envisage the protection of personal data, but not a full and complete protection of privacy—at least in the current state of development of technology—understood in the original sense of the term, as the right to be let alone: the problem that arises «is no longer how to have access to the personal data of the inhabitants of this reality, but only how to manage their treatment» [24].

Even though decentralized identity solutions aim to minimize personal data exposure, certain user actions—like NFT purchases or on-chain voting—are recorded publicly, potentially conflicting with privacy regulations such as the EU's General Data Protection Regulation (GDPR) [23]. Reconciling the transparency of blockchain with the right to data erasure or modification has become a focal point for policy-makers, cryptographers, and platform architects, specifically in relation to Art. 17 of the GDPR that stipulates that the data subject has the right to obtain from the controller the deletion of their personal data without undue delay [25].

In any digital ecosystem, protecting intellectual property (IP) is challenging; in a decentralized metaverse, this complexity multiplies. NFTs minted as representations of digital art must confirm not only the authenticity of the token but also the rightful ownership of the underlying content [2]. IP infringements can be perpetuated by malicious actors who mint tokens tied to stolen or copied artwork. While some NFT marketplaces implement copyright-verification processes, these are often limited in scope.

Furthermore, the boundary between user expression and illegal content becomes murkier in a distributed system lacking a central authority. DAO-based communities might vote to remove certain content or ban a user, but with no unified global standard on censorship or content moderation, such efforts remain ad hoc and context-specific [19]. These issues underline that the evolution of Web 3.0 and the Metaverse is not purely technological, it also demands consensus on governance principles, cultural norms, and robust enforcement mechanisms that do not undermine decentralization.

2.6 *Toward Scalable, Secure, and User-Friendly Architectures*

Addressing scalability and high transaction fees is central to the sustainable growth of blockchain-based platforms [26].

Layer-2 solutions—like rollups or sidechains—bundle transactions off-chain and settle them periodically on the main blockchain, reducing congestion. These approaches also enhance user experience by accelerating confirmation times and curbing gas costs. For Metaverse environments that demand near-instant interactivity, advanced layer-2 frameworks can facilitate real-time gameplay, dynamic NFT creation, and in-world micropayments.

As different blockchains specialize in particular functionalities—ranging from high-speed transactions to privacy-focused operations—interoperability protocols become indispensable [11]. Cross-chain bridges enable token and data transfers between distinct networks but also introduce technical and security complexities. A single point of failure in a bridging contract could jeopardize the liquidity of multiple networks, underscoring the necessity for rigorous audits and multi-layer safeguards. Additionally, bridging solutions must align user experiences by allowing consistent asset naming, metadata standards, and user interfaces to reduce friction.

Although Web 3.0 technologies can, in theory, increase user sovereignty, they remain daunting for individuals unaccustomed to managing private keys, paying gas fees, or comprehending transaction confirmations. The complexity of crypto wallets, the risk of irreversible errors, and the unfamiliar language of cryptography can deter wide-scale adoption. Human-centered design principles—intuitive user interfaces, clear signposting, integrated educational modules, and fail-safe transaction flows—are therefore crucial to bridging the gap between technological potential and mainstream acceptance.

Moreover, ensuring accessibility for users with disabilities becomes even more urgent in 3D virtual environments. Screen readers or alternative input devices must be smoothly integrated so that Metaverse does not replicate or exacerbate digital divides. Projects that invest in inclusive design and robust documentation can differentiate themselves and attract broader, more diverse user communities, ultimately reinforcing the decentralized ethos of open participation.

2.7 *Outlook and Convergence with Other Technologies*

Web 3.0's potential for radical disruption will likely intensify as it converges with other fast-evolving domains, notably artificial intelligence (AI) and edge computing [19]. AI-driven smart contracts may autonomously optimize pricing in real time based on demand signals. Decentralized AI marketplaces might allow users to sell or collectively govern data sets and machine-learning models [4]. Meanwhile, edge computing solutions—where data processing occurs closer to end users—can reduce latency for VR/AR applications critical to a rich Metaverse experience [7].

However, these integrations also raise novel challenges. Algorithmic transparency and accountability remain debated even in centralized contexts, and decentralization can compound the difficulty of regulating biased AI outputs [3]. Additionally, reliance on hardware at the network edge complicates upgrade cycles, given that new cryptographic standards, consensus algorithms, or zero-knowledge proofs might require specialized computing [6]. In short, the intersection of Web 3.0, AI, and edge computing promises both synergy and increased technical complexity.

3 Web 3.0 and the Metaverse

Web 3.0 is a pivotal shift in how the internet functions, primarily marked by the emergence of decentralization, user ownership of data, and a growing reliance on blockchain-enabled mechanisms [1, 3]. In contrast to Web 2.0, where dominant platforms leverage user-generated content and data for profit, Web 3.0 aspires to distribute power and economic benefits more evenly among participants. This evolution is propelled by enabling technologies and governance models such as cryptocurrencies, Non-Fungible Tokens (NFTs), Decentralized Autonomous Organizations (DAOs), and peer-to-peer networks [4]. Taken together, they form an ecosystem poised to reshape digital interactions and create new pathways for economic value creation [12].

3.1 Transition from Web 2.0 to Web 3.0

The rise of social media, streaming platforms, and content-sharing sites under Web 2.0 demonstrated the immense potential of user-generated content and global connectivity. However, this user-centric revolution came with drawbacks, notably the concentration of power in large, centralized entities that collect and monetize personal data [2]. Web 3.0 seeks to address these concerns through a paradigm predicated on decentralization and self-sovereign identity. In practical terms, blockchain technology provides tamper-evident ledgers, and smart contracts automate trust-based transactions, while user-owned tokens furnish participants with unprecedented levels of control and transparency [1, 4].

Decentralization and New Business Models

The ethos of decentralization is evident in the decentralized applications (DApps) that constitute the backbone of Web 3.0. Here, blockchains replace centralized servers, and automated smart contracts can govern interactions without the need for intermediaries. Novel business models have arisen: decentralized finance (DeFi) solutions enable peer-to-peer lending, remittances, and asset exchanges, bypassing legacy institutions [7]. Similarly, DAOs empower communities to make collective decisions about project funding, governance, and resource allocation [19].

Furthermore, tokens, both fungible and non-fungible, underpin many of these models. Fungible tokens can serve as the native currency of a decentralized platform or as an incentive mechanism that rewards user contributions. NFTs represent unique assets, providing verifiable authenticity and ownership of digital (or digitized) creations such as art, music, or real estate deeds. In this environment, the concept of user-centricity transitions from a buzzword to a tangible economic reality, where individuals can hold and trade digital assets without ceding authority to centralized gatekeepers [11].

Despite its promise, Web 3.0 adoption remains in an early, formative stage. Scalability is a critical obstacle: most blockchain infrastructures struggle to match the transaction throughput demanded by mainstream consumer applications. Regulatory uncertainties add further complexity; nation-states and international bodies grapple with how to classify and govern cryptocurrencies, token-driven ecosystems, and transnational data flows. Moreover, the reliance on cryptographic protocols can lead to security vulnerabilities if implementations are rushed or poorly vetted. Consequently, organizations aiming to integrate Web 3.0 tools must balance their technical ambitions with robust legal and risk-management frameworks [1, 4].

3.2 *Metaverse: Convergence of Virtual Worlds*

Alongside the rise of Web 3.0, the Metaverse represents a parallel evolution: a persistent, immersive, and interconnected domain of virtual environments [20]. Though early virtual worlds or online games have existed for decades, today's Metaverse extends far beyond a single platform or developer ecosystem. Instead, it seeks to unite multiple 3D worlds and real-time interactive experiences, enabling users to seamlessly move between entertainment, education, commerce, and social activities. By tapping into blockchain-based ownership and interoperable standards, the Metaverse aims to replicate—and sometimes improve upon—the physical world's commerce, social structures, and cultural institutions [3, 15].

Technological Foundations

Metaverse integrates VR, AR, and mixed reality experiences with distributed ledger technology. Virtual land can be represented and sold as NFTs in projects like Decentraland or The Sandbox, and ownership transfers are recorded on the blockchain. Cryptocurrencies facilitate in-world transactions, whether for digital art, avatar upgrades, or real-estate leasing. Smart contracts provide trustless mechanisms that automatically execute terms agreed upon by participants, such as royalty payments to creators each time a virtual collectible is resold. Additionally, 5G and edge computing solutions can boost the real-time interactivity and scalability needed to maintain large, persistent digital worlds.

From a user standpoint, wallets and identity solutions form critical building blocks. Instead of logging in through centralized credentials, individuals can create decentralized identities tied to cryptographic keys. This approach ensures that digital possessions and in-world reputations remain within the user's control rather than being controlled by a single corporate platform [22].

Metanomics: the economy of the Metaverse

"Metanomics" refers to the evolving virtual economy that thrives at the intersection of Web 3.0 and the Metaverse [20]. Unlike traditional e-commerce, wherein intermediaries hold user data, these new ecosystems prioritize user-generated content, tokenized interactions, and direct user-to-user transfers of goods or value.

Researchers suggest that the Metaverse has the potential to transform industries ranging from tourism to healthcare, leveraging immersive simulations to create compelling user experiences [27, 28]. A tourist, for example, could virtually explore a historical landmark recreated in 3D and later purchase NFT-based souvenirs that serve as proof of having visited the digital replica [20]. Similarly, an immersive healthcare application might use digital twins of medical devices or hospital layouts to improve diagnostic accuracy or facilitate remote treatment [29].

Moreover, luxury brands and entertainment companies are increasingly experimenting with digital collectibles, partnerships, and branded experiences in the Metaverse [1]. These approaches can open new revenue streams based on digital scarcity, exclusive membership programs, or even play-to-earn models in the gaming sector [15]. Yet the question of how best to govern these economies remains open—stakeholders must define fair policies on taxation, intellectual property rights, and content moderation within decentralized virtual spaces [3].

3.3 Interplay and Remaining Challenges

The convergence of Web 3.0 and the Metaverse amplifies both opportunities and complexities [12]. On one hand, decentralization aligns naturally with immersive virtual worlds, enabling user governance and secure asset ownership [30]. On the other hand, the overlapping technological stacks require sophisticated integration and cross-platform interoperability. For instance, bridging a user's NFT collection from one virtual world to another demands standardized protocols and consensus on metadata. The potential for unauthorized duplication of digital assets or exploitation of privacy vulnerabilities further highlights the need for robust cybersecurity and legal frameworks [6].

In addition, user adoption of fully decentralized metaverse environments remains uncertain. While certain segments of the population may be drawn to the ideals of digital sovereignty, others may prefer the convenience or familiarity of centralized, "Web 2.5"–style experiences. Consequently, it is not guaranteed that every Metaverse instance will share the same decentralized ethos that has defined blockchain communities so far. Many early adopters believe that sustained user engagement will hinge on tangible value propositions—more engaging gaming experiences, real ownership of assets, or genuine opportunities for content creators to earn income. Overcoming the novelty factor alone requires a confluence of user-friendly interfaces, seamless onboarding, and mainstream awareness about the benefits of decentralized virtual worlds [12].

Lastly, the regulatory environment for Metaverse projects remains uneven. In some jurisdictions, tokenized assets fall under securities laws, whereas in others, they occupy an unregulated gray area. Questions about digital consumer protection, data governance, and intellectual property infringement add to the complexity. Achieving large-scale acceptance will likely involve finding equilibrium between grassroots innovation and policy frameworks that safeguard participants from fraud, privacy breaches, and algorithmic manipulation.

4 Use Cases and Value Generation

The metaverse represents a multifaceted digital ecosystem where immersive virtual worlds, decentralized technologies, and user-centric governance intersect [3, 15]. Although the concept initially gained traction in entertainment and gaming, its applicability has rapidly extended to numerous sectors—including healthcare [28], education, and real estate (HLB, 2023)—offering new avenues for creating, delivering, and capturing value. This section explores key metaverse use cases and the mechanisms through which they generate tangible benefits for individuals, businesses, and broader communities.

4.1 Metaverse and Web 3.0 Use Cases

Below, we explore some of the most prominent use cases of the Metaverse and the transformative value they bring to various sectors. From virtual real estate and education to entertainment, healthcare, and commerce, the Metaverse is unlocking new possibilities for businesses, individuals, and society at large. Each use case highlights the potential of this technology to revolutionize industries, enhance user experiences, and drive economic growth. As we delve into these applications, it becomes clear that the Metaverse is not just a technological advancement but a paradigm shift with far-reaching implications for the future.

Virtual Real Estate and Architecture

One of the most prominent use cases of the Metaverse is in the realm of virtual real estate and architecture. Companies and individuals can purchase, develop, and monetize virtual land, creating immersive environments for various purposes such as virtual offices, retail spaces, and entertainment venues. The value generation here is twofold: firstly, it provides a new revenue stream for developers and investors, and secondly, it offers businesses a novel way to engage with customers. For instance, virtual real estate platforms like Decentraland and The Sandbox have seen significant transactions, with some virtual plots selling for millions of dollars [31]. The architectural design within the Metaverse also allows for innovative building techniques and materials that are not constrained by physical laws, enabling architects to experiment with new forms and structures [32].

Education and Training

The Metaverse has the potential to revolutionize education and training by providing immersive, interactive learning environments. Virtual classrooms and training simulations can offer students and professionals hands-on experience in a controlled, risk-free setting. For example, medical students can practice surgeries in a virtual operating room, while engineers can simulate complex machinery operations. This use case generates value by enhancing the quality of education and training, reducing costs associated with physical resources, and improving learning outcomes [33]. Moreover, the Metaverse can democratize access to education by making high-quality learning experiences available to a global audience, regardless of geographical location [34].

Entertainment and Social Interaction

Entertainment is a significant use case for the Metaverse, offering new ways for people to consume media, attend events, and socialize. Virtual concerts, movie theaters, and gaming environments are increasingly popular, providing users with immersive experiences that surpass traditional forms of entertainment. The value generation here lies in creating new revenue streams for content creators and enhancing user engagement. For instance, virtual concerts hosted by artists like Travis Scott in Fortnite have attracted millions of attendees, showcasing the potential for large-scale virtual events. Additionally, a special use case is characterized by tourism and museum ecosystems, which can benefit from integrating the Metaverse and Web 3.0 to improve user experience and create new business opportunities [35].

Healthcare and Therapy

The Metaverse also holds promise for healthcare and therapy, offering innovative solutions for patient care and treatment. Virtual reality (VR) environments can be used for pain management, physical rehabilitation, and mental health therapy. For example, VR has been shown to effectively treat conditions such as PTSD, anxiety, and phobias by providing controlled exposure therapy in a safe environment [36]. The value generation in this use case is evident in improved patient outcomes, reduced healthcare costs, and the potential for remote treatment delivery. Furthermore, the Metaverse can facilitate telemedicine, allowing healthcare providers to offer consultations and treatments to patients in remote or underserved areas [37].

Commerce and Retail

The Metaverse is poised to transform commerce and retail by creating virtual marketplaces where users can browse, purchase, and interact with products in a fully immersive environment. Virtual stores and showrooms can offer a more engaging shopping experience, while digital twins of physical products can allow customers to visualize and customize items before making a purchase. The value generation here includes increased sales, enhanced customer satisfaction, and the ability to reach a global audience. For example, brands like Gucci and Nike have already established a presence in the Metaverse, offering virtual goods and experiences that complement their physical products [38]. Additionally, the use of blockchain technology in the Metaverse can enable secure and transparent transactions, further enhancing the trust and efficiency of virtual commerce [39].

The Metaverse offers a wide range of use cases that have the potential to generate significant economic and social value. From virtual real estate and education to entertainment, healthcare, and commerce, the Metaverse is poised to transform various industries and create new opportunities for innovation and growth. As the technology continues to evolve, it will be crucial for businesses, policymakers, and society at large to understand and harness the potential of the Metaverse to maximize its benefits and address any challenges that may arise.

4.2 Use Cases, Value Generation and Business Development: The GUEST Methodology

Efforts to harness Web 3.0 and the Metaverse for tangible value creation frequently encounter a crucial gap: although theoretical principles and enabling technologies have been discussed extensively, few practical frameworks systematically guide how organizations can identify their key stakeholders, articulate their strategic objectives, and devise comprehensive implementation plans. In particular, there is a notable gap in scholarly examinations of how these trends intersect with established business models and theories of value generation. To address this gap, we anchor our framework in foundational research that defines business models as the architecture for creating, delivering, and capturing value [40, 41]. Within this context, emerging technologies, whether Web 3.0 platforms, decentralized applications, or NFTs, do not inherently yield economic or social value. Instead, value arises from the interplay among organizational capabilities, collaborative ecosystems, and market adoption [42].

Against this theoretical backdrop, we aim to propose a structured roadmap that organizations can adopt at the outset of their Metaverse initiatives. Metaverse remains in an early phase of development, provoking legitimate questions about whether a deep transformation will ultimately occur. Precisely due to this nascent stage, we argue that a standardized roadmap can mitigate hype-driven or speculative implementations, as has been demonstrated for other emerging technologies [3]. By guiding stakeholders to articulate value propositions, outline stakeholder relationships, and evaluate both organizational and technological readiness, our approach is grounded in established business model principles (e.g., a clear focus on value creation and capture) to ensure feasibility and relevance. A standardized methodology also supports comparability and iterative learning across multiple pilots, in line with business model research that emphasizes cross-case analysis to identify drivers and barriers to value generation.

One of the leading methodologies for digital transformation use cases is the GUEST methodology. It has been developed by a team of researchers at Politecnico di Torino [43, 44] as an operational framework aiming to provide a user-friendly process applicable throughout decision-making stages [45, 46]. The GUEST methodology is based on two primary theoretical streams: digital transformation frameworks concerning technology adoption [42, 47] and socio-technical system theories related to decentralization and blockchain [4]. GUEST has already been successfully utilized in sectors such as automotive, information technology, and telecommunications, aiding firms in defining clear value propositions, stakeholder roles, and implementation strategies [48–51].

Therefore, our contribution extends beyond one-time consultancy, enhancing the scholarly debate on how emerging technologies are reshaped by evolving business model paradigms. In aligning our roadmap with established research, we draw on insights from business model innovation [40], value co-creation [42], emerging digital ecosystems [52], and business model failure analysis [53]. While business models have historically been framed as systems for generating, delivering, and capturing value [41], digital settings increasingly envision value creation as a collaborative, ecosystem-focused process. Building on these perspectives, we present the GUEST methodology as a structured guide for organizations grappling with the initial complexities of Web 3.0 and metaverse ventures.

We begin with the baseline GUEST model, originally developed for blockchain-specific applications [4]. Through involvement in various European projects spanning automotive, IT, hardware manufacturing, and telecommunications, we introduced GUEST to over 150 experts and practitioners [48–51]. These stakeholders were asked whether GUEST facilitated business modelling, exploitation planning, and the setting of tangible objectives for each use case. Feedback indicated that GUEST enabled rapid articulation of value propositions, development of coherent business models, and clear action planning within just a few weeks. Participants also reported that seminar-style workshops and flipped-classroom activities helped them converge on an executive project plan in under two months. Such favourable responses validate our assertion that GUEST is accessible and well-suited to early-stage technology adoption. Across all cited initiatives, GUEST served as the principal method for exploitation, business modelling, and defining use cases.

The GUEST methodology controls the process from the original idea to its implementation, providing conceptual and practical tools to the different actors involved, enabling them to communicate their vision, difficulties, and opportunities within the same structure. The methodology is articulated in five steps:

- GO—Actor Identification, Roles, and Gains;
- UNIFORM—Consolidating the System Description and Sketching Out Use Cases;
- EVALUATE—Technology Assessment and Feasibility Analysis;
- SOLVE—Building the Execution Plan (Solution Canvas);
- TEST—Piloting, Monitoring, and Benchmarking Results.

Each step allows the actors to monitor their projects and grants standardization of documents and tools that should be used to evaluate ideas, successes, actions, and results. Although each phase serves distinct goals, they interlock to form a continuous feedback loop [4]. In practice, the methodology does not always proceed in a strictly linear fashion, certain steps may be revisited if new information or requirements emerge. This iterative characteristic mirrors the fluidity of Web 3.0 ecosystems, where emergent technologies, changing community sentiments, and evolving regulatory conditions can significantly alter project trajectories.

When applied to a Metaverse use case, the GUEST is described as follows.

Phase 1: GO—Setting the Vision and Identifying Actors

In the GO phase, organizations define the overarching vision of their decentralized solution and identify all key actors who influence or are influenced by the project [4]. Actors can be diverse: they might include end users, governance token holders, NFT creators, or node operators who provide network validation. This initial mapping is vital because Web 3.0 ecosystems typically operate more openly than legacy systems, incorporating communities, open-source developers, and independent validators.

The methodology's hallmark tool, the Value Ring, is commonly employed here, as a graphical tool that promptly illustrates the genuine needs of the actors. By clearly delineating these aspects, the Value Ring helps managers and developers pinpoint areas where value can be enhanced and where strategic interventions are needed. This integrated overview facilitates more informed decision-making, ensuring that every stakeholder's interests are aligned with the overall value generation strategy of the platform (see Fig. 1).

Phase 2: UNIFORM—Consolidating the System View

The UNIFORM phase transforms scattered ideas into a unified system blueprint. Put differently, a wide range of prospective functionalities—like NFT-based certificates, tokenized reward systems, or user profiling—come together in a coherent system architecture [4, 15]. Within GUEST, this phase addresses the "how" behind the project's goals.

The project team enumerates high-level features (e.g., a fungible token used as the system currency, a user interface for experience customization, or a set of backend services to manage KYC procedures).

The methodology encourages stakeholders to outline key transactions [2]. For instance, if a user purchases an NFT representing a bottle of wine, which smart contract is triggered, how ownership is recorded, and how the user claims a physical product? Making these steps explicit helps avoid logical gaps and ensures consistent user experiences across the platform.

A frequent design question is deciding which data resides on the blockchain (immutability at the cost of potential overhead) and which remains off-chain (faster access, potentially less trust) [7]. The UNIFORM phase systematically addresses this trade-off.

In many real-world Web 3.0 pilots, teams discover that certain processes, initially presumed on-chain, may be more efficiently handled off-chain with periodic state

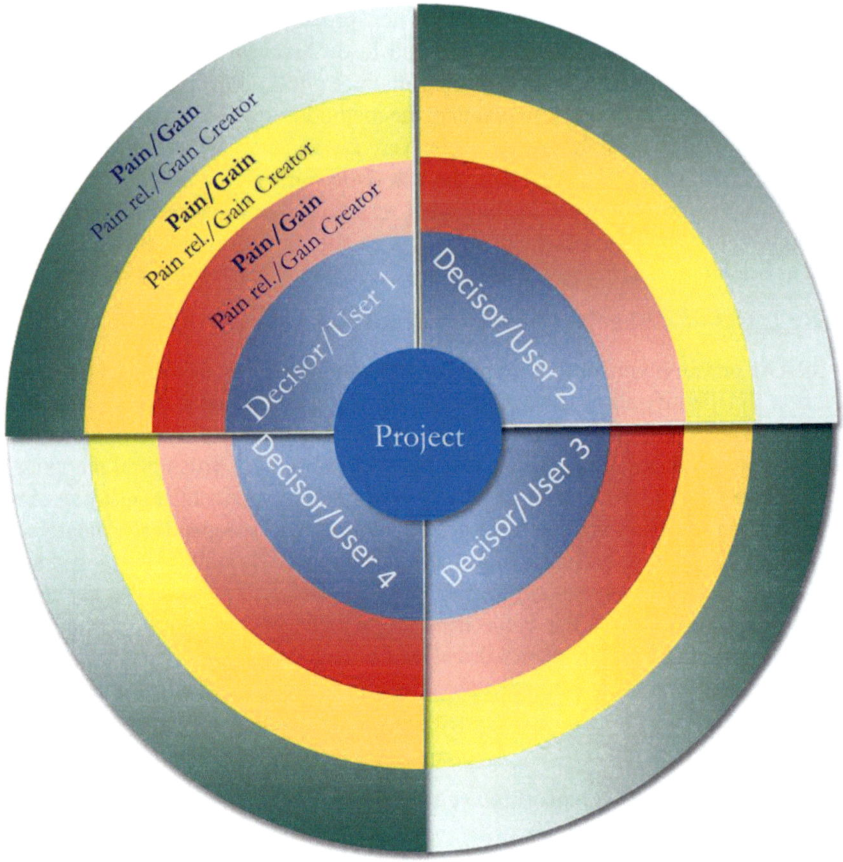

Fig. 1 Value ring

hashing on a public blockchain [6]. The GUEST framework's structure mitigates confusion by promoting clarity about which design decisions optimize value generation for all stakeholders.

Phase 3: EVALUATE—Technical Feasibility and Risk Assessment

Once the system blueprint has taken shape, teams proceed to the EVALUATE phase. This is where technological choices are rigorously tested against scalability, security, usability, and compliance criteria. The EVALUATE phase commonly includes:

- Platform Selection: Public or private blockchain? Ethereum or another chain with higher throughput? The GUEST approach prompts organizations to compare candidate solutions side by side, analyzing compatibility with project requirements and the ecosystem's maturity.
- Consensus Mechanisms: For instance, a project needing extremely rapid transactions might favor delegated proof-of-stake (DPoS), while a supply-chain pilot that

values broad trustlessness might opt for Ethereum's proof-of-stake environment [7]. GUEST encourages teams to weigh throughput, transaction costs, security, and even environmental impact, aligning these metrics with the 'gains' identified earlier.

- Legal and Regulatory Analysis: Web 3.0 solutions often face legal uncertainties around token classification (security versus utility), data privacy laws (e.g., GDPR or local equivalents), and potential liabilities for decentralized governance structures [20]. By incorporating these dimensions, the EVALUATE step provides a holistic risk map, suggesting compliance strategies and clarifying lines of responsibility.
- Human-Centric Design and UI/UX: The evaluation is not purely technical. A Web 3.0 solution may falter if its user onboarding process is complex or if wallet integrations alienate non-expert users [15]. Therefore, GUEST emphasizes assessing user interface prototypes, collecting feedback from pilot testers or focus groups, and ensuring the system can accommodate a broad user base.

Phase 4: SOLVE—The Execution Plan via the Solution Canvas

Following a successful feasibility analysis, the SOLVE phase converts the project blueprint into an actionable roadmap [4]. This stage is centered around the Solution Canvas, a condensed, visual template that captures key decisions, constraints, resources, channels, and performance objectives (see Fig. 2):

- Decisions and Constraints: Organizations detail the technical, operational, or regulatory choices that structure the entire platform. For instance, they must decide whether to issue a native utility token or to rely solely on an existing cryptocurrency. Constraints often relate to the limited throughput of a chosen blockchain or the legal requirements for user identification [3].
- Information and Resources: Teams specify the data sets, cryptographic libraries, or specialized developer skill sets required to implement each solution component [6]. The GUEST methodology underscores that resource planning should be accompanied by cost analyses, ensuring that potential economic benefits justify the project's complexity [19].
- Users and Channels: A distinct feature of GUEST's Solution Canvas is the clarity in how users learn about and interact with the system [4]. This channel definition reduces friction points. For example, if end users discover and purchase NFTs on a third-party marketplace, the solution plan must define how that marketplace integrates with the project's blockchain back end.
- KPI Specification: Teams identify Key Performance Indicators (KPIs) that represent progress in user adoption, transaction volume, or cost reduction. The GUEST methodology encourages each KPI to be traceable to a stakeholder's "gain" or a solution objective, fostering accountability [2].

By consolidating these elements into a single, consistently updated artifact, the Solution Canvas streamlines cross-functional collaboration. It helps the team pivot swiftly when unexpected constraints emerge, such as changes in transaction fees,

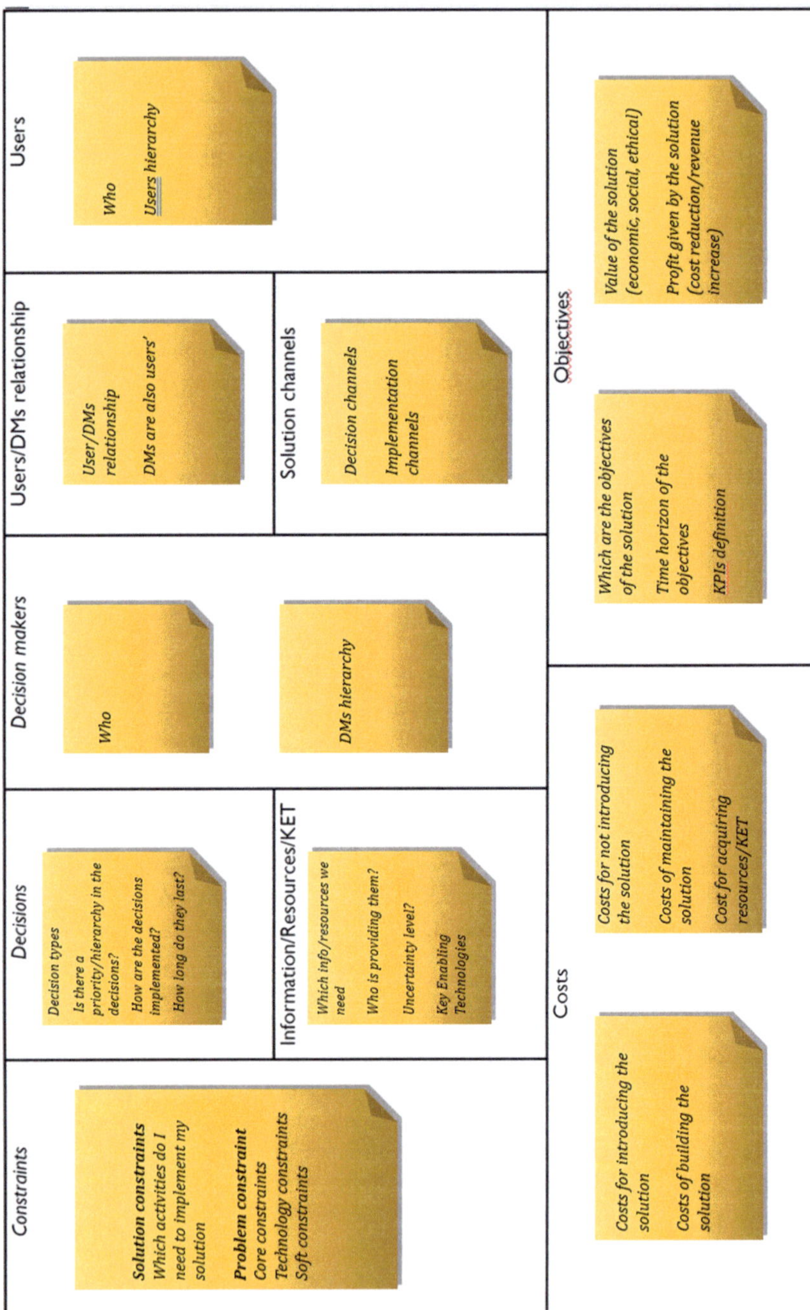

Fig. 2 Solution canvas

newly enacted regulations, or shifting user preferences. Compared to tools such as Lean Canvas and Business Model Canvas, it provides a more practical vision of the use case and makes it easier to develop a proper action plan.

To effectively implement the methodology, the initial step involves defining the various actors engaged in the process. This entails identifying the specific gains each actor seeks and the pains associated with their tasks. Subsequently, prioritizing the gains and pains for each actor allows for highlighting the most critical or urgent ones. This prioritization is visualized using the Value Ring.

Then, the Solution Canvas is introduced. It is a concise, visual framework that captures and organizes all key aspects of a solution in one place. It details critical decisions, constraints, required resources, stakeholder roles, interaction channels, and performance metrics, helping teams align their vision and strategies while ensuring that every component of the solution supports value generation and operational feasibility. Compared to other tools such as Lean Canvas and Business Model Canvas, it provides a more practical vision of the use case and makes it easier to develop a proper action plan (see Fig. 2).

Phase 5: TEST—Monitoring the Results, Refinement, and Scaling

Although the final step in GUEST is titled TEST, it arguably initiates a new cycle of iteration, bridging project pilots to full-scale deployment:

- Pilot Implementation: A limited rollout with a subset of users or network nodes reduces exposure to large-scale failure. This pilot helps identify friction in user experiences, unexpected legal hurdles, or insufficient network capacity.
- KPI Tracking: The project's success metrics—adoption rates, transaction frequency, average handling time, and operational cost savings—are gathered from on-chain analytics or integrated dashboards. This evidence-based approach highlights which features yield the highest returns and which require retooling.
- Scalability and Migration Path: Based on pilot performance, teams may decide to expand node count, switch to another blockchain, or shift certain off-chain components on-chain for heightened transparency. The GUEST framework assists by retaining the structured solution blueprint, enabling a strategic pivot without losing alignment with stakeholder objectives.
- Community Engagement: Web 3.0 and Metaverse solutions often rely on community input for refining token economics and platform governance. In the TEST phase, user feedback loops guide the next series of improvements.

4.3 *How GUEST Facilitates Value Generation*

Bridging Tech and Business Model Innovation

One of the challenges with emerging technologies like NFTs, DAOs, or DeFi lies in ensuring they serve a clear, validated use case. GUEST stands out by embedding each design choice, such as introducing token rewards or verifying transactions on-chain, into a broader business rationale. The methodology requires that every introduced feature solve a specific stakeholder problem or exploit a demonstrated market opportunity [6].

The focus here is to transform what might otherwise be an abstract pilot into a viable proposition for value generation. Rather than sprinkling tokenomics arbitrarily onto existing workflows, organizations following GUEST systematically design how tokens influence user behavior, how decentralized ledgers reduce costs, and how smart contracts can govern intangible assets. In turn, the final deliverable is more likely to resonate with end users, investors, and regulatory bodies.

Achieving Sustainable Network Effects

Web 2.0 success stories (e.g., major social media platforms) thrived on network effects, but also fostered monopolistic data ownership. Web 3.0 reimagines network effects by emphasizing shared governance and value distribution. Through GUEST, participants can identify 'choke points' such as platform fees or data integration friction and design token-based incentives to encourage collaboration.

Building Reputational and Regulatory Confidence

No matter the project's aspirations, regulatory scrutiny is inevitable if tokens or large user communities are involved. By establishing a formal approach to stakeholder identification, risk assessment, and solution documentation, GUEST outputs can serve as reference material for external audits, investor due diligence, and compliance checks. Particularly in regulated industries—finance, healthcare, or supply chain tracking of food and pharmaceuticals—this methodical approach to design fosters confidence among regulators and market participants alike.

Furthermore, by specifying how personal data, payments, or user identities are managed, GUEST ensures that privacy considerations are embedded in the architectural design rather than treated as afterthoughts. In this respect, the methodology supports "privacy by design," aligning well with established data-protection regulations.

5 Managerial Insights and Future Developments

This section introduces a holistic managerial perspective that addresses both the disruptive challenges and strategic opportunities presented by decentralized technologies. It underscores how shifting from centralized control to distributed governance requires new leadership skills and adaptive organizational structures. Managers who can balance these evolving dynamics with operational efficiency and regulatory compliance will be best positioned to drive innovation and unlock new avenues for value co-creation across their digital ecosystems.

5.1 Toward a Holistic Managerial Perspective

From blockchain-based supply chains to fully immersive Metaverse experiences, the evolving Web 3.0 landscape presents an array of strategic challenges and opportunities for managers. The decentralized ethos underpinning these technologies forces leaders to reevaluate traditional decision-making processes, risk management, and organizational structures. In decentralized networks, functions such as user authentication, trust-building, or financial transactions must be distributed across multiple actors in the ecosystem. Consequently, top-down control may recede in importance, replaced by governance systems that rely on smart contracts and token-based incentives [4, 6].

This transformation, although disruptive, can yield strategic advantages. By redistributing ownership of data and assets, organizations may unlock new forms of innovation and value co-creation with customers, suppliers, and even competitors. For instance, a brand that tokenizes its loyalty program enables consumers to trade or gift points, turning what was a closed, top-down system into a user-driven, liquid market for brand engagement. However, implementing such changes requires managerial foresight in shaping an ecosystem that balances user autonomy with a consistent, secure, and user-friendly experience.

5.2 Managerial Themes for Web 3.0 and the Metaverse

Governance and Organizational Restructuring

In decentralized contexts, governance often moves beyond traditional hierarchical schemes to include mechanisms like token-based voting and smart contract-driven decision-making [6]. For managers, adopting these models can be both empowering and disorienting. On one hand, distributing governance among employees, customers, or community members taps into the collective intelligence of stakeholders. On the other hand, establishing effective protocols to handle disputes, define voting quorums, and enforce accountability can be elusive when no single authority holds ultimate sway. Leaders must therefore develop new skill sets, including the capacity to steward decentralized autonomous organizations (DAOs), facilitate multiparty collaborations, and articulate transparent governance guidelines. Firms might opt to start with ‘hybrid’ organizational structures: partially centralized and partially token-governed, mitigating risk while testing distributed decision-making. Case-by-case experimentation, monitored through a robust set of KPIs, will reveal how well such structures align with a firm’s strategy, culture, and consumer expectations [12].

Scalability and Operational Readiness

Entering a distributed, token-driven marketplace entails adjustments to operational workflows. For example, if a company deploys an NFT-driven supply chain, traditional departments must integrate automated smart contracts, ensuring each transaction triggers accurate data entries and ledger updates [4]. This requires familiarity with Web 3.0 tooling, from on-chain analytics to wallet integrations, and the capacity to troubleshoot novel forms of risk, such as smart contract exploits. Managers must also anticipate potential bottlenecks, particularly when high network traffic inflates transaction fees or slows the finalization of blockchain transactions [6]. To mitigate scalability concerns, some organizations employ layer-2 solutions or sidechains to improve throughput and reduce fees. This involves forging partnerships with specialized providers or blockchains, effectively requiring 'vendor management 2.0,' where the vendor's platform availability, code governance, and future technical roadmaps become critical variables.

User Experience (UX) and Onboarding

In the Metaverse, the quality of virtual experiences can make or break the success of a platform. Beyond immersive visuals, managers should recognize that many consumers remain unfamiliar with concepts like wallet custody, token transactions, and private key management. User onboarding, particularly for mainstream audiences, must be frictionless, combining intuitive sign-up flows with clear guidance on how to secure digital assets. From a design standpoint, bridging the complexity of distributed applications with a welcoming interface often requires multi-faceted collaboration among front-end designers, blockchain specialists, cryptographers, and marketing experts. The emergent nature of the Metaverse also underscores the importance of performance optimizations. High-fidelity, low-latency experiences anchored in real-time interactions and transactions can be immensely resource-intensive, prompting the need for advanced infrastructure solutions such as edge computing or custom sidechains.

Regulatory and Compliance Strategy

The decentralized ethos of Web 3.0 often collides with regional laws governing data privacy, securities, and consumer protection. Failure to address these regulatory factors can lead to costly compliance breaches, user mistrust, and truncated product rollouts. Hence, proactive engagement with policymakers and industry consortiums is no longer a mere afterthought; it should be an integral part of strategic planning. Token classification—whether a digital token is deemed a utility, commodity, or security—remains a hot-button topic. Legal teams must scrutinize tokenomics models carefully, establishing disclaimers and governance structures that minimize liabilities [4]. Additionally, cross-jurisdictional complexities can arise: a DAO with token holders spanning multiple countries may be subject to contradictory regulatory environments. Ultimately, an adaptable compliance strategy that anticipates evolving legal frameworks will position companies for long-term stability amid a fast-changing regulatory sphere [23].

5.3 *Future Developments and Research Directions*

Interoperable Ecosystems

Although current metaverse projects frequently operate as siloed environments, future success may hinge on genuine interoperability. Whether transferring NFTs across platforms or combining DeFi protocols from multiple blockchains, bridging solutions are critical for delivering a seamless user journey. As multiple chains adopt distinct consensus algorithms and data structures, managers must decide how (or if) to integrate bridging protocols. This necessitates heightened security measures, multi-signature frameworks, and potentially orchestrated partnerships among platform developers [13, 14].

In a fully connected future, a user's digital identity, assets, and reputation could remain consistent across different Metaverse realms. If realized, such cross-platform continuity stands to unleash new user experiences and revenue pathways. For instance, loyalty points gained in one ecosystem might be redeemed in another, or a brand's digital collectible might seamlessly appear in different virtual stores.

AI-Driven Decentralized Platforms

The intersection of AI and Web 3.0 is a nascent but burgeoning field. On one hand, AI can streamline decentralized governance by detecting anomalies, predicting user preferences, or automating repetitive decisions within a DAO. On the other hand, decentralization can bolster AI development by creating shared data markets where producers remain in control of their data yet collectively train machine-learning models. This synergy could reshape analytics-driven fields such as logistics, healthcare, and finance by injecting greater transparency and inclusiveness into model-building processes [4].

However, merging AI with public ledgers raises ethical and technical questions about data sovereignty, algorithmic bias, and real-time inference costs. Tools like zero-knowledge proofs and secure multiparty computation may partially address privacy concerns, but full solutions will require deeper regulatory and technical collaboration to guard against flawed or discriminatory AI outputs [54].

Legal Standardization and Norm Formation

Despite incremental progress, the legal landscape for tokens, DAOs, NFTs, and cross-border digital transactions remains fragmented. A major leap forward could occur if international bodies converge on uniform standards, allowing businesses to reliably forecast compliance costs and user obligations. Until then, managers might need to craft patchwork solutions, selectively implementing geofencing for certain features, forging alliances with specialized law firms, or adopting 'regulatory sandbox' statuses.

In the near term, policymakers and industry coalitions may accelerate standardization of technical facets (e.g., smart contract security certifications) and definitional clarity (e.g., "What is a utility token?"). Such consensus offers the advantage of lowering barriers to entry for new entrepreneurs, thereby fostering more robust competition and market growth. At the same time, standardizing too rigidly risks stifling the innovations that make Web 3.0 appealing in the first place. Hence, the onus is on managers, consortiums, and governments to negotiate the right balance between stable legal frameworks and entrepreneurial flexibility [12].

Bridging the Digital Divide

Although enthusiasm for the Metaverse and token economies is high among certain tech-savvy demographics, large portions of the global population remain underrepresented in these new ecosystems. Whether due to limited internet access, lack of digital literacy, or skepticism toward cryptocurrencies, substantial demographic gaps persist [1]. Forward-looking managers aiming for inclusive growth need to tailor user experiences, documentation, and educational programs that address these divides. Partnerships with non-governmental organizations (NGOs) or public institutions may further help in bridging infrastructure gaps, ensuring that decentralization does not inadvertently replicate existing inequities.

5.4 Concluding Remarks

In moving from theory to practice, managers who embrace Web 3.0 and the Metaverse face a confluence of technological, organizational, and regulatory considerations. The stakes are high: by orchestrating robust ecosystems, harnessing user-centric innovations, and mitigating novel forms of risk, companies and institutions can position themselves at the forefront of next-generation digital transformation. However, success demands a deliberate, multilayered approach that reconciles decentralization with effective governance, fosters user trust, and builds resilient market positions against regulatory headwinds.

Looking ahead, further research and cross-industry collaborations are necessary to refine best practices, from stable on-chain governance structures to advanced interoperability protocols. As the Metaverse and Web 3.0 continue to converge, managers who adopt an agile, forward-thinking stance—prioritizing user empowerment, transparent governance, and inclusive design—will be uniquely poised to catalyze value across physical and virtual domains. By thoughtfully incorporating new technologies into well-crafted business models and frameworks such as the GUEST methodology, organizations can chart a course toward sustained innovation in an increasingly decentralized and immersive digital world.

References

1. Almeida F, Santos JD, Monteiro JA (2014) e-commerce business models in the context of Web 3.0 paradigm, CoRR, abs/1401.6102

2. Potts J, Rennie E (2019) Web3 and the creative industries: how blockchains are reshaping business models. In: A research agenda for creative industries. Edward Elgar Publishing, pp 93–111
3. Guan C, Ding D, Guo J (2022) Web3.0: a review and research agenda. In: 2022 RIVF international conference on computing and communication technologies (RIVF). IEEE, pp 653–658
4. Perboli G, Musso S, Rosano M (2018) Blockchain in logistics and supply chain: a lean approach for designing real-world use cases. IEEE Access 6:62018–62028
5. Zambardino F (2024) The applications of the blockchain technology. A Comparative Law Study, Edizioni Scientifiche Italiane
6. Capocasale V, Perboli G (2022) Standardizing smart contracts. IEEE Access 10:91203–91212
7. Lage O, Saiz-Santos M, Zarzuelo JM (2022) Decentralized platform economy: emerging blockchain-based decentralized platform business models. Electron Mark 32:1707–1723
8. Zheng Z, Xie S, Dai H, Chen X, Wang H (2018) Blockchain challenges and opportunities: a survey. Int J Web Grid Serv 14(4):352–375
9. Casino F, Dasaklis TK, Patsakis C (2019) A systematic literature review of blockchain-based applications: Current status, classification and open issues. Telematics Inform 36:55–81
10. Borroni A (2019) Blockchain: uses and potential value. In: Borroni A (ed) Legal perspective on blockchain theory, outcomes, and outlooks, Pubblicazioni del Dipartimento di Scienze Politiche Jean Monnet dell'Università degli Studi della Campania Luigi Vanvitelli, Edizioni Scientifiche Italiane
11. Zambardino F (2022) La blockchain e la protezione dei dati personali: una tecnologia privacy compliant by design? | Blockchain and the protection of personal data: a privacy compliant by design technology? Euro J Priv Law Technol 2:136–152
12. Borroni A, Perboli G, Pratali S, Vandoni C (2024). Metaverso e web 3.0. In: Borroni A (a cura di) Le nuove frontiere dei beni culturali: il Metaverso percorsi tra diritto, cultura e tecnologia, pp 39–52
13. Capocasale V, Gotta D, Perboli G (2023) Comparative analysis of permissioned blockchain frameworks for industrial applications. Blockchain Res Appl 4(1):100113. https://doi.org/10.1016/j.bcra.2022.100113.
14. Capocasale V, Pedone F, Perboli G (2024) Parallel transaction execution in blockchain and the ambiguous state representation problem. In: 2024 19th European dependable computing conference (EDCC), pp 131–138. https://doi.org/10.1109/EDCC61798.2024.00035
15. Ooi K-B et al (2023) The metaverse in engineering management: Overview, opportunities, challenges, and future research agenda. IEEE Trans Eng Manage 1–8
16. Borroni A (2016) Bitcoins: regulatory patterns, banking and finance law review, pp 47–68
17. Zambardino F (2020) La tokenizzazione del settore immobiliare. In: IANUS—Rivista di diritto e finanza, pp 81–99
18. Capocasale V, Musso S, Perboli G (2022b) Interplanetary file system in logistic networks: a review. In: 2022 IEEE 46th annual computers, software, and applications conference (COMPSAC), pp 1684–1689. https://doi.org/10.1109/COMPSAC54236.2022.00268
19. Sadowski J, Beegle K (2023) Expansive and extractive networks of web3. Big Data Soc 10(1):1–6
20. Ey L (2023) Web 3.0: Metaverso e NFT. https://assets.ey.com/content/dam/ey-sites/ey-com/it_it/generic/genericcontent/report_metaverso.pdf. Accessed Dec 2024
21. Hsieh YY, Vergne JP, Anderson P, Lakhani KR, Reitzig M (2018) Bitcoin and the rise of decentralized autonomous organizations. J Organ Des 7:14. https://doi.org/10.1186/s41469-018-0038-1
22. Park SM, Kim YG (2022) A Metaverse: taxonomy, components, applications, and open challenges. IEEE Access 10:4209–4251. https://doi.org/10.1109/ACCESS.2022.3142053
23. Perboli G, Borroni A, Bruni ME (2025) Developing metaverse use cases: the LETO methodology. In: Borroni A, Perboli G, Piccinelli GM (eds) Metaverses: reshaping law, economy, and society. Springer

24. Borroni A, Zambardino F (2024) The data protection in the metaverse. Roman Bus Law Rev 3:17–32
25. Zambardino F (2021) L'inquadramento giuridico dei token. Un'analisi comparatistica della regolamentazione italiana e sammarinese. In: IANUS—Rivista di diritto e finanza, pp 21–36
26. Abdelhamid M, Sliman L, Ben Djemaa R, Perboli G (2025) A review on blockchain technology, current challenges, and AI-driven solutions. ACM Comput Surv 57(3):1–39. https://doi.org/10.1145/3700641
27. Envols (2023) Tutankhamun: explore the first virtual Egyptian city in the metaverse, taking us to the time of king Tut. https://www.en-vols.com/en/inspirations-en/culture-en/tutankhamun-metatut-virtual-city-egypt-metaverse/#:~:text=%E2%80%9CMetatut%20is%20the%20first%20Egyptian. Accessed Dec 2024
28. Shardeo V et al (2024) Adoption of metaverse in healthcare sector: an empirical analysis of its enablers. IEEE Trans Eng Manage 71:8671–8685
29. Borroni A, Zambardino F (2023) Distributed ledger technologies (DLTs) and right to health: the impact of blockchain in the healthcare sector. Euro J Priv Law Technol 2:93–110
30. Dwivedi YK, Hughes DL, Baabdullah AM, Ribeiro-Navarrete S, Giannakis M, Al-Debei MM, Dennehy D, Metri B, Buhalis D, Wamba SF (2022) Metaverse beyond the hype: multidisciplinary perspectives on emerging challenges, opportunities, and agenda for research, practice and policy. Int J Inf Manage 66:102542. https://doi.org/10.1016/j.ijinfomgt.2022.102542
31. Dionisio JDN, Burns WG III, Gilbert R (2013) 3D virtual worlds and the metaverse: current status and future possibilities. ACM Comput Surv (CSUR) 45(3):1–38
32. Kolarevic B (2003) Architecture in the digital age: design and manufacturing. Taylor & Francis
33. Bailenson JN, Yee N, Blascovich J, Beall AC, Lundblad N, Jin M (2008) The use of immersive virtual reality in the learning sciences: digital transformations of teachers, students, and social context. J Learn Sci 17(1):102–141
34. Freina L, Ott M (2015) A literature review on immersive virtual reality in education: state of the art and perspectives. Int Sci Conf Elearning Softw Educ 1:133
35. Vandoni C, Perboli G (2025) Understanding value drivers in museum NFTs: a multi-phase, technological forecasting & social change, forthcoming. Analysis of consumer preferences and adoption patterns
36. Rizzo AS, Difede J, Rothbaum BO, Johnston S, McLay RN, Reger G, Gahm G, Parsons TD, Graap K, Pair J (2011) VR PTSD exposure therapy results with active duty OIF/OEF combatants. Stud Health Technol Inform 167:277–282
37. Haleem A, Javaid M, Singh RP, Suman R (2021) Telemedicine for healthcare: capabilities, features, barriers, and applications. Sens Int 2:100117. https://doi.org/10.1016/j.sintl.2021.100117
38. Hilken T, de Ruyter K, Chylinski M, Mahr D, Keeling DI (2017) Augmenting the eye of the beholder: exploring the strategic potential of augmented reality to enhance online service experiences. J Acad Mark Sci 45:884–905
39. Tapscott D, Tapscott A (2016) Blockchain revolution: how the technology behind bitcoin is changing money, business, and the world. Penguin
40. Chesbrough H (2010) Business model innovation: opportunities and barriers. Long Range Plan 43:354–363
41. Teece DJ (2010) Business models, business strategy and innovation. Long Range Plan 43(2–3):172–194
42. Venkatesh V, Thong JYL, Xu X (2012) Consumer acceptance and use of information technology: extending the unified theory of acceptance and use of technology. MIS Q Manag Inf Syst 36(1):157–178
43. Perboli G (2016) Guest-OR: linking lean business and OR. In: Workshop of the 28th European conference on operational research, Poznan, Poland
44. Perboli G (2017) The GUEST initiative. http://www.theguestmethod.com
45. Brotcorne L, Perboli G, Rosano M, Wei Q (2019) A managerial analysis of urban parcel delivery: a lean business approach. Sustainability 11:3439

46. Fadda E, Gobbato L, Perboli G, Rosano M, Tadei R (2018) Waste collection in urban areas: a case study. Interfaces 48:307–322
47. Tornatzky LG, Fleischer M (1990) The processes of technological innovation. Lexington Books, Lexington, MA
48. 5G-LOGINNOV Consortium (2021) 5G-LOGINNOV project web site. https://5g-loginnov.eu/. Accessed Dec 2024
49. CONCORDIA Consortium (2020) CONCORDIA project web site. https://www.concordia-h2020.eu/. Accessed Dec 2024
50. INCIT-EV Consortium (2021) INCIT-EV project web site. https://incit-ev.eu/. Accessed Dec 2024
51. Sinfonica Consortium (2023) Sinfonica project web site. https://sinfonica.eu/. Accessed Dec 2024
52. Autio E, Nambisan S, Thomas LD, Wright M (2018) Digital affordances, spatial affordances, and the genesis of entrepreneurial ecosystems. Strateg Entrep J 12(1):72–95
53. Cantamessa M, Gatteschi V, Perboli G, Rosano M (2018) Startups' roads to failure. Sustainability 10(7):2346
54. Perboli G, Tronzano A, Rosano M, Tarantino L, Velardocchia F (2021) Using machine learning to assess public policies: a real case study for supporting SMEs development in Italy. In: 2021 IEEE technology & engineering management conference—Europe (TEMSCON-EUR), pp 1–6. https://doi.org/10.1109/TEMSCON-EUR52034.2021.9488581

Developing Metaverse Use Cases: The LETO Methodology

Guido Perboli, Andrea Borroni, and Maria Elena Bruni

Abstract Metaverse, the concept of a persistent, virtual space shared by users worldwide, has garnered enormous attention among scholars, technologists, and business leaders. However, effectively designing and deploying Metaverse use cases requires a comprehensive framework incorporating multiple dimensions. Unfortunately, existing methodologies typically focus only on technology and operational aspects. This paper introduces the LETO methodology. It provides a holistic view to ensure solutions are compliant, sustainable, and operationally efficient, addressing the legal, ethical (including environmental and social), technological, and operational dimensions. The paper demonstrates how LETO can help stakeholders navigate the complexity of Metaverse through the WineMetaverse Platform, a blockchain-based pilot project enhancing the wine purchasing experience. Furthermore, benefits, limitations, and future research directions of the LETO are discussed.

Keywords Metaverse · LETO Methodology · Sustainable digital innovation

1 Introduction

The rapid evolution of digital technologies has given rise to transformative phenomena reshaping how businesses operate and how consumers interact with digital environments. Among these, the emergence of the Metaverse, a persistent, immersive, and interconnected virtual space, stands out as one of the most promising

G. Perboli (✉) · M. E. Bruni
CIRRELT, Montreal, Canada
e-mail: guido.perboli@polito.it

G. Perboli
Politecnico Di Torino, Turin, Italy

A. Borroni
Università Degli Studi Della Campania Luigi Vanvitelli, Caserta, Italy

M. E. Bruni
Università Della Calabria, Rende, Italy

A. Borroni et al. (eds.), *Metaverses: Reshaping Law, Economy, and Society*, PoliTO Springer Series, https://doi.org/10.1007/978-3-032-08664-8_4

yet complex frontiers of digital society. The Metaverse promises to integrate virtual, augmented, and mixed reality technologies with blockchain-based decentralized systems, enabling new forms of digital ownership, economic activity, and social interaction [1, 2].

Despite the significant attention it has garnered, most of the current research and industry efforts have focused predominantly on the technological aspects of the Metaverse [3, 4]. However, as businesses begin to explore and invest in this new paradigm, it is evident how a successful Metaverse solution must account for more than just technical innovation. It must also address legal uncertainties, environmental and ethical concerns, including the social impact of the solutions, and operational challenges that arise from deploying such disruptive technologies within complex organizational structures [5, 6]. On the other hand, as has already happened in the Blockchain, the current methodologies focus mainly on the technology and operational aspects [7].

This paper aims to fill this gap by introducing the LETO Methodology, a comprehensive framework designed to guide the development of Metaverse use cases across four critical dimensions: Legal, Ethical (including environmental and social impacts), Technological, and Operational. By integrating these dimensions, the LETO Methodology provides a balanced approach that not only ensures the technical viability of Metaverse solutions but also aligns them with regulatory requirements, sustainability goals, and practical business operations. This holistic perspective is essential in an era where digital platforms must navigate complex regulatory environments, ensure data integrity and security, and deliver genuine value to both users and stakeholders [8]. This is because the transformative potential of Metaverse, as many innovations, is not limited to technological disruption. It also extends to redefining business models, value creation processes, and consumer engagement strategies. Businesses that successfully harness the power of the Metaverse can unlock new revenue streams, enhance customer experiences through immersive and personalized interactions, and build decentralized ecosystems that foster innovation and collaboration. However, achieving these benefits requires a well-structured roadmap that addresses the multi-faceted challenges inherent in transitioning from current web paradigms to fully realized Metaverse applications [9, 10]. To illustrate the practical application of this framework, we present a detailed case study of the WineMetaverse Platform, a pilot project that creates an augmented world mixing digital and physical elements in a single Metaverse to transform the wine purchasing experience. This case study not only demonstrates the efficacy of the LETO approach but also provides insights into how businesses can manage risk, ensure compliance, and create sustainable value in a decentralized digital ecosystem.

The remainder of the paper is structured as follows. Section 2 describes the landscape of Metaverse and the current state of the art. Section 3 introduces the LETO methodology, which is then exemplified by the WineMetaverse use case in Sect. 4. Finally, Sect. 5 gives insights about the benefits, the limitations and the future development of the LETO.

2 Background and Related Work

The current state of the art in Metaverse research is characterised by different directions of studies that, while extensive, remain fragmented in their approach. This section outlines how the literature is currently structured and highlights the pressing need for methodologies that address these additional dimensions in an integrated manner. In more detail, we explore the literature along three axes: the definition of Metaverse, the economic models of the Metaverse and the methodologies used in Metaverse for managing the innovation design process.

2.1 Definition of the Metaverse

Early definitions of the Metaverse focused predominantly on immersive digital environments created through virtual reality (VR) and augmented reality (AR) technologies. In those initial visions, the Metaverse was primarily seen as a digital art form or a simulated world—an experience largely divorced from the complexities of the physical environment. While these early definitions emphasized the transformative potential of immersive experiences, they capture only a fraction of what the modern Metaverse entails. Recent advances have broadened our understanding considerably. The modern Metaverse is now more accurately described as an enhanced version of the real world, a dynamic ecosystem where multiple layers of data and interaction converge. Key aspects of this enriched definition include:

- **Integration of Real-World Data**. The Metaverse continuously incorporates metadata from Internet of Things (IoT) networks. Sensors and connected devices collect real-time data about environmental conditions, infrastructure, and human activities. This data layer adds context and dynamism to the digital experience.
- **Advanced Data Processing and AI**. Artificial Intelligence (AI) is employed to analyse, interpret, and manipulate vast amounts of real-time data. AI enables context-aware and personalized experiences that adapt to both user behaviour and changing environmental conditions.
- **Immersive Interfaces Beyond VR/AR Alone**. Although VR and AR remain critical components, they are now part of a broader suite of technologies. Rather than merely simulating physical spaces, immersive technologies in the modern Metaverse enhance real-world experiences by integrating additional data and interactive elements.
- **Decentralized, Web3.0-Based Infrastructure**. Blockchain the foundation of the modern Metaverse is the Web3.0 paradigm. This framework utilises digital ledger-based systems (blockchains) to ensure secure, transparent, and immutable records of transactions and ownership. It supports shared data spaces and decentralized governance, whereby decision-making is distributed among users rather than controlled by a central authority. Federated learning mechanisms further enhance

the system by enabling collaborative AI model training across distributed nodes while preserving data privacy.

This integrated ecosystem is best visualized as a convergence of multiple technological layers that collectively enhance and extend the physical world (see Fig. 1). As Gan et al. [11] explain, the shift from isolated virtual environments to an interconnected digital ecosystem represents the true evolution of the Internet. Similarly, Krishnan (2022) notes that the integration of decentralized technologies and real-world data fundamentally redefines the parameters of eCommerce and digital interaction. In addition, Beniiche et al. [12] highlight that the move toward a "Society 5.0"—where digital and physical realms merge—is essential for realizing the full potential of this new economic paradigm.

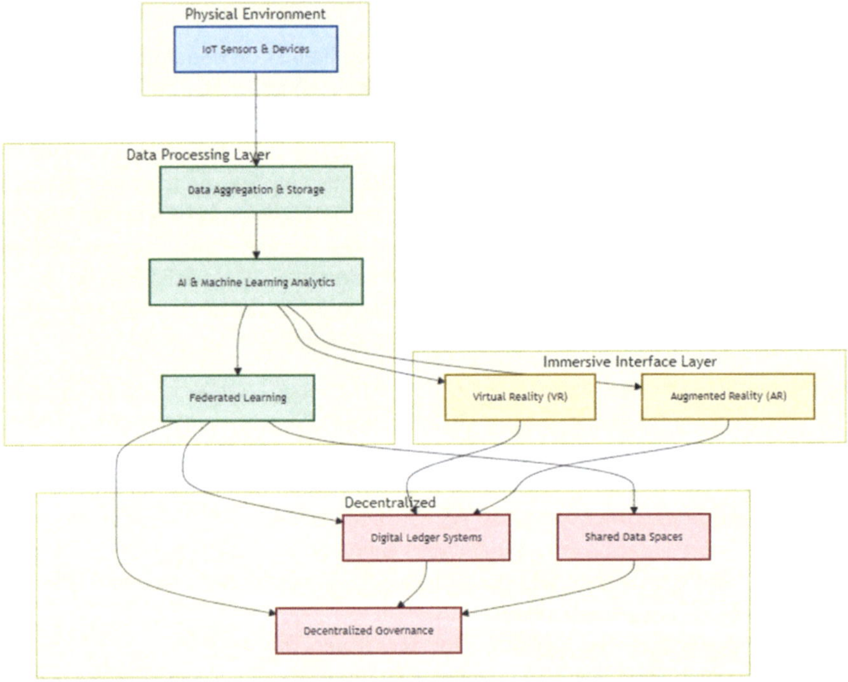

Fig. 1 The Integrated Metaverse Ecosystem: This diagram illustrates how real-world data from IoT sensors is aggregated and processed using AI and federated learning techniques. The processed data enhances immersive experiences through VR and AR, while digital ledger systems, shared data spaces, and decentralized governance underpin a Web3.0 framework that governs the entire ecosystem

2.2 The Economics of the Metaverse

The emergence of Metaverse has introduced a paradigm shift in economic structures by transforming traditional asset ownership, transaction methods, and value creation processes. At the core of this digital economy is the tokenization of assets, which enables both tangible and intangible goods to be represented as digital tokens on blockchain networks. This tokenization not only facilitates secure ownership and transferability but also creates new markets for assets that were previously difficult to exchange or value, such as art, virtual real estate, and even experiences [1].

One of the most transformative aspects of this new economic model is the rise of non-fungible tokens (NFTs). Unlike cryptocurrencies which are fungible and interchangeable, NFTs are unique digital certificates that verify the authenticity and ownership of a particular asset. This uniqueness has catalyzed the growth of digital art markets and has provided creators with novel revenue streams, while simultaneously offering consumers a way to invest in and collect scarce digital goods [2]. NFTs also pave the way for fractional ownership, where high-value assets can be divided into smaller, tradable units, thus broadening access to investment opportunities.

Decentralized finance (DeFi) is another pillar of the Metaverse economy. DeFi platforms leverage blockchain technology to eliminate intermediaries in financial transactions, thereby reducing costs and increasing transaction efficiency. By enabling direct peer-to-peer lending, borrowing, and trading of digital assets, DeFi creates an ecosystem where users have greater control over their financial activities. This shift not only democratizes access to financial services but also challenges traditional financial institutions, forcing them to adapt or risk obsolescence [5].

Moreover, the decentralized nature of Metaverse implies that economic power is redistributed away from centralized institutions towards a more diverse range of participants. In a traditional economy, central banks and regulatory bodies play a pivotal role in controlling monetary policy and ensuring market stability. However, in Metaverse, blockchain technology allows for transparent, immutable transaction records and self-executing smart contracts, which reduce the need for centralized oversight [13]. This change could potentially lower transaction fees, enhance trust among users, and lead to more competitive market dynamics. However, it also introduces regulatory challenges, as existing financial regulations may not be fully equipped to manage a decentralized and borderless economy [6].

The economic implications of these developments are profound. For example, industries such as real estate and entertainment are experimenting with virtual property markets and digital collectables, thereby creating entirely new revenue streams. Furthermore, the ability to verify the provenance and authenticity of digital goods through blockchain technology enhances consumer trust and may drive higher valuations for digital assets. Yet, these opportunities are tempered by challenges such as market volatility and the lack of standardized valuation metrics for many digital assets.

The economics of the Metaverse are characterized by innovation in asset tokenization, the emergence of NFTs, and the growth of decentralized finance, all of which contribute to a more accessible, efficient, and democratized digital economy. However, these advancements also come with challenges that require robust frameworks and regulatory adaptations to ensure sustainable growth and consumer protection in this rapidly evolving landscape.

2.3 Methodologies for Innovation Design

Innovation design methodologies have long provided the tools and frameworks necessary to transform ideas into viable products and services. Traditional approaches, such as those outlined by Chesbrough [14] and Teece [9], emphasize that sustainable business models emerge from the alignment of technological capabilities with strategic business practices. These models stress the importance of value creation and capture by integrating internal competencies with external market opportunities.

A widely recognized framework in this field is the Technology–Organization–Environment (TOE) model [15]. The TOE framework examines the factors that influence technology adoption by considering not only the technological attributes but also the organizational readiness and the external environment in which the business operates. This comprehensive approach has been applied in numerous studies to explain how companies integrate new technologies and transform their business models.

Another perspective comes from Service-Dominant Logic, as proposed by Vargo and Lusch [10]. This paradigm shifts the focus from products to services, emphasizing that value is co-created through interactions between providers and consumers. In the context of digital transformation, this logic encourages organizations to design innovations that leverage collaborative networks and ecosystems rather than relying solely on a one-way value delivery system.

Despite the strengths of these methodologies, much of the literature in innovation design has primarily focused on technological implementation and product development. In recent years, especially with the rise of blockchain and decentralized platforms, research has largely concentrated on the technical feasibility of systems such as secure distributed ledgers, smart contracts, and immersive user interfaces [1]. These studies provide detailed insights into the technological mechanisms that underpin Metaverse but tend to overlook other critical dimensions.

For example, while the TOE model robustly addresses the internal and external factors affecting technological adoption, it does not explicitly cover the legal, environmental, ethical, and operational challenges. Similarly, while Service-Dominant Logic provides a framework for understanding value co-creation, it often does not address the regulatory uncertainties or the environmental impact associated with digital innovations in the Metaverse [6].

Lean business approaches, popularized by the Lean Startup method, emphasize iterative development, rapid prototyping, and validated learning. Cantamessa et al. [16] discuss how lean methods can reduce the risks associated with overinvestment in unproven technologies by fostering an environment of continuous improvement and adaptability in rapidly changing digital markets. Furthermore, the GUEST methodology has been developed specifically for blockchain-centric projects and now extends to Metaverse use case design [17, 18]. This methodology integrates lean principles with operations research to provide a structured, step-by-step process for articulating value propositions, defining coherent business models, and mapping out actionable implementation plans. The GUEST approach demonstrates how iterative design, combined with rigorous evaluation, can help organizations navigate the uncertainties of decentralized systems [19, 20]. Additionally, socio-technical systems theory offers an alternative lens for understanding innovation. Tornatzky and Fleischer [21] highlight that technological innovation is a process deeply intertwined with organizational dynamics. This perspective is particularly relevant for Metaverse, where the interplay between decentralized technology and organizational readiness is critical. It underscores that successful innovation requires not only the development of robust technologies but also the alignment of those technologies with organizational structures and strategic goals.

In response to these challenges, there is a growing recognition of the need for comprehensive methodologies that encompass the full design process of Metaverse use cases. While existing frameworks such as TOE, Service-Dominant Logic, and lean methods provide robust tools for technology adoption and product development, they primarily focus on technological and operational aspects, often neglecting the integration of legal frameworks, ethical considerations, and environmental impacts—factors that are particularly critical in decentralized, blockchain-powered ecosystems. To bridge this gap, a more holistic approach is required, one that integrates traditional innovation design frameworks with additional dimensions to ensure that new solutions are not only technologically robust but also sustainable, compliant with regulatory requirements, and strategically aligned with broader business objectives [8, 22]. By incorporating these elements, organizations can mitigate risks associated with fragmented efforts and drive digital transformation initiatives that are both comprehensive and future-proof. This integrated methodology is essential for translating technological potential into practical, sustainable implementation within the emerging Metaverse landscape.

3 The LETO Methodology

In the rapidly evolving landscape of digital transformation, particularly with the advent of decentralized virtual environments such as the Metaverse, a holistic design approach is essential. The LETO Methodology is an integrated framework that addresses the full spectrum of challenges associated with developing Metaverse use cases. This comprehensive approach bridges the gap between emerging technologies and real-world business implementation, enabling organizations to harness the transformative potential of the Metaverse.

In more detail, the LETO Methodology provides a comprehensive framework that integrates four key dimensions—Legal, Environmental & Ethics, Technology, and Operational—to guide the development of sustainable and scalable Metaverse use cases. This multidimensional approach ensures that innovations are not only technologically advanced but also compliant, sustainable, and operationally sound.

3.1 Legal

The legal dimension of LETO focuses on establishing clear regulatory frameworks, ensuring intellectual property protection, and defining contractual relationships within decentralized digital ecosystems. With emerging technologies—such as blockchain and smart contracts—traditional legal paradigms are being re-examined. Scholars such as Lessig [23] and Finck [24] argue that law must evolve to address the challenges of digital ownership, data sovereignty, and cross-jurisdictional disputes. Impact includes:

- **Digital Asset Ownership**. In environments where non-fungible tokens (NFTs) represent physical or digital assets, legal clarity on ownership is critical. A well-structured legal framework can ensure that each NFT has a traceable provenance, reducing the risk of fraud and disputes [2].
- **Compliance with Data Protection Regulations**. As digital platforms collect vast amounts of personal data, ensuring compliance with regulations such as the General Data Protection Regulation (GDPR) is vital. Legal frameworks must outline how data is collected, processed, and stored while maintaining user privacy [25], more recently revisited in debates by Zuboff [26].
- **Smart Contract Enforcement and Legal value**. The use of self-executing smart contracts raises questions about enforceability in traditional legal systems. Establishing standards and legal interpretations for smart contracts can help bridge the gap between code-based agreements and conventional contract law [24].
- **Digital Well-being**. The design and governance of digital platforms should consider their impact on mental health, reducing digital addiction and ensuring a balanced user experience. Ethical UX design principles can help mitigate the negative effects of excessive screen time and algorithm-driven engagement loops.
- **Transparency and Accountability**. Establishing transparent governance mechanisms and clear accountability structures is essential to maintaining user trust. Open-source initiatives and explainable AI models contribute to ensuring that digital platforms operate in an ethically responsible manner (Zambardino, 2024).
- **Fair Labor Practices**. The rise of gig economy platforms and remote work necessitates ethical labor standards, ensuring fair wages, worker protections, and equitable treatment. Ethical technology deployment should also consider how automation impacts job displacement and economic inequality (Signorini and Zambardino, 2024).
- **Cybersecurity and Digital Rights**. Protecting users from cyber threats, misinformation, and digital surveillance is crucial to upholding ethical standards in

digital ecosystems. Ethical cybersecurity practices should ensure data sovereignty, encryption, and the right to digital privacy.

3.2 Ethical (Including Environmental and Social)

This dimension addresses the ethical aspects along different axes, including the environmental and the social impacts in deploying large-scale digital platforms. The debate over the energy consumption of blockchain technologies, for instance, has been widely discussed in the context of "green computing" [27]. At the same time, ethical considerations—ranging from algorithmic fairness to digital inclusivity—are central to building trust in these systems. Impacts include:

- **Sustainable Technology Choices**. Sustainable Technology Choices refer to the deliberate selection, development, and implementation of technologies that prioritize environmental stewardship, social equity, and economic viability over their entire lifecycle.

4 Ethical AI and Data Use

- Ethical frameworks for AI ensure that data processing respects individual privacy and prevents bias. This is particularly relevant in systems where AI is used to analyze IoT data and tailor user experiences [28].
- **Social Responsibility**. Metaverse platforms can drive social change by promoting digital inclusivity and supporting local communities. For example, platforms designed with decentralized governance models often include mechanisms to ensure that marginalized voices are represented in decision-making processes [29].

4.1 Technology

The technology dimension encompasses the technical infrastructure and tools necessary for building robust Metaverse applications. This includes blockchain networks, smart contracts, IoT integration, AI analytics, and immersive interfaces. The literature on digital transformation [30] emphasizes that technology is the enabler of innovation—but its success depends on the seamless integration of diverse systems including:

- **Hybrid Blockchain Architecture**. Modern applications benefit from combining private ledgers (for confidentiality) with public blockchains (for transparency). This hybrid approach addresses scalability and security concerns and has been discussed in research on enterprise blockchain solutions [31].

- **Standardized Smart Contracts**. Leveraging proven, standard templates for smart contracts reduces vulnerabilities and builds trust among users. This is critical in high-stakes environments such as decentralized finance (DeFi) [32].
- **Integration of IoT and AI**. Using IoT devices to collect real-time data and applying AI to process this data enhances the responsiveness of digital systems. For example, in agricultural applications or supply chain management, sensor data combined with AI-driven predictive analytics leads to improved operational efficiency [33].
- **Immersive Technologies**. VR and AR are no longer standalone novelties; they have become integral to creating engaging hybrid experiences that augment the physical world. Their evolution is supported by advances in hardware and software that are well documented in technology adoption literature [34].

4.2 Operational

The operational dimension of LETO focuses on the practical aspects of deploying and managing digital platforms. Scholars such as Hammer and Stanton [35] have demonstrated that efficient operational frameworks are critical for translating technological innovation into business value. The operation dimension includes:

- **User Onboarding and Interface Design**. Simplified and intuitive interfaces enhance user adoption. For instance, incorporating step-by-step tutorials and integrated wallet services can reduce friction in user onboarding processes [36].
- **Decentralized Governance Models**. Operational success in the Metaverse depends on clear governance structures. Implementing decentralized autonomous organizations (DAOs) allows stakeholders to participate in decision-making, ensuring that the platform remains agile and responsive [37].
- **Performance Monitoring**. Establishing clear Key Performance Indicators (KPIs) such as system uptime, transaction throughput, and user engagement metrics is essential for continuous improvement. In particular, the balanced scorecard approach, as discussed by Kaplan and Norton [38], can be adapted to monitor digital platforms effectively.
- **Scalability and Maintenance**. Routine security audits, regular system updates, and scalability testing are vital to maintain operational integrity. The principles of lean operations ensure that resources are allocated efficiently, and feedback loops are established for continuous optimization [26].
- **Optimization and Decision Support Systems**. Robust optimization and decision support systems are essential for streamlining operations and making data-driven decisions. These systems integrate advanced analytics and modeling techniques to evaluate performance in real time, simulate various operational scenarios, and optimize resource allocation. By providing actionable insights and facilitating rapid, informed decision-making, they support strategic planning and risk management, thereby enhancing overall operational efficiency.

5 Building Metaverse Use Cases: The WineMetaverse Platform

The WineMetaverse Platform represents a pioneering approach to reimagining the wine industry in the Langhe region through the lens of the Metaverse. By creating an immersive digital ecosystem that mirrors and enhances real-world experiences, the platform seeks to transform traditional wine commerce. It offers a comprehensive environment where producers, distributors, and consumers interact via digital representations, unlocking new opportunities for engagement, transparency, and value creation.

This project is a pilot study for a larger initiative on developing a comprehensive Italian food metaverse starting in 2025 with the Italian Ministry of Enterprises and Made in Italy.

Within the WineMetaverse platform, which is based on Web 3.0 technology, complete user interface customization is a distinctive feature. Utilizing Web 3.0 technology, the user interface can be adapted and personalized to display the most relevant and interesting content, experiences, and realities for each individual.

This platform leverages user and attractor profiling, analyzing their behaviors, preferences, and interactions to create an environment where value is generated from aligning attractors' offerings with user needs. This approach fosters the creation of a more engaging and interactive ecosystem where users can find precisely what they desire, while attractors reach an audience more aligned with their offerings.

The underlying vision of the WineMetaverse Platform is to offer a unique experience to the community of wine enthusiasts while creating a new platform to engage and educate customers on sustainable and conscious wine preparation and consumption. It also serves as a valuable marketing channel for the global agri-food industry.

The ultimate goal is to create a system that not only measures interactions but also allows one to comprehend the impacts resulting from implemented changes, such as policy adjustments, and the introduction of new actors, products, or services.

5.1 Methodology

The design of innovative digital solutions has long benefited from structured methodologies that guide organizations from ideation to implementation. Two such frameworks that have emerged in recent years are the GUEST methodology and the LETO Methodology. While both approaches aim to bridge the gap between technological innovation and business implementation, they differ in scope and focus.

The GUEST methodology was developed primarily for blockchain-centric projects. GUEST provides a step-by-step process to articulate and implement innovative ideas in a structured manner. It is composed of distinct phases—commonly identified as GO, UNIFORM, EVALUATE, SOLVE, and TEST—which serve to:

- Define and prioritize key actors and value propositions: GUEST begins by identifying the various stakeholders and understanding their gains and pains.
- Structure the use case design: It outlines the system's functioning, identifies the primary use cases, and assesses the technological infrastructure needed.
- Implement the solution through iterative feedback: GUEST employs tools such as the Value Ring and Solution Canvas to refine the business model and operational strategy.

This methodology has proven effective in guiding innovation projects in multiple sectors, helping practitioners articulate value propositions and structure actionable plans. It has been used in more than 200 companies and 20 international projects [39–41]. However, its primary focus is on operational and technological aspects. As such, while GUEST successfully addresses the mechanics of innovation and deployment, it may not fully capture the broader contextual factors necessary for emerging digital ecosystems like the Metaverse. For this reason, it has recently hybridized with the LETO Methodology, which can extend GUEST to consider the other dimensions, i.e., legal, ethical, environmental, and social.

5.2 *WineMetaverse Platform*

At its core, the WineMetaverse Platform is designed to enhance transparency, foster immersive experiences, and empower stakeholders. The platform leverages blockchain technology to record every stage of wine production, ensuring a permanent, tamper-proof record that enhances consumer trust. By integrating immersive technologies, the platform creates a virtual environment that allows users to explore vineyards, cellars, and production facilities, thereby deepening their connection to wine. Finally, the platform enables wine producers to showcase their heritage and unique products while offering consumers personalized experiences and authentic digital certificates that validate product origin and quality. On the Metaverse side, the platform bridges the physical and digital worlds. It emphasizes the convergence of real-world data—such as sensor information from vineyards—with virtual experiences, enabling users to interact with a digital reflection of the physical wine production process.

These objectives not only address long-standing challenges in the wine industry—such as counterfeiting and opacity—but also harness the potential of the Metaverse to create a new, enriched wine purchasing experience.

A critical step in designing the WineMetaverse Platform is the identification and segmentation of key actors. The approach distinguishes between two groups:

1. **Core Technological Actors**:

    - **Wallet Services**: Facilitate secure storage of digital tokens and ensure smooth financial transactions.

- **Network Validators**: Maintain the integrity of digital records by validating and recording transactions.
- **Developers**: Create and maintain digital infrastructure, including smart contracts and system protocols.
- **Community Managers**: Foster an engaged user community and promote the platform's offerings.

2. **WineMetaverse-Specific Actors**:

- **Platform Managers (PM)**: Oversee the entire digital ecosystem, managing roles and ensuring the system functions seamlessly.
- **Mappers (M)**: Objectively assess and score the value offered by each wine producer and service provider.
- **Attractors (A)**: Comprise local wine producers, wineries, restaurants, and cultural attractions that contribute distinctive products and experiences.
- **Users (U)**: Wine enthusiasts and consumers who seek an immersive, interactive, and personalized wine purchasing experience.

The Value Ring in Fig. 2 maps the benefits and priorities of each main actor. *Platform Managers* benefit from enhanced oversight and regulatory compliance; *Attractors* gain increased exposure and the opportunity to connect directly with consumers; and *Users* enjoy personalized experiences enriched by detailed product provenance and interactive virtual tours.

The WineMetaverse platform is then declined into three main use cases.

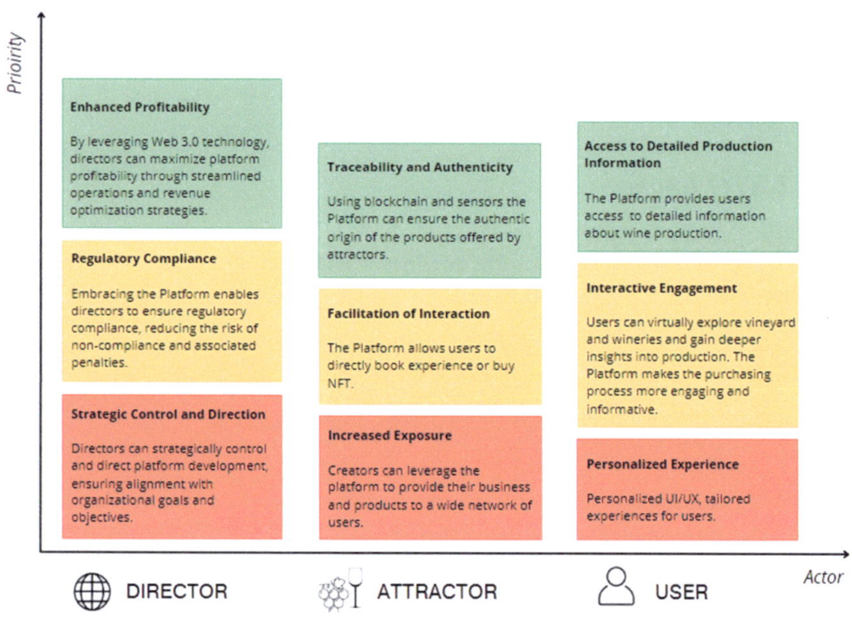

Fig. 2 The value ring of the proposed platform

- **Enhanced Wine Experience through Immersive Digital Certification**. In this use case, the Metaverse is leveraged not merely as a virtual replica of reality but as an augmented environment where real-world data and digital innovation converge to create a factual solution for the wine industry. The platform provides users with an immersive experience that integrates verified production data with interactive digital features. By merging real-world sensor data with immersive virtual experiences, this use case demonstrates how the Metaverse can provide a factual solution that enhances transparency, ensures authenticity, and deepens consumer engagement.
- **Transparent Wine Production Chain through Enhanced Digital Traceability**. This use case illustrates the transformative power of the Metaverse in delivering a fully transparent, data-driven solution that tracks the entire wine production process. Here, the Metaverse is utilized as an enhanced overlay of the real world, where every stage of production is enriched with digital data that verifies and augments physical processes. This use case exemplifies how the Metaverse, when viewed as an enhanced version of the real world, can deliver a factual solution that not only ensures product integrity but also transforms the consumer experience by combining verified real-world data with immersive, interactive digital environments.
- **Digital Heritage and Immersive Vision for Cultural Food Heritage**. Food companies have long recognized the value of preserving and showcasing their heritage. Many such companies establish museums—as exemplified by the Lavazza Museum—and maintain historical buildings, including castles and villas, as part of their brand identity and cultural legacy. In this use case, the WineMetaverse Platform's underlying architecture is adapted to create immersive digital environments that not only preserve this heritage but also enhance it through interactive and data-driven experiences (Fig. 3).

5.3 LETO Impact

The LETO methodology ensures that the WineMetaverse Platform is developed on a robust, multi-dimensional foundation. In this section, we detail the impact across its four key areas.

Legal Impact

To build a trustworthy and compliant platform, legal aspects are integrated from the onset:

- **GDPR Compliance**. The platform strictly adheres to the General Data Protection Regulation (GDPR), ensuring that all personal data collected through user interactions and IoT sensors is processed, stored, and transmitted in a secure manner.

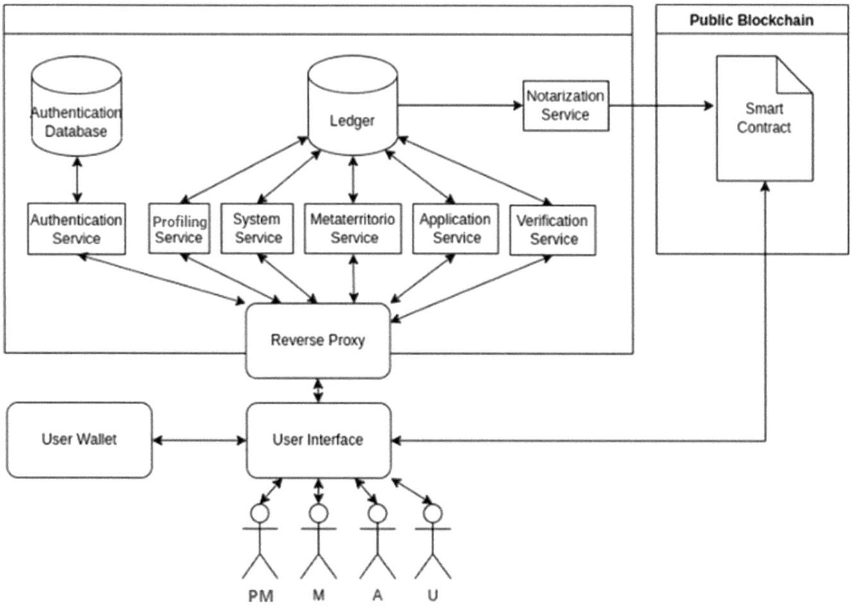

Fig. 3 The WineMetaverse architecture

- **European Data Act**. Data shared in collaborative data spaces are managed under the European Data Act, which promotes transparent, equitable data sharing among stakeholders while safeguarding data sovereignty.
- **Regulatory Blockchain Standards**. The platform complies with emerging blockchain standards to ensure that all transactions are secure, verifiable, and auditable.
- **AI Act (for Artificial Intelligence)**. For its AI-driven components, the platform anticipates and aligns with the requirements of the forthcoming AI Act, ensuring that automated processes are transparent, fair, and accountable.

Ethical (including environmental and social)

Sustainability and ethical considerations are central to the platform's design.

- **CO_2 Impact Calculation**. Continuous monitoring and computation of the carbon footprint—especially associated with blockchain operations—enable the platform to adopt energy-efficient practices and reduce its environmental impact.
- **Sustainable Purchasing Analysis**. Advanced analytics track consumer purchasing patterns, enabling the identification and promotion of sustainable buying behaviors.
- **Decentralized Governance**. A decentralized governance model ensures that decisions regarding environmental and ethical practices are made collaboratively by

a diverse group of stakeholders, ensuring that sustainability remains a core focus of the platform's evolution.

Technology Impact

The technological foundation is critical to delivering a seamless, immersive experience.

- **Hybrid ledger**. The platform employs a hybrid ledger architecture that utilizes a dual-layer approach, where sensitive production data is maintained in a private ledger and periodic state updates are recorded on a public ledger for enhanced transparency (see Fig. 1)
- **Smart Contract Automation**. Smart contracts automate essential functions such as the issuance of digital certificates, transaction processing, and ownership transfers. This reduces manual intervention and minimizes errors.
- **User Profiling and Personalization**. Advanced profiling mechanisms allow the platform to tailor experiences to individual preferences, ensuring that users receive content and recommendations that are both relevant and engaging.
- **Immersive Interfaces**. By integrating VR and AR technologies, the platform creates a rich, interactive digital environment that enables users to virtually explore vineyards, cellars, and production facilities.
- **Shared Data Spaces**. The platform incorporates shared data spaces that facilitate the secure exchange of information among stakeholders, ensuring that all data processing complies with stringent regulations like the GDPR and the European Data Act.

Operational Impact

Efficient operational processes are essential to the platform's success:

- **Intuitive User Interface**. A user-centric interface ensures that all interactions are seamless and engaging, which is critical for widespread adoption.
- **Decentralized Management and Monitoring**. The system includes tools for continuous performance monitoring, security audits, and scalability testing to maintain operational efficiency as the user base grows.
- **Comprehensive Service Integration**. The platform integrates key services—such as token exchange, authentication, and transaction signing—to deliver a smooth, end-to-end user experience. These operational elements work in conjunction with the technological backbone to ensure the system remains robust and reliable.
- **User-centric optimization**. The system includes tools for optimizing the performances of the user experience, by determining relevant patterns in the system and proposing to each user the option better fitting his/her behaviours.

6 LETO: Benefits, Limitations and Future Directions

The insights derived from our earlier discussion and practical application in the WineMetaverse Platform suggest that a holistic, multidimensional approach is critical for navigating the complexities of digital transformation. The comprehensive framework provided by LETO—encompassing legal, environmental, technological, and operational dimensions—offers a strategic advantage by mitigating risks, ensuring sustainability, and enhancing overall system resilience.

This section presents some recommendations based on these insights. The following subsections will explore practical implementation considerations, highlight potential challenges, and propose strategies to facilitate the adoption of Metaverse solutions in diverse organizational contexts.

The LETO Methodology offers several key benefits for organizations venturing into the Metaverse. By simultaneously considering the Legal, Environmental & Social, Technological, and Operational dimensions, LETO helps organizations avoid one-dimensional "tech-only" approaches. This comprehensive perspective ensures that solutions are robust not only from a technical standpoint but also in terms of compliance, sustainability, and practical execution. The mitigation of the risks is another key point. In particular, the early identification of legal and organizational pitfalls is crucial. By addressing these issues at the design stage, LETO reduces potential financial and reputational pitfalls. Finally, including environmental and social criteria fosters ethical adoption of Metaverse technologies. This approach encourages the use of sustainable, energy-efficient technologies and ensures that platforms are designed with considerations for data privacy, social inclusion, and corporate social responsibility.

While the LETO Methodology is designed to be comprehensive and robust, several limitations should be noted. Although LETO provides a structured framework, it still heavily relies on the expertise of practitioners. The qualitative nature of some assessments may result in outcomes influenced by the subjective judgment of experts, potentially leading to inconsistencies in implementation. Second, small organizations with limited resources may struggle to implement all the four dimensions of LETO thoroughly. The integrated approach requires significant investment in terms of time, expertise, and capital, which might not be feasible for every enterprise. Finally, Metaverse is in continuous development and Metaverse-related technologies, such as AR/VR hardware and blockchain protocols, are evolving constantly. The rapid technological advancements and shifting standards may require frequent updates to the methodology to remain current and effective.

To further enhance the LETO framework and its application in diverse contexts, new features should be studied and incorporated. The development of standardized KPIs for measuring the success of Metaverse initiatives in each of the LETO dimensions is essential. This would enable more objective evaluation and continuous improvement. The LETO ecosystem should also incorporate some elements related to the tracking of how Metaverse use-case perform over multiple years. This would provide insights into how organizational cultures and practices evolve with digital transformation and inform adjustments to the methodology over time.

In conclusion, designing and validating Metaverse solutions calls for a holistic approach that transcends mere technological feasibility. The LETO Methodology provides a practical, iterative framework that integrates the Legal, Ethical, Technological, and Operational dimensions. By considering these four perspectives simultaneously, organizations can better navigate challenges such as regulatory uncertainty, sustainability concerns, and internal resistance to change. LETO offers a means for harnessing the transformative potential of the Metaverse in a manner that is safe, sustainable, and strategically sound.

Acknowledgements Partial funding for this project comes from the NOUS project, call HORIZON-CL4-2023-DATA-01, contract number 101135927.

References

1. Iansiti M, Lakhani KR (2017) The truth about blockchain. Harvard Business Review, Jan–Feb 2017
2. Sinfonica Consortium (2023) Sinfonica Project Website. Retrieved from: https://sinfonica.eu/. Accessed Dec 2024
3. Gan W, Ye Z, Wan S, Yu PS (2023) Web 3.0: the future of internet. arXiv preprint arXiv:2304.06032
4. Ooi K-B et al (2023) The metaverse in engineering management: overview, opportunities, challenges, and future research agenda. IEEE Trans Eng Manage 1–8
5. Nissenbaum H (2010) Privacy in context: technology, policy, and the integrity of social life. Stanford University Press, Stanford
6. Paik M et al (2020) ImmuDB: a lightweight, performant immutable database. Available at: https://codenotary.s3.amazonaws.com/Research-Paper-immudb-CodeNotary_v3.0.pdf. Accessed Dec 2024
7. Potts J, Rennie E (2019) Web3 and the creative industries: how blockchains are reshaping business models. A research agenda for creative industries. Edward Elgar Publishing, Cheltenham, pp 93–111
8. GUEST (2017) The GUEST initiative. Retrieved from: http://www.theguestmethod.com Accessed Dec 2024
9. Tornatzky LG, Fleischer M (1990) The processes of technological innovation. Lexington Books, Lexington, MA
10. Warren SD, Brandeis LD (1890) The right to privacy. Harv Law Rev 4(5):193–220
11. Guan C, Ding D, Guo J (2022) Web3.0: a review and research agenda. In: 2022 RIVF international conference on computing and communication technologies (RIVF). IEEE, pp 653–658
12. Beniiche A, Rostami S, Maier M (2022) Society 5.0: internet as if people mattered. IEEE Wirel Commun 29:160–168
13. Capocasale V, Perboli G (2022) Standardizing smart contracts. IEEE Access 10:91203–91212
14. Chesbrough H (2010) Business model innovation: opportunities and barriers. Long Range Plan 43:354–363
15. Baker J (2012) The technology–organization–environment framework. In: Dwivedi YK, Wade MR, Schneberger SL (eds) Information systems theory: explaining and predicting our digital society. Springer, New York, pp 231–245

16. Cantamessa M, Gatteschi V, Perboli G, Rosano M (2018) Startups' roads to failure. Sustainability 10(7):2346
17. Hammer M, Stanton S (1999) How process enterprises really work. Harvard Business Review, Jul–Aug 1999
18. Perboli G, Musso S, Rosano M (2018) Blockchain in logistics and supply chain: a lean approach for designing real-world use cases. IEEE Access 6:62018–62028
19. Brotcorne L, Perboli G, Rosano M, Wei Q (2019) A managerial analysis of urban parcel delivery: a lean business approach. Sustainability 11:3439
20. Fadda E, Gobbato L, Perboli G, Rosano M, Tadei R (2018) Waste collection in urban areas: a case study. Interfaces 48:307–322
21. Vargo SL, Lusch RF (2008) Service-dominant logic: continuing the evolution. J Acad Mark Sci 36(1):1–10
22. Guan C, Ding D, Guo J, Teng Y (2023) An ecosystem approach to Web3.0: a systematic review and research agenda. J Elect Bus Digit Eco 2(1):139–156
23. Manyika J et al (2011) Big data: the next frontier for innovation, competition, and productivity. McKinsey Global Institute
24. Finck M (2018) Blockchains and data protection in the European Union. Eur Data Prot Law Rev 4(1):17–35
25. Westerman G, Bonnet D, McAfee A (2014) Leading digital: turning technology into business transformation. Harvard Business Review Press
26. Zuboff S (2019) The age of surveillance capitalism: the fight for a human future at the new frontier of power. PublicAffairs, New York
27. Böhme R, Christin N, Edelman B, Moore T (2015) Bitcoin: economics, technology, and governance. J Econ Perspect 29(2):213–238
28. Crawford K (2021) Atlas of AI: power, politics, and the planetary costs of artificial intelligence. Yale University Press, New Haven
29. NOUS Consortium (2023) NOUS Project Website. https://nous-project.eu/. Last Accessed Dec 2024
30. Womack JP, Jones DT (2003) Lean thinking: banish waste and create wealth in your corporation. Free Press, New York
31. INCIT-EV Consortium (2021) INCIT-EV project web site. Retrieved from: https://incit-ev.eu/. Accessed Dec 2024
32. Catalini C, Gans JS (2016) Some simple economics of the blockchain. NBER Working Paper Series, p 22952
33. Money (2022) How banks can make money in the metaverse, future of finance. Retrieved from: https://futureoffinance.biz/how-banks-can-make-money-in-the-metaverse-transcript/. Accessed Dec 2024
34. Brynjolfsson E, McAfee A (2014) The second machine age—work, progress, and prosperity in a time of brilliant technologies. W. W. Norton & Company, New York
35. Huang H, Wu J, Zheng Z (2023) From blockchain to Web3 & metaverse. Springer
36. Davenport TH, Ronanki R (2018) Artificial intelligence for the real world. Harvard Business Review, Jan–Feb 2018
37. Buterin V (2014) DAOs, DACs, DAs and More: an incomplete terminology guide. Ethereum Blog. Retrieved from: https://blog.ethereum.org/2014/05/06/daos-dacs-das-and-more-an-incomplete-terminology-guide. Accessed Dec 2024
38. Lessig L (2006) Code: and other laws of cyberspace, Version 2.0. Basic Books, New York
39. 5G-LOGINNOV Consortium (2021) 5G-LOGINNOV project website. Retrieved from https://5g-loginnov.eu/. Accessed Dec 2024
40. Kaplan RS, Norton DP (1996) Using the balanced scorecard as a strategic management system. Harvard Business Review, Jan–Feb 1996
41. Teece DJ (2010) Business models, business strategy and innovation. Long Range Plan 43(2–3):172–194

Banks in the Metaverse: The Legal Framework Between Evolution and Revolution

Andrea Borroni

Abstract The relentless advancement in technology has profoundly reshaped the nature of the world economy, transitioning from a primarily international model to a transnational one where national borders are perceived as a factual obstacle. Against this backdrop, cryptocurrencies, underpinned by blockchain technology and offering secure, immutable, and transparent transactions without the need for financial intermediaries, have emerged as an alternative to fiat currency—a development that has catalyzed the rise of a so-called decentralized finance (DeFI) dimension. Nevertheless, States and financial institutions are not passive observers; likewise, leading banks are proactively exploring the potential of the evolving digital ecosystem and, more specifically, of the metaverse as a new frontier for financial innovation and customer engagement. The integration of virtual assets like cryptocurrencies and NFTs within this space necessitates a profound transformation of the banking sector and prompts urgent regulatory considerations regarding transparency, data protection, accountability, and the enforceability of blockchain-based agreements. This study is aimed at critically examining the evolving nexus between the metaverse and banking institutions through a multidisciplinary approach, identifying current issues and future trajectories.

Keywords Metaverse banking · Fintech · Artificial Intelligence · Cryptocurrencies · Blockchain · Decentralized Finance (DeFi)

A. Borroni (✉)
University of Campania, "Luigi Vanvitelli", Caserta, Italy
e-mail: andrea.borroni@unicampania.it

A. Borroni et al. (eds.), *Metaverses: Reshaping Law, Economy, and Society*, PoliTO Springer Series, https://doi.org/10.1007/978-3-032-08664-8_5

1 The Metaverse: Defining Features and Concrete Manifestations

There is no denying how the popularity of the metaverse, as well as the study of its possible applications in a wide variety of contexts, has grown exponentially in the last period. Giving a strong impetus to this trend was, without a doubt, Facebook's announcement of its change of corporate name to Meta [1], coupled with its declaration that it wanted to facilitate connections between people through the discovery (and, in some ways, heteroinduced emergence) of new communities and the subsequent development of related businesses [2].

Since its first mention in the famous 1992 science fiction novel Snow Crash [3], the term "metaverse" has gradually evolved to accommodate manifestations of increasingly immersive virtual experiences, circulating in domains of knowledge and research with an increasingly rapid pace [4]: in a nutshell, it now stands for a persistent virtual space in which multiple interoperable platforms converge [5] and, therefore, are shared by users who can enjoy them through their avatars while maintaining a continuum in terms of digital identity and proprietary assets [6].

From a regulatory point of view, access to the synthetic domain is driven by the rules of the game, the so-called terms of service [7]; acceptance of them, together with the payment of registration fees and periodic fixed costs, constitute for users the *conditio sine qua non* for access to the online services provided by the owning company—including the software and server that support them, access to messaging platforms, the viewer that allows them to view the simulated environment, etc. [8].

On the technical side, however, to ensure the most authentic and immersive experience possible, the metaverse takes advantage of all the latest technologies functional to its purpose—in particular, those related to virtual reality, such as special visors [9]; an experience that is not limited solely to the area of digital entertainment but embraces a potentially very wide range of dynamics including the possibility of traveling, working, socializing and, above all, carrying out economic transactions such as, for example, transactions of goods and services, cryptocurrency exchanges, investments, purchase of NFTs or virtual real estate [10].

In particular, in relation to the latter, we are witnessing the emergence of a new economic ecosystem in which the ownership of digital assets exchanged through cryptocurrencies or NFTs guarantees users of the metaverse a high degree of personalization of their experience [11].

It still follows that, since they are indispensable to the concrete realization of a virtual reality so understood, it is necessary for States to invest in applications and technologies that facilitate the smooth circulation of massive amounts of data while managing, at the same time, to guarantee the security of the same [12].

For what is of interest here, with specific reference to the banking and *lato sensu* financial sector, the metaverse represents, finally, the place where money "meets" virtually but also, and above all, concretely [13]. In this regard, "banking is necessary, but banks are not" [14] is a popular and exceedingly topical quote: it reflects the recent paradigm shift in the interaction between customers and lending institutions

[15] stating that the former are expressing an increasingly widespread preference for accessing banking services through online platforms [16] or virtual branches, and a palpable interest in a wide range of experiences "ranging from augmented reality account management, virtual financial education, and virtual reality property tours to mixed reality banking applications" [17].

In light of this, the metaverse represents a virtual environment within which interaction with advisors can be instantaneous and in which transactions such as transferring funds, making payments, or managing one's accounts themselves can be done through intuitive gestures [18]. It also stands as a significant investment opportunity for States, since—much like the derivatives market exceeds that of its underlying assets—the metaverse is seen by stakeholders as a promising market whose virtual economy "will be much, much bigger than the economy of the physical world. You're going to have more cars build and designed in virtual worlds, you'll have more buildings, more roads, more houses—more hats, more bags, more jackets" [19].

Therefore, it's becoming increasingly evident how necessary for banks is to adapt their own business models to the new economic and social context, incorporating a greater degree of sensitivity to the needs emerging from technological progress in their product and service offerings and ensuring a more pronounced involvement of customers, even in "remote" interactions [20]. From the considerations just outlined, the plausibility of a potential shift in the traditional concept of banking follows: in analyzing a phenomenon that is still in the making, the question arises, in particular, whether the emergence of "meta-reality" may trigger a revolution in the processes and mechanisms of these institutions or, rather, a form of adaptation and evolution of established patterns.

2 The Technological Impact on the Financial Sector: Blockchain and Decentralized Finance. Virtual Money and Real Economy in a Multiverse World

The relentless process of technological, transportation systems, and media development has profoundly changed—and continues to transform—the distinctive features of the world economy: the economic boom of the years between 1950 and 1970, and, specifically, factors such as the emergence of multinational corporations, the new international division of labor, and the rise of tax and financial havens, contributed to the concretization of the transition from an essentially international economy to one with traits of transnationality, according to which the borders of States were no longer merely a structural factor, but a real obstacle [21].

Added to this, then, is the fact that with the financial crisis that occurred between 2007 and 2008, two of the main points on which traditional finance is based, namely trust and regulation, were undermined [22]: trust in credit institutions began to break down following the default of Lehman Brothers while, as far as the regulatory aspect is concerned, the crisis resulted in the need for increased controls and, consequently, the adoption of particularly strict measures in this regard [23].

In this complex scenario, the opportunities offered by constant technological advancement have had a major impact on changes in the financial sector, making the existing regulatory substrate inadequate to cope with new realities and in constant need of updating. In particular, one of the innovations that has emerged is the area of so-called digital banking, which has responded to the need for flexibility imposed by the coronavirus pandemic by allowing customers to take advantage of banking services "with little to no assistance, or 'touch,' from registration, opening new bank accounts, conducting several banking transactions (transfers, payments), closing accounts, to other related services such as financial advising, investment, e-commerce transactions, etc." [24].

Contributing to the conditioning of traditional schemes, then, has been the phenomenon of FinTech companies, which combine the provision of financial services and products with the most innovative technologies (mainly information and communication technologies, ICTs); these companies promise to reduce costs, improve quality, and contribute to financial inclusion by offering such services "to the Bottom of the Pyramid unbanked people" [25].

These firms can, potentially, take the form of entities that form partnerships with traditional banks, or—trespassing more into the latter's areas of expertise—as true fintech banks. There is, at the moment, no comprehensive definition of the latter type of reality [26]; nevertheless, their relevance on today's financial landscape is such that it has prompted the European Central Bank (ECB) to publish the "Guide to assessments of fintech credit institution license applications" [27] which, while not entailing substantial changes to the current regulatory framework as a non-binding instrument, clarifies the meaning of the expression under consideration [28]: in the text, in fact, the term "fintech" does not identify a type of banking institution, but is understood as purely descriptive of the peculiar business model of those activities that carry out banking functions using, to a greater extent or *in toto*, innovative technological finds which, because of the cybernetic risks that underlie them, entail the need for *ad hoc* supervisory activity [29].

In addition to this, under the impetus of globalization, the advent of cryptocurrencies [30], real keys for accessing the new phenomenon of the metaverse, has been particularly popular: described as "digital gold" [31] in that they are anchored in limited quantities of virtual assets, they are based on blockchain technology, a shared digital ledger in which transactions are permanently recorded and entries are contained in concatenated blocks in chronological order managed through cryptography [32]. Data management, in particular, is described as decentralized, in that it is not done by a Central Bank but through validation by all nodes in the network [33].

In this regard, people have begun to talk about, precisely, decentralized finance (or DeFi) [34] which represents "neither a legal nor a technical term. It is nonetheless increasingly used in the context of discussions about the future evolution of finance and its regulation" [35]; in its most common meaning, it refers to "an open, permissionless, and highly interoperable protocol stack built on public smart contract platforms, such as the Ethereum blockchain" [36]. DeFi reproduces the core functions of traditional financial services while, however, the involvement of intermediaries [37] or centralized authorities. Instead, contractual obligations are executed through code, transactions are transparent and verifiable, and authorized changes are immutably recorded within the blockchain architecture [38]; the result is "an immutable and highly interoperable financial system with unprecedented transparency, equal access rights, and little need for custodians, central clearing houses, or escrow services, as most of these roles can be assumed by 'smart contracts'" [39].

According to the thinking of part of the scholarly writings, therefore, there is an ongoing phasing out of traditional financial institutions which, relying on a system of fees and intermediaries, may no longer be the most effective and efficient means for the individual to complete monetary transactions [40].

In spite of this, it should not be forgotten how control over the circulation of money has always been one of the fundamental prerogatives of the state and how, despite the fact that transactions in the metaverse are linked to the sphere of the virtual, they give rise to problems with a strong impact on the sphere, on the other hand, real that, at least at present, escape adequate regulation: for example, particularly difficult is the management of enormous masses of personal data, which exceed in value and detail those generated for traditional banking; moreover, of undoubted moment is the issue of international money laundering [41], linked to the anonymity of users in the context of crypto-asset exchanges, to the absence of supervisory activities by public authorities and to the transnational nature of the exchanges themselves [42].

Ultimately, while it is tangible how we see a shrinking role for them in this field, the interest of lending institutions in the metaverse and the tools it offers is far from absent. As mentioned earlier, in a world where foreign exchange, international payments, and financial assets are increasingly connected to the virtual and individuals continue to invest in cryptocurrencies and, for now, NFTs, a metamorphosis of the banking sector is necessary: investing in the metaverse-and at the same time making it suitable for ensuring certainty and security of transactions, so that lending institutions can experiment with innovative ways to strengthen the relationship and communication with their customers in addition to new tools in the field of contracting-is likely to become a necessary step for these institutions.

2.1 *Central Banks Responses to Cryptocurrencies: The Introduction of CBDCs*

The idea of a currency not issued by central banks has roots far back in time: as early as the 1970s, F.A. Hayek hypothesized the creation of a "denationalized" monetary system, asserting that "[t]here is no justification in history for the existing position of a government monopoly of issuing money. It has never been proposed on the grounds that government will give us better money than anybody else could" [43].

The author's suggestion was, therefore, to provide for issuing institutions, or banks, located in different regions of the world, operating in a state of "healthy competition [...] resulting in positive economic returns throughout the system" [44]. At the same time, this would not have translated into a total suppression of the role of the state: the government would still be left with some room for maneuver, as "new rules of law [would be] needed to provide an appropriate legal framework within which the new banking practices could successfully develop" [45].

The current approach of central banks, however, seems to be leaning toward a greater degree of state intervention that, in parallel with technological development, sees a metamorphosis of the very nature of the national currency: concerns arising from the advent of cryptocurrencies, are resulting in the eventual introduction of the so-called Central Bank Digital Currency (CBDC), tool whose possible uses in banking and finance as well as possible synergies with blockchain technology are still being tested.

As reported in a well-known paper by the International Monetary Fund (IMF), "CBDC could be the next milestone in the evolution of money" [46]; it remains to be seen, however, how the various jurisdictions will decide to regulate this phenomenon. Indeed, one of the key legal issues arising from the issuance of CBDCs is certainly that of the regulatory regime applicable to them, as their assimilability to legal tender does not appear to be universally agreed upon [47].

Generally speaking, however, the doctrine has identified the following distinguishing characteristics of CBDCs: (i) being convertible on par with cash; (ii) being easy to use, just like cash (which should include the possibility of offline transactions); (iii) being secure against cyber attacks; (iv) guaranteeing immediate or near-immediate payments; (v) being interoperable; and (vi) complying with regulatory and prudential standards [48].

Notable in this regard are the experiments of the PRC, which is already using its own digital yuan (e-CNY) [49], and India, which announced, in December 2022, the first pilot project to launch the digital rupee, based on distributed ledger technology, which "will test the robustness of the entire process of digital rupee creation, distribution and retail usage in real time" [50]. Similar to the Chinese project, the Indian project is enjoying a fair amount of success, the target of one million daily digital rupee transactions in December 2023 having been reached, thanks to the provision of incentives by Indian banks [51].

Similarly, countries such as Japan, South Korea, Sweden, Australia, Russia, South Africa, and Ghana have begun the pilot phase; having a fully operational CBDC to date, however, include Jamaica, the Bahamas, Eastern Caribbean Central Bank member countries, and Nigeria [52]. Finally, the evaluative phase of the possible introduction of its own Central Bank Digital Currency still persists, for countries such as the United States [53], Canada [54], the United Kingdom [55] and the European Union. The latter in particular seems to have taken more concrete steps toward the adoption of the digital euro, having concluded the investigation phase and initiated the preparatory phase of the project related to it; in particular, the ECB's legal opinion on the Commission's June 2023 proposal for a Regulation [56] is awaited in order to move forward.

In this regard, several Articles of the proposal take up the definition of CBDC outlined earlier. In detail, under Article 7, it is stipulated that the Digital Euro is legal tender, which implies "the obligation to accept it, at full face value, with the power to extinguish a payment obligation. [...] By virtue of the obligation to accept the Digital Euro, the payee shall not reject the proposed Digital Euro in payment to fulfill this obligation. [...]" [57]; Article 12, then, stipulates that this currency is convertible at par. Finally, there is provision in Article 30 for both online and offline payment transactions, which are settled instantaneously.

That being said, it is highlighted that the impact of the digital euro, due to its nature as a medium of exchange, is felt more by the payment company sector than the banking sector [58], as the main function of credit institutions is to collect deposits. As evidence of this, it has been observed how, between 2016 and 2022, positive statements by central bankers on the possible introduction of the European CBDC had significant repercussions on the stock market: in particular, it has been estimated that European payment companies (on all of them, Worldline, Nexi, Edenred, and Adyen), experienced an increase in value of $23 billion in the stock market; conversely, their U.S. counterparts, such as Mastercard, Paypal, or Visa, suffered a total loss of $127 billion in the same time frame [59]. Therefore, considering that the latter play a key role in the European payments' scenario, it has been pointed out that there is a real geopolitical dimension to CBDC research, related to the increasing autonomy on the part of local payment service providers at the expense of foreign ones [60].

3 The Contribution of AI: Advancing Innovation and Transformation in the Banking Sector

As noted, among the implications arising from the advent of the metaverse is the development of a new financial ecosystem, which arises separate from the one existing in physical reality and is able to offer new opportunities for commercial banks [61].

While the importance of typical products of the metaverse has been grasped, starting with the forerunner blockchain [62], the field in which more concrete steps have been taken is that of customer interaction. In general terms, in wishing to sketch the interface that the metaverse is able to configure in this field, artificial intelligence tools come to the fore, whose path, starting from the first hints found in the works of Alan Turing and John McCarthy, is in constant evolution: their main function is that of data collection for the purpose of data processing aimed at the execution of specific operations—more or less automated, depending on the degree of human intervention required. The activity of such tools, being even predictive in nature, raises important questions in relation to the algorithmic operators' capacity to regulate the machine learning process. in particular, in the financial sector, the use of chatbots and voice assistants whose operation is linked to AI and machine learning mechanisms is increasingly common in front offices: they are used to simulate real-life employees, improve customer relations and suggest personalized services to them, provide users with calculated recommendations and assist in everyday financial decisions as well.

Moreover, AI robots like Jiao Jiao, introduced by the Bank of Communications of China in its branches, have been employed as lobby managers and are "able to welcome guests, provide service guidance and check account information for customers using speech recognition, touch interaction and body language" [63].

Even if these tools are meant to simplify the client experience and alleviate the operational burden placed on human personnel [64], they also mark a broader shift towards automation [65]. The impetus driving banks toward a widespread adoption of AI is not fundamentally different from the motivations influencing other economic actors: the main goal is to maintain a competitive position in the market and to pursue greater cost-effectiveness. Financial institutions are thus deepening their investments not only in the implementation phase but also in the research and study of technological innovation solutions [66].

Aside from customer support services, in fact, other functions attributable to AI solutions relate to good governance and management, fraud prevention, regulatory compliance and, above all, to the lending sphere—specifically, risk assessment in lending operations [67]. Digital lending, especially in the form of digital mortgages, now stands as one of the most promising fields: moving beyond the mere digitization of traditional processes, recent developments point toward the emergence of fully integrated ecosystems wherein applicants might be able to interact with AI virtual agents or avatars, and be guided through documentation upload, identity verification, and contract execution via qualified electronic signatures, blockchain-based authentication, and smart contract deployment. While traditional mortgage lending has shown signs of strain—due to rising interest rates, inflationary pressure, and a climate of economic uncertainty—short-term credit products and Buy Now, Pay Later (BNPL) solutions have gained larger popularity thereby fostering innovation in the banking sector as well [68]. However, such a transformation unfolds within a broader financial context, where the regulatory, temporal, and risk-related complexity of mortgage financing cannot be equated with those of consumer goods payment deferral. Mortgage approval processes continue to require third-party interventions—most notably that of the property appraiser—whose assessments remain human-dependent (albeit increasingly supported by technology) to not undermine legal certainty [69].

Beyond operational optimization, however, AI—and in Particular Generative Artificial Intelligence (GenAI)—is now being explored for its transformative potential in the design, structuring and delivery of financial services themselves: as noted, "GenAI can enhance environmental contributions by optimizing operations to reduce resource usage and energy consumption at data processing centers. […] companies using GenAI for green finance initiatives can develop products that support environmentally responsible investments, thereby aligning innovation strategies with ESG principles" [70].

Lastly, as AI redefines the products that financial institutions have to offer, it might also completely change the way they interact and function internally. When the rise of immersive technologies converges with that of the Metaverse, the contemporary banking scenario confronts a familiar legal dilemma across legal traditions: that of what constitute "presence" for legal purposes. It is likewise—even imminent—that a board member might attend meetings via a hyper-realistic avatar, enabled by AR or VR technologies, whose gestures, expressions, and interactions could be near-indistinguishable from those of their physical self. In the case of such avatars, the rational mind may reasonably acknowledge that the individual is physically elsewhere, yet what a person may *believe* could actually be in contrast with what they may *alieve* (or perceive): a dissonance between cognition and perception that reinforces the legal debate surrounding the subject. This blurring of physical and virtual presence is further intensified by the advent of AI-enabled tools, like the smart glasses which already allow real-time translation of speech across multiple languages. In cross-border banking governance, such tools could enable instant multilingual interaction without interpreters, reinforcing the illusion of co-presence during board meetings and thus requiring a recalibration of the legal definition of "presence".

Legal systems have long diverged in the treatment of this concept across various domains—some attaching more weight to corporeal proximity, others to functional participation. In recent years, then, the digital turn in corporate governance has encouraged a growing convergence toward models that embrace technological facilitation [71]. Yet, a deeper layer emerges when an avatar is infused with A.I. and can act autonomously based on algorithms trained on that board member's preferences and decision-making patterns: while a semblance of functional participation may still be preserved, the attribution of liability becomes increasingly fuzzy. In such scenarios, the avatar may not merely represent the individual but effectively deliberate, influence, or even cast votes, raising legitimate questions on the legitimacy of non-human decision-making entities and their legal personhood—a particularly contentious issue in the banking sector, where institutional trust is paramount, and lapses in judgment can have systemic repercussions.

It is evident that the banking sector is undergoing a widespread transition toward the integration of AI. But while exhibiting great potential, these systems are also bearers of major risks that must be carefully pondered by regulators. First of all, AI systems are only as reliable as the data on which they are trained, that could be incomplete, unrepresentative, or embedded with prejudices (bias) and could consequently produce discriminatory or unjust outcomes—for instance, concerning loan approvals. Another challenge that must be addressed is the misinformation spread by these technologies—especially the generative ones that can be used to produce synthetic data (meaning realistic yet entirely false information) which, again, if incorrectly or maliciously exploited can severely undermine the integrity of the financial ecosystem as a whole. Moreover, in this evolving scenario, when board participation occurs through immersive technology, or when avatars driven by AI proxies stand in for directors, not only questions of agency, consent, and liability become increasingly difficult to demarcate, but the quality of deliberation might be affected (if not diminished), too: technical interruptions could hinder real-time interactions, sensitive financial information discussed in such settings could be exposed to data leaks, unauthorized access, or algorithmic misuse, and the mechanics of electronic voting and its revocation should be well addressed from a procedural point of view.

These problems are symptoms of the imperative need for a thorough and exhaustive regulatory framework and for the eventual set-up of a robust detection mechanism capable of identifying false information and biases. Also, respect for the basic principles of fairness, transparency, and inclusiveness is of the utmost importance and must govern the whole process from the design to the implementation and monitoring of AI systems and solutions, to ensure that innovation can flourish efficiently in a competitive environment while also holding on to those values that matter equally to any society.

4 Metaverse Banking. Opportunities and Experiments by Financial Institutions. Revolution or Evolution?

The customer contact phase has been totally revolutionized by the development of so-called metaverse banking applications [72] or VR banking platforms that are capable of enabling customers to visit virtual branches, participate in meetings and, more generally, access services using their own personal computers or appropriate equipment. In addition to this, banks are already considering concrete uses of the spaces offered by the metaverse in order to introduce procedures to facilitate transactions in the virtual world, safeguard and guard their customers' digital assets or, again, deliver training courses for their employees [73].

Progressively, in order to meet the challenges arising from so-called "disruptive innovations" [74], take advantage of existing opportunities, and increase their competitive advantage, key players in the business and financial sector have begun to implement new operational strategies: among them, the U.S. bank J. P. Morgan was the first financial institution to access the metaverse through the opening of a virtual office (Onyx Lounge) in 2022 on the Decentraland platform in the opulent Metajuku shopping district [75].

Similarly, in Germany, Deutsche Bank [76] has announced the upcoming opening of a new 3D lounge, also in Decentraland [77], where visitors—using their own avatars—will have the opportunity to explore the virtual environment, observe forecasts of Stephan Ulrich, Chief Investment Officer of Deutsche Bank for corporate and private clients, and use the clothes and accessories for their virtual alter egos that can be exchanged with special tokens issued by the credit institution; this is a phase that could be called "exploratory", a first step through which Deutsche Bank intends to understand "what people are interested in in virtual worlds and what skills we need to offer them" [78].

In Asia, a notable development is the establishment by South Korea's KB Kookmin Bank of a fully virtual financial city, that's been operational since July 2021 and features a financial and commercial hub. This digital environment includes a public relations and recruitment booth, as well as an auditorium and a teleworking center designed to enhance communication and collaboration between remote and in-office employees [79].

Again, in 2022, Standard Chartered Bank (Hong Kong) Limited ("SCBHK") entered a partnership with The Sandbox, in whose Mega City neighborhood it purchased virtual land to use, as stated by Mary Huen, Chief Executive of the bank, with the aim of reimagining "our relationship with existing and potential clients on this new platform and our approach to enhance client journeys. Having acquired virtual land in Mega City, a natural choice for the Bank given its distinctive Hong Kong theme, perfectly fits with our promise of strengthening our continued presence in Hong Kong, whether physical or virtual" [80].

Finally, in Italy, particular attention has attracted the initiative of the Sella group which launched the Metaverse 4 Finance Accelerator, the first Italian acceleration program that aims identifying and supporting the growth of startups for the development of solutions and innovative, economically sustainable, inclusive, and safe technologies, within the metaverse applied to finance [81].

Ultimately, at present, the virtual environments created by commercial banks are primarily aimed at interaction and communication with the customer [82]; for the moment, no real structural change has happened in the credit institutions. Therefore, various technological [83] and, above all, legal limits will have to be overcome so that these institutions can concretely provide services in the metaverse such as the exchange between cryptocurrencies and traditional currencies or the provision of loans and financing: on the other hand, managing the intersection between real and virtual finance will require new efforts and new skills, also and above all on the part of the jurist who, wanting to regulate a phenomenon in constant and rapid development such as the one analyzed so far, runs the risk of developing a legal *datum* soon intended to obsolescence.

5 Concluding Remarks: New Horizons and Challenges in the Banking Sector

Law and information technology—the latter of which has literally revolutionized the representation of legal knowledge [84]—unfold through a complex relationship from which new problems and new choices arise.

The network accentuates the detachment of commercial transactions from the territorial dimension and deterritorialisation and the digital era discolour the model which sees the State as the main subject authorized to set the rules, now instead binding individuals to the rules of the virtual community, independent of the physical place in the which they are located or the geopolitical entity to which they belong. The rapid evolution of web 3.0 and, more generally, of the Internet make any legal controls and prohibitions considerably more problematic: the authorities struggle to keep up with innovation, while the decentralization of the underlying technologies and the very nature of the intended metaverse as a system of interconnected platforms they complicate the definition and application of a complete regulatory framework as well as the creation of a centralized control system.

As for the evolution of the banking sector, it too is inevitably conditioned by the blurring of the demarcation between the physical and digital domains: the structure of the metaverse represents fertile ground for the digital transformation of credit institutions but, in order for them to clutch the opportunities that arise, an in-depth understanding of the dynamics that take place there is necessary [85] and a strategic ability to adapt to them [86]. In this sense, the plausibility of superimposing existing operational models on new paradigms represents a challenge of considerable relevance for both operators and trainers [87].

In particular, when contemplating the concept of "navigating the digital frontier", used in the more general context of AI to convey the idea of exploring a new territory [88], we must ask ourselves whether the innovations that are involving all facets of banking represents a long journey or a temporary deviation from traditional banking patterns.

Without a doubt, in recent decades, the timing and methods of financial activities have faced a sometimes radical change; as we have seen, the recent emergence of FinTech realities which is implying a wide-ranging archipelago of tools, techniques, and approaches to the use of increasingly elaborate technologies, so much so that it is now devoid of a univocal connotation [89], is accompanied by the affirmation of DeFi and AI tools as a further challenging element for the legislator. And equally varied is the literature that analyzes the implications, risks and implications at a prudential level of phenomena of this magnitude, as well as the reasonableness of any standardization processes regarding procedures and mechanisms.

Certainly, a future in which decentralized finance can take total precedence over traditional finance is still difficult to imagine; in fact, albeit slowly, the counter-reaction of the State and Central Banks to phenomena such as that of cryptocurrencies is starting to manifest itself—and, moreover, it was unrealistic to hypothesize total *inertia* and staticity in the face of the rapid progress of a reality which in some ways was still blurry.

At the same time, a sudden and absolute metamorphosis of the legislative framework is not yet to come, which upsets traditional legal categories: as is known, the law adapts, transforms and evolves progressively together with the context in which it is developed; at least for the moment, the main references for national states still seem to be the currently operational instruments and the comforting municipal legal solutions. The response to the challenges posed by the new horizons that are emerging, although still uncertain, can only consist of a synergistic approach between banking institutions and regulators; at the same time, the adaptation of pre-existing structures to new cases can only lead to the adoption of measures aimed at guaranteeing some key principles, such as interoperability, data and transaction security and, it would seem, a certain degree of supervision by part of the central banks.

References

1. Which means 'beyond'; the juxtapositions of 'meta' and 'universe', from which the name 'metaverse' derives, is well indicative of the sense with which this new reality was conceived

2. Meta (2021) Introducing meta: A social technology company. https://about.fb.com/news/2021/10/facebook-company-is-now-meta/. Last accessed 2025/04/10
3. Stephenson N (1992) Snow crash. Bantam Books, New York
4. The metaverse of today is, by Meta staff's definition, a hybrid space consisting of the set of current social experiences available online "sometimes expanded into three dimensions or projected into the physical world. It will let you share immersive experiences with other people even when you can't be together—and do things together you couldn't do in the physical world. It's the next evolution in a long line of social technologies […]". Meta, Introducing Meta: A Social Technology Company… cit
5. Achieving effective interoperability in relation to banking and intermediaries is made more complicated by the widely held belief that users cannot create a single profile granting access to most or all of the metaverse platform. This issue is further exacerbated by accessibility challenges across various portals, bowsers, mobile apps, and the different entry points. Additionally, another significant obstacle to interoperability lies in the difficulty of integrating diverse digital assets and services within the metaverse, which are developed by different systems and creators. Leo S, Panetta IC (2023) Il posizionamento degli intermediari nel metaverso: a metà strada tra presente e futuro. In: Cardella PL, della Valle E, Paparella F (eds.), Tributi, economia e diritto nel metaverso, p. 57. Pacini Giuridica, Pisa
6. The definitions of metaverse are numerous and vary for several reasons, ranging from the number of analytical profiles required, to the experience and background of those formulating them. An additional element of temporary instability arises from the circumstance that this notion is still in its infancy, to the extent that a shared and consolidated definition of the metaverse—particularly within the legal field, which is characteristically reactive—has yet to emerge. Among the many definitions found in the literature, recent ones include: Bocchini R (2023) Nuovi beni digitali e mondi dematerializzati. I metaversi. In: Bocchini E (ed.) Manuale di Diritto Privato dell'Informatica. Edizioni Scientifiche Italiane, Napoli; Cerrina Feroni G (2023) Il metaverso tra problemi epistemologici, etici e giuridici. Media Law 1; Giovanardi E (2022) Il metaverso e le nuove frontiere del mercato disintermediato. In: Deotto D, De Simone D, Giovanardi E, Rampone F, NFT e Metaverso. Cosa sono, come funzionano, caratteristiche e aspetti tecnici. Guida operativa alle regole giuridiche e fiscali, Il Sole 24 Ore Professional. Although the mainstream seems to be moving towards accepting the approach given by Ball M (2022) The Metaverse: And how it will Revolutionize Everything. Liveright, New York
7. Bocchini R, Nuovi beni digitali e mondi dematerializzati... cit., p 170
8. *Ibidem.* It should be added that the praxeological formant within this framework has already responded with the preparation (albeit informal and limited to insiders) of mostly standard contracts that are beginning to circulate and spread among legal practitioners
9. The metaverse five "core technologies" are considered to be: Augmented Reality (AR), Virtual Reality (VR), Mixed Reality (MR), Extended Reality (XR) and Artificial Intelligence (AI). See Robinson J, Whyte J, Segura C (2022) Exploring Metaverse and The Digital Future, GSM Association, p. 7. Available at https://www.gsma.com/about-us/regions/asia-pacific/wp-content/uploads/2022/02/270222-Exploring-the-metaverse-and-the-digital-future.pdf, last accessed 2025/04/10. Moreover, thanks to the advent of such new technologies, the metaverse is expected to be characterized by: "interoperability across networked platforms; immersive, three-dimensional user experience; real-time network access and the spanning of the physical and virtual worlds". Rosati E (2023) The localization of IP infringements in the online environment: from Web 2.0 to Web 3.0 and the Metaverse. Journal of Intellectual Property Law & Practice 18(10), p. 721
10. Clearly, these activities give rise to challenges related to taxation and transaction tracking, as well as to the verification of the possession of the necessary authorisations by operators who must, in any case, offer investors adequate levels of asset guarantees. With specific reference to the taxation of cryptocurrency transactions, a preliminary regulatory input for the Italian system originates from the Italian Revenue Agency, which, through Ministerial Resolution No. 72/E of September 2, 2016, proposes the following interpretative solution: with reference to a company which is conducting the activity of purchasing and selling bitcoin on behalf of its clients, it should deem it appropriate, for the purpose of direct taxation, to consider the

realized capital gains (or capital losses) as revenues and costs of its core business activities, thereby including them, along with other incurred costs and revenues, in the determination of the financial statement result and the taxable base for IRES and IRAP. Furthermore, if the Company holds bitcoin at the end of the fiscal year, it shall be required to provide adequate disclosure in its financial statements, taking into account the fair value or current value (as is the case with traditional foreign currencies). In such instances, unrealized gains or losses from exchange rate fluctuations will not have fiscal relevance unless and until they are effectively realized. On this topic, see also Celotto A (2022) Tema del "metaverso" e delle sue implicazioni per l'ordinamento giuridico. GiustAmm.it 7

11. It is no coincidence that one of the most profitable sectors in this area is thehin virtual real estate market. Epicoco IO, Pierro M, Sarzana Di Sant'Ippolito F (2022), Il diritto del metaverso. NFT, DeFi, GameFi e privacy. Giappichelli, Torino, p 12
12. *Ibidem*
13. As stated, by gaining access to the new synthetic world, the individual becomes *homo economicus* and operates within this synthetic world using virtual money, which is converted into real currency. Bocchini R, Nuovi beni digitali e mondi dematerializzati... cit., pp. 169–170
14. This is a well-known quote from Bill Gates, founder of Microsoft, dating back to 1994. Naturally, Gates' statement has been the subject of extensive critical discussion, particularly by supporters of the crypto world
15. Gandhi R, Deputy Governor of the Reserve Bank of India, during his farewell speech in 2016, pointed out that "[t]he plethora of technological products and services, such as the World Wide Web, mobile phones and Apps, have helped emergence of FinTech companies who offer lower cost services for traditional services, such as e-payments and online trading". The transcription of the speech is available at https://www.bis.org/review/r160822b.pdf. Last accessed 2025/04/10
16. *Ibidem.*
17. Nguyen LT, Duc DTV, Dang T-Q, Nguyen DP (2023) Metaverse banking service: are we ready to adopt? A deep learning-based dual stage SEM-ANN analysis. Human Behavior and Emerging Technologies, p 2
18. In this regard, it can be stated that the "Metaworld" consists of two essential elements: the intrinsic, albeit sometimes unconscious, human need for a "better-than-real world" and the technology capable of connecting this ideal world with the real one. In other words, "[t]he goal of merging a permanent virtual environment with a virtual, augmented physical reality is to create a seamless bridge between the virtual and actual worlds". *Ivi*, pp. 2–4. On this topic, see also Zainurin MZL, Haji Masri M, Besar MHA, Anshari A (2023) Towards an understanding of metaverse banking: a conceptual paper. Journal of Financial Reporting and Accounting 21(1), pp. 178–190
19. As declared by Nvidia founder and CEO Jensen Huang to Yahoo Finance. The interview is available at https://finance.yahoo.com/news/nvidia-ceo-the-metaverse-will-be-much-much-bigger-than-the-physical-world-174256652.html. Last accessed 2025/04/03
20. According to a report published by PwC in 2014, five "megatrends" underpin the transformation of the banking sector: "demographic and social change; shifts in global economic power; rapid urbanisation; climate change and resource scarcity; and technological breakthroughs. While these all are relevant to banking, the most influential will arguably be the first and last of these megatrends: demographic and social change, creating new customer demands and stakeholder expectations; and technological breakthroughs changing everything from customer relationships to business model". PwC (2014) The future shape of banking. Time for reformation of banking and banks? 2014. Available at https://www.pwc.com/im/en/publications/assets/pwc-the-future-shape-of-banking-v4.pdf. Last accessed 2025/04/10
21. Hobsbawm EJ (2014) Il Secolo breve, 1914–1991, p 326. Rizzoli, Milano
22. Epicoco IO, Pierro M, Sarzana Di Sant'Ippolito F, Il diritto del metaverso… op cit., p 113
23. In general terms, as it was observed, each State represents a distinct realty, and at times decisive intervention on its part is deemed as necessary, while at other times it is more efficient to allow room for self-regulation. *Ibidem.* More specifically, forms of deregulation, combined with the

process of economic internationalization, can trigger what have been described as mechanisms of competition between legal systems (such as the "choice of law" through arbitration), regulatory sectors, or individual institutions of different States. The unfavourable activation of such mechanisms is perceived by States as a loss of economic sovereignty and constitutional legitimacy; thus, consequently, the typical response can be twofold, involving either the introduction of repressive measures or, conversely, a form of adaptation to the regulation trend aimed at enhancing their appeal in order to attract more investments and resources. di Plinio G (2008) Sulla Costituzione economica. Contributo per una teoria degli effetti costituzionali dell'economia. Il Risparmio LVI (1), pp. 48–50. Among the extensive literature on these phenomena, please refer in particular to Di Plinio G (2005), Il common core della deregulation. Dallo Stato regolatore alla costituzione economica sovranazionale. Giuffrè, Milano; Monateri PG (1989) Deregulation. Dig. disc., priv., Sez. civ., vol. V, p. 302 ff. Utet Giuridica, Torino; Zoppini A (2004) La concorrenza tra ordinamenti giuridici. Laterza, Roma-Bari

24. Anggara MRH, Davie MR, Margani M (2022) The Presence of Commercial Banks in Metaverse's financial ecosystem: Opportunities and tisks. J Central Banking Law Insts 1(3):406. More precisely, financial institutions can offer services such as online banking, mobile banking and/or virtual banking, in addition to online payment systems. Nguyen LT, Duc DTV, Dang T-Q, Nguyen DP, Metaverse Banking Service… cit., pp. 4–5 provide a definition for each of these categories, identifying online banking services as those offered by banks to their client via the Internet (e-banking), accessible at any time and designed to replace paper-based transactions. The term mobile banking, on the other hand, refers to the provision of banking services through applications available on mobile devices, such as smartphones and tablets, or via websites that have been optimized for such devices. The distinction between the two categories lies in "devices, portability, user experience, features" as well as the security measures employed—for instance, mobile banking apps may implement biometric authentication systems. Finally, the expression virtual banking refers to a mode of delivering banking services that, as is readily apparent, requires the use of virtual reality devices (like headsets) and dedicated avatars to provide a highly personalized experience in accessing the more traditional forms of digital banking described above. On this topic, see also Saeed MA, Reehan A (2021) Using online banking among COVID-19 pandemic: a systematic literature review. Research Journal of Finance and Accounting 12(18), pp. 1–5; Teo AC, Tan GWH, Ooi KB, Hew TS, Yew KT (2015) The effects of convenience and speed in m-payment. Industrial Management and Data Systems 115 (2); Lee MC (2009) Factors influencing the adoption of internet banking: an integration of TAM and TPB with perceived risk and perceived benefit. Electronic Commerce Research and Applications 8(3):130–141
25. Moro-Visconti R, Cruz Rambaud S, López Pascual J (2020) Sustainability in FinTechs; An explanation through business model scalability and market valuation. Sustain 12(24)
26. See Lener R (2024) Regolazione Fintech e Testo unico bancario. Quaderni di Ricerca Giuridica della Consulenza Legale della Banca d'Italia 100, according to which Fintech is a generic and multifaceted term. See also Id. (2021) La «digitalizzazione» della consulenza finanziaria. Appunti sul c.d. robo-advice. In: Id. La consulenza digitalizzata. Quaderni di Minerva Bancaria; Lanteri A, Esposito M, Tse T (2021) From fintechs to banking as a service: global trends banks cannot ignore. LSE Business Review; Lüth K (2021) Banks and FinTechs: Friends or foes? What it takes for banks and FinTechs to successfully cooperate. Journal of Digital Banking 6
27. Full text is available at https://www.bankingsupervision.europa.eu/ecb/pub/pdf/ssm.201803_guide_assessment_fintech_credit_inst_licensing.en.pdf. Last accessed 2025/04/10
28. As has been stated, "A fintech company could grant credits (not collect deposits, pursuant to art. 11, paragraph 2, TUB), one of the typical activities of the bank, without however being a bank [...]" (author's translation); if, however, the offer of a bank's products and services occurs mainly through technological tools (above all, apps replacing branches and algorithms), then it could be called a fintech bank. In light of this, the interpreters will have to establish the existence or otherwise of a type of fintech bank subject to the traditional existing regulations on banking supervision and supervision. Among the most difficult activities to regulate is the practice, which has spread in recent times, of issuing virtual IBANs, which do not fall within the scope

of the SEPA Regulation and can take the form of two main models: the first consists in the offer, by the provider of the payment service, of the opening of an account (master account) identified by the traditional IBAN and, at the same time, of secondary accounts (sub-accounts) connected to the virtual IBAN and aimed at the automation of accounting; the second model, which could be less similar to the regulated cases, provides that the payment service provider allows another person carrying out the same activity to access, indirectly, the payment systems of a different jurisdiction: for example, in the case of a bank with a European passport authorized in a country other than Italy, this could to resolve the well-known issue of "IBAN discrimination", which arises when a debtor—typically, but not exclusively, an employer—does not agree to pay off the debt if the creditor requests payment into an account that has an IBAN from another country in the Union. This is done by making sure that its Italian customers can obtain a "recognizable" Italian IBAN, which starts with the letters IT. Or, through an intermediary, a virtual IBAN starting with IT can be associated with the "real" IBAN. Lener R, Regolazione Fintech e Testo unico bancario… cit., pp. 61–68. See also Grabowski M (2022) Virtual IBAN as a Service in the Law of the European Union and Poland. Journal of Risk and Financial Management 15(12)

29. With specific reference to the supervisory activity required to address technological risks associated with the fintech sector, the EU Regulation known as the Digital Operational Resilience Act (DORA) for the financial sector comes into focus. This regulation, applicable starting from January 17, 2025, establishes a regulatory framework requiring all companies to ensure their ability to withstand and respond to all types of threats stemming from information and communication technologies (ICT). See Annunziata F (2024) Fintech: tra infrastrutture tecnologiche e intelligenza artificiale. Quaderni di Ricerca Giuridica della Consulenza Legale della Banca d'Italia 100, p. 272. Within this framework, the debate on the concepts of RegTech (Regulatory Technology) and SupTech (Spervisory Technology) has emerged. Both being considered as distinct segments of the phenomenon under analysis, the former refers to the use of technology by regulated entities to optimize compliance and risk management activities; SupTech, on the other hand, involves tools employed by financial authorities, which similarly focus on data collection and analysis. However, the activities of regulatory authorities in this domain often struggle to keep pace with the frequent market developments: since market dynamics tend to increase informational asymmetries between regulators and regulated, it is believed that better integration of RegTech and SupTech would have the positive effect of enhancing the effectiveness of supervision in a context where traditional legal categories are fuzzy and new risks and complexities are stemming from digitalization and datafication processes. *Ivi*, p. 278
30. As it was already mentioned, the 2008 financial crisis resulted in a loss of trust "in regulators, banks, and other financial institutions, and this may have helped pave the way for a more widespread acceptance of alternative means of payment, such as crypto-currencies, because of their decentralised and transnational nature, and, at least originally, the fact they are supposed to be immune from the control of public authorities". Borroni A, Seghesio M (2016) Hayek and Bitcoins: Which Governance for an International Currency. In: Gimigliano G (ed.) Money, Payment Systems and the European Union, p. 104. Cambridge Scholars Publishing, Newcastle upon Tyne. For further insights on this topic, please refer to Borroni A, Seghesio M (2019) Bitcoin and Blockchain: Un'analisi comparatistica dalla nascita alla potenziale regolamentazione. Ianus 19; Borroni A (2016) Bitcoins: Regulatory Patterns. Banking & Finance Law Review 32(1); Cian M (2019) La criptovaluta—alle radici dell'idea giuridica di denaro attraverso la tecnologia: spunti preliminari. Banca borsa e titoli di credito 3, p. 339 ff.; Gambino AM, Bomprezzi C (2019) Blockchain e criptovalute. In Finocchiaro G, Falce C (eds.) Fintech: diritti, concorrenza, regole, Zanichelli, Bologna; Vardi N (2015) Criptovalute e dintorni: alcune considerazioni sulla natura giuridica dei bitcoin. Il diritto dell'informazione dell'informatica 3, p. 443 ff
31. It has been stated that Bitcoin will likely be considered appealing to those "who like gold-backed currencies because its value depends on the availability of a limited (albeit virtual) resource rather than discretionary actions by central bankers". Grinberg R (2012) Bitcoin: An innovative alternative digital currency. Hastings Sci Tech Law Journal 4(1):168

32. In other terms, blockchain is "a distributed database of records, a public ledger of all transactions or digital events executed and shared among participating users. Each transaction in this database is verified through the consensus of the participants in the blockchain, because, once the information has been entered into the system, then it cannot be erased unless the changes are approved by the majority of the computing power of the entire blockchain. If someone tries to modify the register, the change will be blocked because the 'consent' between the nodes—computers connected to the blockchain—that keep the blockchain intact would be lost. Therefore, the blockchain contains various immutable records of every single transaction made". Borroni A (2019) Blockchain: Uses and Potential Value. In: Borroni A (ed.) Legal Perspective on Blockchain. Theory, Outcomes, and Outlooks. Edizioni Scientifiche Italiane, Napoli. See also Di Ciommo F (2023) Smart Contract e blockchain. In: Bocchini R (ed.), Manuale di Diritto Privato dell'Informatica, Edizioni Scientifiche Italiane, Napoli; Poncibò C (2022) Il Diritto Comparato e la Blockchain. Edizioni Scientifiche Italiane, Torino; Wright A, De Filippi P (2015) Decentralized Blockchain Technology and the Rise of Lex Cryptographia. Available at https://papers.ssrn.com/sol3/papers.cfm?abstract_id=2580664. Last accessed 2025/04/10
33. As regards the need for regulation of the crypto-assets markets, we note the recent adoption of EU Regulation 2023/1114, known as the "MiCA" Regulation (Markets in Crypto-Assets) aimed at protecting consumers and focused, mainly, on two cases: tokens linked to assets - stablecoins issued by banks based in the EU, subject to stringent monitoring and potentially capable of replacing money—and electronic money tokens—equated, in fact, to electronic money and issued by banks and EMIs. Furthermore, the Regulation regulates various other services that can be provided by legal entities or supervised intermediaries subject to authorization. For further information, see. de Tomasi F (2024), Sistema dei pagamenti. Evoluzione della moneta bancaria ed euro digitale. Quaderni di Ricerca Giuridica della Consulenza Legale della Banca d'Italia 100, pp. 255–256
34. DeFi seeks to eliminate the inefficiencies of a flawed and centralized system, through the decentralization of some financial dynamics which occurs through blockchain technology. New technologies have the potential to significantly reduce the risks inherent in centralized systems of traditional finance, can reduce transaction costs, expand the scope of transactions and enhance peer-to-peer transactions, stimulating a new wave of innovation. Epicoco IO, Pierro M, Sarzana Di Sant'Ippolito F, Il diritto del metaverso… cit., p. 113. Decentralized service provision means, on the other hand, "provision by multiple participants, intermediaries, and end-users spread over multiple jurisdictions, with interactions facilitated, and often in fact enabled in the first place, by technology". Zetzsche DA, Arner DW, Buckley RP (2020) Decentralized Finance. Journal of Financial Regulation 6(2), pp. 173–174
35. Zetzsche DA, Arner DW, Buckley RP Decentralized finance… cit., p 173
36. Schär F (2021) Decentralized Finance: On Blockchain—and Smart contract-based financial markets. Federal Reserve Bank of St. Louis Rev 103(1):153
37. In a broader sense, the fundamental goal of the development of digital commerce is the removal of middlemen and the establishment of a decentralized system. In web 1.0, consumers could only passively visit websites to obtain information; in web 2.0, however, we saw the emergence of large social platforms, which allowed users to transform from passive readers to active creators of content, generating revenue for both the platforms and businesses; and in web 3.0, an economy was established that transfers information directly from the creator, who has complete control over what he posts online, to the person who consumes it. Bocchini R, Nuovi beni digitali e mondi dematerializzati... cit., pp.167–168
38. The concept of DeFi differs from both that of Traditional Finance (Trad Fi) and that of Centralized Finance (CeFi); the first expression indicates the traditional financial ecosystem, dominated by traditional banking infrastructures, main players who enjoy public trust. In Trad Fi "[b]anks play a major role in moving financial markets, through receiving funds from people who have funds and channeling funds to people in need of credit through various financial services" while technology and digitalisation play a minority role. The expression CeFi, however, refers to that financial ecosystem which embodies the transition between Trad Fi and DeFi; it is characterized by the gradual decline of the dominance of banking as the primary source of

financial transactions, "replaced by technology-based institutions such as Fintech, centralized crypto exchanges. The increasing popularity of electronic money, digital wallets, and other technology-based forms of payment is another characteristic of CeFi, where even in the case of electronic currencies, the financial ecosystem still relies on several financial institutions but has shifted to financial institutions other than established commercial banks". It is hypothesized that the financial ecosystem of the metaverse (Metaverse Finance or Meta FI) will arise from the combination of DeFi, Trad Fi and CeFi. Anggara MRH, Davie MR, Margani M, The Presence of Commercial Banks in Metaverse's Financial Ecosystem... cit., pp. 412–413

39. Schär F, Decentralized Finance: On Blockchain—and smart contract-based financial markets... cit., p. 153. On this topic, see also Doop J (2022) Decentralized finance. Georgetown Law Tech Rev 6(1):373–385 and Johnson KN (2021), Decentralized Finance: Regulating Cryptocurrency Exchanges. William & Mary Law Rev 62(6):1911–2002
40. Shapiro J (2022), Il futuro delle criptovalute è una scommessa. Domino 9. This appears even more probable if we consider the so-called smart contracts, which provide for the automatic transfer of sums of money if objective parameters indicative of the fulfillment of the contract are met
41. In a 2014 report titled Virtual Currencies Key Definitions and Potential AML/CFT Risks, the Financial Action Task Force (FATF) has stated that "virtual currencies are the wave of the future for payment systems; and [..] provide a powerful new tool for criminals, terrorist financiers and other sanctions evaders to move and store illicit funds, out of the reach of law enforcement and other authorities". The report is available at https://www.fatf-gafi.org/content/dam/fatf-gafi/reports/Virtual-currency-key-definitions-and-potential-aml-cft-risks.pdf, last accessed 2025/04/10. Similarly, on the use of so-called virtual currencies, the Bank of Italy stated: the virtual currency network may lend itself to being used for transactions related to criminal activities, including money laundering; although virtual currency transactions are visible, in fact, the owners of the electronic wallets and, more generally, the parties involved can generally remain anonymous; this could make it necessary for the authorities to intervene to close the exchange platforms, preventing access or use of any funds held on them. The Communication, dating back to 30 January 2015, is available at https://www.bancaditalia.it/compiti/vigilanza/avvisi-pub/avvertenza-valute-virtuali/index.html, last accessed 2025/04/10. For further information on the Bank of Italy's position on cryptocurrencies, see also PasquinoT (2023) Le monete virtuali. In: Bocchini R (ed.), Manuale di Diritto Privato dell'Informatica, Edizioni Scientifiche Italiane, Napoli, pp. 233–235
42. With specific reference to the Italian regulatory framework on the subject, we note, in particular, the publication of the Decree of the Ministry of Economy and Finance on 17 February 2022 on the "modalities and timing with which service providers relating to use of virtual currency and digital wallet service providers are required to communicate their operations on the national territory as well as forms of cooperation between the Ministry of Economy and Finance and the police forces". The decree, in art. 3, establishes the obligation for service providers relating to the use of virtual currency and digital wallet services to register in the special section of the register managed by the OAM (Body for the management of the lists of agents in financial activity and mediators credit) in order to be able to provide these services on the territory of the Italian Republic. Furthermore, for the OAM to confirm the registration, it is necessary to meet the requirements set out in the art. 17-bis, paragraph 2, of the Legislative Decree. 13 August 2010, n. 141, namely: "(a) for natural persons: Italian citizenship or citizenship of a European Union State or of a different State according to the provisions of article 2 of the consolidated text of the provisions concerning the regulation of immigration and rules on the of the foreigner, pursuant to Legislative Decree 25 July 1998, n. 286, and domicile in the territory of the Republic; (b) for entities other than natural persons: registered and administrative headquarters or, for EU entities, permanent establishment in the territory of the Republic" (author's translation). The text of the Decree issued by the Ministry of Economy and Finance is available at https://www.gazzettaufficiale.it/eli/id/2022/02/17/22A01127/sg, last accessed 2025/04/10. As has been observed, this obligation meets the need to ensure that the Ministry of Economy and Finance acquires information relating to the size and operation of the market for

services relating to the use of virtual currency. See Lombardi M (2023) Compliance in materia di tutela dei dati personali e di antiriciclaggio. In: Piccinali M, Puccio A, Vasta S (eds.), Il metaverso. Modelli giuridici e operativi, pp. 159–162. Giuffrè, Milano

43. Hayek FA (1979) Toward a Free Market Monetary System. Journal of Libertarian Studies 1(3), p. 7. Among the most authoritative definitions of virtual currency, we remind to, above all, that of the ECB (2015), Virtual currency schemes—a further analysis, according to which the virtual currency represents "a digital representation of value, not issued by a central bank, credit institution or e-money institution, which in some circumstances can be used as an alternative to money" and that of the Bank of Italy, Warning on the use of so-called "virtual currencies", 2015 which talks about "digital representations of value, used as a means exchange or held for investment purposes, which may be transferred, stored and traded electronically". More generally, on the subject of electronic money, v. De Tomasi F, Sistema dei pagamenti. Evoluzione della moneta bancaria ed euro digitale… cit.; Mattassoglio F (2023) Moneta e tecnologia. Come intelligenza artificiale e DLT stanno trasformando lo strumento monetario. Giappichelli, Torino; Motroni R (2023) I pagamenti non monetari nella finanza digitale europea. La prospettiva italiana. Cacucci editore, Bari; Sica S, Stanzione P, Zeno Zencovich V (2006) La moneta elettronica: profili giuridici e problematiche applicative. Giuffrè, Milano as well as the commentaries on Articles 114-*bis*/114-*quinquies* in Capriglione F (2018) Commentario al Testo Unico delle leggi in materia bancaria e creditizia. Cedam, Milano
44. Borroni A, Seghesio M Hayek and Bitcoins: Which governance for an international currency… cit., p 103
45. Hayek FA (1990) Denationalisation of money: The argument refined (An Analysis of the Theory and Practice of Concurrent Currencies Series). Coronet Books, London, p 129
46. Bossu W, Itatani M, Margulis C, Rossi A, Weenink H, Yoshinaga A (2020) Legal aspects of central bank digital currency: Central bank and monetary law considerations. IMF Working Paper. https://www.imf.org/en/Publications/WP/Issues/2020/11/20/Legal-Aspects-of-Central-Bank-Digital-Currency-Central-Bank-and-Monetary-Law-Considerations-49827. Last accessed 2025/04/10
47. de Vauplane H (2019) Cryptocurrencies and Central Banks. In: Madir J (ed) Fintech—law and regulation, p 122. Edwar Elgar Pub, Cheltenham
48. As reported in BIS (2020) Central bank digital currencies: foundational principles and core features. https://www.bis.org/publ/othp33.pdf. Last accessed 2025/04/10
49. For further insights on the Chinese experience with the e-CNY, reference is made to the constribution of Paone EM, Evaluating the Potential of the Metaverse: The Chinese Perspective, in this volume
50. Reserve Bank of India (2022) Operationalisation of central bank digital currency, https://www.rbi.org.in/Scripts/BS_PressReleaseDisplay.aspx?prid=54773. Last accessed 2025/04/10
51. Kalra J (2024) India's digital currency transactions top 1 mln/day in Dec-sources, Reuters, https://www.reuters.com/business/finance/indias-digital-currency-transactions-top-1-mlnday-dec-sources-2024-01-04/. Last accessed 2025/04/10
52. The Central Bank of the Bahamas was the first in the world to recognize the legal tender of the Sand Dollar, its CBDC. With particular reference to Nigeria, however, the eNaira CBDC was introduced with a dual purpose: to combat the so-called underground economy and encourage foreign money remittances. Bursi M (2023) Euro Digitale: una questione di sovranità. Federalismi.it, n. 24, p. 9 ff. As is well known, Nigeria is part of the West African Monetary Zone (WAMZ), together with Gambia, Ghana, Guinea, Liberia and Sierra Leone, whose ultimate goal is the creation of a new common currency. However, as this has not yet been achieved, it remains to be seen what implications this could have on locally processed CBDCs. For further information on the role of Nigeria in the WAMZ and on the current state of the level of integration of the geographical area of reference, please refer to Borroni A, Petrucciano F (2023) The West Africa: seeking the path to monetary stability. In: Borroni A, Karaboue K, Zambardino F (eds.) Percorsi di comparazione nell'Africa occidentale, p. 177 ff. Artetetra Edizioni, Capua
53. In more details, the U.S. Federal Reserve has stated: "[w]hile the Federal Reserve has made no decisions on whether to pursue or implement a central bank digital currency, or

CBDC, we have been exploring the potential benefits and risks of CBDCs from a variety of angles, including through technological research and experimentation. Our key focus is on whether and how a CBDC could improve on an already safe and efficient U.S. domestic payments system". Board of Governors of the Federal Reserve System (2024), Central Bank Digital Currency (CBDC), https://www.federalreserve.gov/central-bank-digital-currency.htm, last accessed 2025/04/10. Furthermore, it should be noted that the introduction of the Digital Dollar is currently the subject of constant debate on the political scene. For example, Florida governor Ron DeSantis has stated several times that he wants to ban the use of CBDCs, both foreign and federal. See https://www.flgov.com/eog/news/press/2023/governor-ron-desantis-announces-legislation-protect-floridians-federally-controlled. Last accessed 2025/04/10

54. Bank of Canada, Digital Canadian Dollar. https://www.bankofcanada.ca/digitaldollar/. Last accessed 2024/01/18
55. Bank of England, The digital pound. https://www.bankofengland.co.uk/the-digital-pound. Last accessed 2025/04/10
56. For further information, please refer to https://finance.ec.europa.eu/digital-finance/digital-euro_en?prefLang=it#policy-making-timeline. Last accessed 2024/01/18
57. In derogation to the last point, Article 9 provides that the beneficiary has the right to refuse the digital euro under the following circumstances: "a) where the payee is a an enterprise which employs fewer than 10 persons or whose annual turnover or annual balance sheet total does not exceed EUR 2 million, or is a non-profit legal entity as defined in in Article 2, point (18), of Regulation (EU) 2021/695 of the European Parliament and of the Council 44, unless it accepts comparable digital means of payment; (b)where a refusal is made in good faith and where such refusal is based on legitimate and temporary grounds in line with the principle of proportionality in view of concrete circumstances beyond the control of the payee; (c)where the payee is a natural person acting in the course of a purely personal or household activity; (d)where, prior to the payment, the payee has agreed with the payer on a different means of payment, subject to Article 10. For the purposes of point (b), the burden of proof to establish that legitimate and temporary grounds existed in a particular case and that the refusal was proportionate shall be on the payee". The text of the Proposal is available at https://eur-lex.europa.eu/legal-content/EN/TXT/?uri=CELEX:52023PC0369. Last accessed 2025/04/10
58. As can be seen from a series of simulations conducted by the Bank of Italy, the impact of CBDCs on the banking system is fundamentally dependent on the specific features that they will ultimately possess: their total demand as well as their degree of substitutability with banknotes and traditional bank deposits, will be shaped by a range of variables, including "the legal tender status, the off-line functionality, the security and confidentiality safeguards, the remuneration, and the introduction of holding or transaction limits". More specifically, under standard conditions, a non-interest-bearing CBDC is expected to function in a manner akin to banknotes, without exerting substantial effect on banks' balance sheets. Moreover, the overall impact is projected to be limited and manageable, particularly if appropriate caps are placed on the amount of digital euro that can be held in deposit. For further details, see Auer A, Branzoli N, Ferrero G, Palazzo F, Rainone E (2024) CBDC and the banking system. Questioni di Economia e Finanza (Occasional Papers) 829, https://www.bancaditalia.it/pubblicazioni/qef/2024-0829/QEF_829_24.pdf, last accessed 2025/04/10. As regards, however, the impact on the payments system, it is perceived as predominant in light of the irreversible nature of the digitalisation of economic processes: in particular, online commerce, implying an almost total replacement of cash transactions with electronic ones, plays a significant role in the redefinition of the concept of money, as the legal tender currency appears to be, from a structural point of view, unsuitable in this context. It remains to be seen how, under the impetus of technological advancement, the regulation and structure of payment systems will continue to evolve. See de Tomasi F, Sistema dei pagamenti... cit., pp. 258–262
59. Berg T, Keil J, Martini F, Puri M (2024) CBDCs, payment firms, and geopolitics. https://papers.ssrn.com/sol3/papers.cfm?abstract_id=4402740. Last accessed 2025/04/10
60. *Ibidem.* The issue was also raised by Piero Cipollone, member of the ECB executive committee and head of the digital euro project; in particular, in his recent speech "Innovation, integration

and independence: taking the Single Euro Payments Area to the next level", he highlighted how "[a]ccording to the most recent data, international card schemes account for close to two-thirds (64%) of all electronically initiated transactions with cards issued in the euro area". In fact, due to the fragmentation found in the non-cash payments sector on the European scene, the problems related to the competitive aspect are exacerbating: "cross-border transactions within the euro area have become dependent on a very small number of non-European market players. This hampers competition, innovation and resilience". The transcription of the speech is available at https://www.ecb.europa.eu/press/key/date/2024/html/ecb.sp240424~12ecb60e1b.en.html. Last accessed 2025/04/10

61. The metaverse also offers opportunities to the other major branch of the financial system, the insurance one. Firstly, as for banks, its use can be aimed at greater involvement of customers, employees and other stakeholders (for example, one could think of employee training or customer assistance); secondly, the possession of new forms of digital assets such as NFTs, avatars and virtual real estate is inevitably linked to the emergence of new types of risks and the consequent need for protection and coverage from them. See PwC (2023), The metaverse and insurance: A new frontier in user experience and risk coverage, https://www.pwc.com/us/en/tech-effect/emerging-tech/the-metaverse-and-insurance.html, last accessed 2025/04/10. For further information on the relationship between the insurance sector and the metaverse, see also Manta O, Folcu O, Militaru I (2023) Artificial Intelligence, Integrity, and Opportunity in Insurtech. Journal of Information Systems & Operations Management 17.1; Saraf K, Mohapatro S, Reddy YK, Kumari N (2023) The Metaverse Technology in Revolutionizing the Insurance Industry. In: Hemachandran K, Rodriguez RV (eds.) The Business of the Metaverse. Routledge, New York, Routledge. In terms of applications of virtual reality products in the insurance field, the use of a model known as digital twin is becoming increasingly popular, i.e. "an integrated multiphysics, multi-scale, probabilistic simulation of a complex product, e.g. a machine, and uses the optimal available physical models, sensor updates, etc., to mirror the life of its corresponding twin". Ansari F, Nixdorf S, Sihn W (2020) Insurability of Cyber Physical Production Systems: How Does Digital Twin Improve Predictability of Failure Risk? IFAC-PapersOnline 53(3), p. 297. In particular, it is possible to mention the so-called failure-based insurance policies, i.e. "a type of product where an insurer and insured parties can continuously respond to shifts in underlying risks or behaviors in (near-to) real-time, using digital twin technology for continuous learning of machines or equipment's health score based on the interdependency between cyber-physical failure sources, estimating associated remaining useful-time (RUL), and subsequently adjusting premium as well as coverage". In other words, these policies use digital twin technology for a continuous assessment of the state of the machines or equipment, later fine-tuning the amount of the premium and coverage. The digital twin solution, therefore, helps ensure greater risk predictability. *Ivi*, p. 298
62. For example, Goldman Sachs Group Inc. has been actively exploring blockchain technology and has launched its own cryptocurrency exchange platform, known as Circle Trade. The aforementioned J.P. Morgan, in partnership with the startup Digital Asset Holdings, which deals with blockchain, is also developing a platform for trading digital currencies. Dubey V, Mokashi A, Pradhan R, Guta P, Walimbe R (2022) Metaverse and Banking Industry—2023 The Year of Metaverse Adoption. Technium. 4 (10), pp. 70–71. Furthermore, it seems useful to mention the creation of the R3 consortium, in which around a hundred entities participate, including banking institutions (for example, one of these is the Italian bank Intesa Sanpaolo). Aiming to speed up the "digital transformation" of a system that is still too slow today because it is primarily "paper-based", that is, entrusted to the exchange and verification of paper documents, projects that promote international trade in a broad sense by introducing into the ecosystem one or more networks that are shared and accepted by all parties involved and that allow the digital exchange of data and documents (etc.) are promoted in this context. Di Meo A (2019) Crediti documentari, Wolters Kluwer, Milano
63. Lu L (2023) Fintech: technology-enabled financial innovation for digital trade. In: Collins D, Geist M (eds) Research handbook on digital trade, p 310, Edward Elgar, Cheltenham
64. Lu L Fintech: technology-enabled financial innovation for digital trade… cit.

65. As evidenced by institutions such as Ping'an Bank, where AI-driven systems have effectively replaced 80% of its customer service employees, the progressive substitution of human roles with AI systems is exacerbating job insecurity and diminishing opportunities for entry-level employment. See Zungu NP, Amegbe H, Hanu C, Asamoah ES (2025) AI-driven self-service for enhanced customer experience outcomes in the banking sector. Cogent Business Manag 12(1):1–21
66. Notably, JPMorgan has established an AI Research team based in New York, focusing on developing 'solutions that are most impactful to the firm's client and business' and are based on advanced machine learning and cryptography technologies. See https://www.jpmorgan.com/technology/artificial-intelligence/research-agenda. Last accessed 2025/04/02
67. Annunziata F Fintech: tra infrastrutture tecnologiche e intelligenza artificiale… cit., p 270
68. According to Deloitte, Experian, Cetif—Università Cattolica del Sacro Cuore (2025) Osservatorio Digital Lending 2025, the generational shift is playing a central role in the digital transformation of financial services in Italy: while approximately 75% of Generation Y (Millennials) actively use digital banking channels, this figure drops to around 50% among Baby Boomers. The pandemic has further reinforced this acceleration, particularly for low-complexity, high-frequency products such as BNPL. In contrast, digital adoption in more complex products such as mortgage lending remains modest. Due to structural barriers, the digital disbursement of mortgages is expanding at a slower pace, with a growth of just 18% between 2019 and 2024; as of 2024, only 15% of the total applications are processed entirely online. Nonetheless, the integration of AI and remote property valuation is progressively reshaping the workflow in this field. The report is available at https://www.deloitte.com/it/it/services/consulting/research/digital-lending-la-nuova-frontiera-del-credito.html. Last accessed 2025/07/06
69. CETIF (2024) Il digital lending è un laboratorio per mutui e prestiti del futuro. See https://www.cetif.it/about-us-they-say/il-digital-lending-e-un-laboratorio-per-mutui-e-prestiti-del-futuro. Last accessed 2025/07/06
70. Zada M, Khan S, Megmood S, Nicolás Contreras-Barraza N (2024) Generative artificial intelligence in FinTech: Applications, environmental, social, and governance considerations, and organizational performance: The moderating role of ethical dilemmas. Oeconomia Copernicana 15(4):1303–1347
71. In Italy, for instance, both the corporate law reform and the implementation of the Shareholder Rights Directive (SHRD) have aimed at enhancing shareholder participation by reducing its logistical and financial burdens through digital tools and pre-meeting disclosures—an approach also driven by the increasing internationalization of corporate structures, which demands more flexible, border-transcending mechanisms for participation. See, *passim*, Bianchini M, Gasparri G, Resta G, Trovatore G, Zoppini A (eds.) (2022) Gli sviluppi tecnologici del diritto societario. Quaderno giuridico Consob 23. Comparatively, common law jurisdictions such as Delaware have taken a more expansive stance, allowing corporations—subject to their own bylaws—to hold shareholder meeting entirely via remote communication platforms. In contrast, Chinese law continues to mandate the physical convening of such meetings, albeit supplemented by digital voting rights and remote attendance option. See Hopt KJ (2022) Company Law and the Corporate Governance Effects of the Pandemic—The New German Law on Virtual General Meetings of Stock Corporations and the Effect of the Pandemic on the Board. Orizzonti del Diritto Commerciale 2:563-584; Ainuson K (2025) Shareholder Rights in a Digital Age: The Ghana Experience. Journal of Law, Policy and Globalization 147: 16-29; Xi C (2022) Shareholder Voting and COVID-19: The China Experience. The Chinese Journal of Comparative Law 125: 129–130
72. Metaverse banking apps allow users to access their traditional bank accounts and digital assets even in the virtual environment. For instance, The Sandbox has entered into a partnership with South Korean commercial bank KEB Hana Bank to establish a virtual branch in the Sandbox metaverse, offering basic banking services while simultaneously promoting investment opportunities. The Sandbox (2022) The Sandbox partners with KEB Hana Bank to build global business model, https://sandboxgame.medium.com/the-sandbox-partners-withkeb-hana-bank-to-build-global-business-model-d7acc452c0ee. Last accessed 2025/04/10

73. A notable example of this last function can be found in the case of Bank of America, which adopted immersive technologies (specifically, VR ones) to train employees across more than four thousand US financial centers: "[t]his innovative training technology will allow approximately 50,000 employees to practice a range of routine to complex tasks and simulate client interactions through a virtual environment". Bank of America (2021) Bank of America is First in Industry to Launch Virtual Reality training Program in Nearly 4,300 Financial Centenrs, https://newsroom.bankofamerica.com/content/newsroom/press-releases/2021/10/bank-of-america-is-first-in-industry-to-launch-virtual-reality-t.html. Last accessed 2025/04/10
74. The expression refers to "a shift in the technological paradigm and business routines due to the introduction of new technologies, which allow for subverting an economic sector, increasing its competitiveness, redefining pre-existing business models and changing the consumers' behavior in the approach to products and services". Elia G, Stefanelli V, Ferilli GB (2023) Investigating the role of Fintech in the banking industry: what do we know? European Journal of Innovation Management 26(5), p. 1365. See also Bower JL, Christensen CM (1995) Disruptive technologies: catching the wave. Harvard Business Review 1, p. 53 according to which: "[n]o matter the industry, a corporation consists of business units with finite life spans: the technological and market bases of any business will eventually disappear. Disruptive technologies are part of that cycle. Companies that understand this process can create new businesses to replace the ones that must inevitably die. To do so, companies must give managers of disruptive innovation free rein to realize the technology's full potential-even if it means ultimately killing the mainstream business. For the corporation to live, it must be willing to see business units die. If the corporation doesn't kill them off itself, competitors will"
75. J.P. Morgan launched the Onyx Lounge after seeing an increase in its customers' interest in Ethereum-based services. The bank's goal is to be able to offer, in the future, the same banking services offered by its physical branches in the metaverse. At the same time, the bank published a report analyzing the market opportunities offered by the metaverse, which can be summarized as follows: (i) economic transactions—since around 54 billion dollars are spent on virtual goods every year, nearly twice the global spending on digital music; (ii) social interaction—as exemplified by platforms such as Roblox, where users exchange around 60 billion messages a day); (iii) content creation—as demonstrated by the monetization of user-generated content on platforms like Second Life; (iv) ownership of NFTs; (v) experimentation and new partnerships. Furthermore, the report states that "[t]he success of building and scaling in the metaverse is dependent on having a robust and flexible financial ecosystem that will allow users to seamlessly connect between the physical and virtual worlds. […] We believe the existing virtual gaming landscape (each virtual world with its own population, GDP, in-game currency and digital assets) has elements that parallel the existing global economy. […]. We are building and scaling new emerging technologies to modernize infrastructure and business models including but not limited to tokenization and digital identity, as we strive for perpetual innovation and better ways to organize financial transactions and payments in the decentralized web". J.P. Morgan (2022) Opportunities in the metaverse. How businesses can explore the metaverse and navigate the hype versus reality. https://www.jpmorgan.com/content/dam/jpm/treasury-services/documents/opportunities-in-the-metaverse.pdf. Last accessed 2025/04/10
76. As stated on their official website: "[b]anks too, recognise the metaverse as an opportunity: for them, it is about accompanying their customers in the new digital universe". Deutsche Bank (2023) Why the metaverse could change banking. https://www.db.com/what-next/digital-disruption/Metaverse/banking/index?language_id=1&kid=wn.stageGHP.Metaverse.en. Last accessed 2025/04/10
77. In reality, Deutsche Bank had already opened a first 3D lounge, also in Decentraland, where visitors could view video content and read articles; the old lounge will be combined with the new one, in order to create an environment in which it is possible to interact in a new way with its customers. According to what is reported on its official website, Deutsche Bank could, in the future, use the metaverse as a space to provide advice on products. *Ibidem*
78. As stated by Ufuk Avci, who is managing the project on behalf of Deutsche Bank. *Ibidem*

79. Barnes B (2022) Extended reality and the metaverse are leading financial services into a new realm of possibility. International Banker. https://internationalbanker.com/technology/extended-reality-and-the-metaverse-are-leading-financial-services-into-a-new-realm-of-possibility/. Last accessed 2025/04/10
80. See https://www.sc.com/en/press-release/weve-partnered-with-the-sandbox-to-create-metaverse-experience/. Last accessed 2025/04/10
81. As many as 117 startups from 26 different countries took part in the Sella group's initiative, which envisages an overall investment budget of three million euros. See ANSA (2023), Incubatore dpixel del gruppo Sella investe in quattro startup, https://www.ansa.it/ansa2030/notizie/finanza_impresa/2023/07/03/incubatore-dpixel-del-gruppo-sella-investe-in-quattro-startup_c8a5fdb2-4a80-4523-a216-24e29ac25119.html. Last accessed 2025/04/10
82. *Breviter,* as has been stated, the involvement of banks and financial intermediaries in the metaverse occurs fundamentally along three lines: (1) using the metaverse to carry out marketing plans, primarily as a medium for communication, consumer engagement, and integrative distribution; (2) using the metaverse to enhance internal procedures; (3) the development of fresh goods and services. The first point relates to enhancing customer interactions and reaffirming your brand; the second is about optimizing processes such as employee management, customer onboarding, costs, and contracts (including smart ones) through data collection and digitalization; and the third is about the area of development where there is currently the most uncertainty: Although the phenomena of CBDCs appears to be gaining traction, this cannot yet be stated for the provision of less conventional services such 1. the creation of CDs, with the exception of occasional tests. Ad hoc payment services for the metaverse; 2. the metaverse-wallet offer; 3. The metaverse-ATM. Leo S, Panetta IC, Il posizionamento degli intermediari nel metaverso... cit., pp 48–56
83. From a technical standpoint, certain physiological limitations must be addressed before the banking sector can fully leverage the opportunities presented by the metaverse. In fact, it is well established that extended use of VR devices may lead to adverse effect such as dizziness, disorientation, and a sense of detachment from the physical world; moreover, as stated in the case of Deutsche Bank, "technical limitations prevented the team from writing ideas on a virtual whiteboard or working on shared documents". Deutsche Bank, Why the metaverse could change banking… cit. A further obstacle to the large-scale diffusion of metaverse banking services is, undoubtedly, linked to the costs that would burden end users in terms of acquiring the devices necessary to access them; however, it cannot be excluded that the evolution of these technologies could lead, together with their greater accessibility, also to a reduction in the costs involved in the production process and, therefore, in market prices. Nguyen LT, Duc DTV, Dang T-Q, Nguyen DP, Metaverse Banking Service: Are We Ready to Adopt?... cit., pp 5–6
84. Alpa G, Guarneri A, Monateri PG, Pascuzzi G, Sacco R (1999) I mezzi di informazione, Le fonti non scritte e l'interpretazione, vol. II of Sacco R (directed by) Le fonti del diritto italiano, del Trattato di diritto civile, p 553 ff. Torino. For a particularly in-depth analysis, see Pascuzzi G (2001) Il diritto fra tomi e bit. Generi letterari e ipertesti, Cedam, Padova
85. As it was stated, the regulations pertaining to consolidation, banking activity, and supervisory authority must be readable by the future banking supervisor from a legal perspective. However, it must also be able to confirm, from the perspective of the market, if and how a collection of technological-financial firms, or fintech, should be regulated. Additionally, it will need to be able to keep an eye on corporate activities that, more or less intentionally, aim to circumvent the regulatory system. To better explain this concept, the author uses the example of a fintech group that does not include any bank, but is made up of a payment institution and a consumer credit company: none of these entities, considered individually, carries out deposit collection and credit disbursement activities (typical banking activity); however, the situation changes if the group is considered in its unitary dimension. In summary, for the letter of the Consolidated Banking Act to remain current, it is essential to operate with a spirit of adaptation, favoring,

where necessary, an approach making substance prevailing over the form in the legal qualification of reserved activities or, in any case, in the ones subject to authorization. Lener R, Regolazione Fintech e Testo unico bancario… cit., p. 70

86. In particular, the doctrine has highlighted the problem of the supervision of banking activity in the metaverse and, therefore, of the applicable law in this context. In this regard, two theories in particular have emerged: the Data Center Location Theory and the Law of the Metaverse Theory. According to the first, specifically, the applicable law would coincide with that of the country in which the data collection center relevant for that metaverse has its registered office; this is a theory originating from the predisposition that can be recognized in the countries in which the GDPR is applied to determine the law applicable to goods or services by making it correspond to that of the domicile of the company or bank concerned. According to the second theory, however, the jurisdiction of the metaverse is a *sui generis* jurisdiction, separate from those currently existing; therefore, no positive law can be applied to the cases that arise there, making it necessary to develop a specific and detailed regulatory framework. Anggara MRH, Davie MR, Margani M, The Presence of Commercial Banks in Metaverse's Financial Ecosystem… cit., pp. 423–426. In a critical sense, see also Leo S, Panetta IC, Il posizionamento degli intermediari nel metaverso… cit., pp. 58–59, who, while recognizing the simplicity of application of the first theory, underline that it fails to account for the possibility that the metaverse may be distributed across multiple data centers, each potentially located in a different jurisdiction. With reference to the second theory, however, it is underlined that it has the advantage of guaranteeing greater legal certainty for those who operate in the metaverse; at the same time, however, a development in this sense would entail the burden of identifying a suitable subject to regulate and supervise such a vast subject as a whole, probably making the data center theory the only viable path—as of now, at least. Another relevant aspect in this sense is that linked to customers' trust in banks: "customers are more inclined to participate in banking service transactions in the metaverse if banks demonstrate kindness, integrity, and expertise". Ibidem. And, in fact, establishing a relationship of trust in the context of the metaverse can be a rather difficult undertaking where the perception one has of the most innovative aspects is not purely positive, despite the interaction being more immediate and personalized—avatars, in fact, are often associated with the world of video games and, therefore, not necessarily perceived as "professional". Finally, the issue of security, as well as privacy, of users comes to the fore: data breaches, loss of sensitive information, unauthorized access to accounts, theft of virtual assets, represent threats that financial institutions must prevent through implementing the appropriate cybersecurity measures, by providing clear guidelines and instructions for customers and by developing detailed regulatory compliance solutions. See Nguyen LT, Duc DTV, Dang T-Q, Nguyen DP, Metaverse Banking Service: Are We Ready to Adopt?... cit., p. 6
87. On the other hand, in general terms, the Internet effect has accelerated the speed of data circulation (and therefore also of rules, institutions and models), further detached law from territorial space and state ties, exacerbated the crisis of criteria traditional methods of determining the law applicable to instantaneous and in every respect globalized relationships. Borroni A (2014) I contratti di scommessa: osservazioni di diritto comparato. In: Mazzei G, Espartero Casado J (eds.) Problematiche giuridiche e ruolo sociale dello sport. Problemática jurídica y papel social del deporte, p. 150. Edizioni Scientifiche Italiane, Napoli
88. KPMG (2023) Navigating the digital frontier: the role of digital transformation and Artificial Intelligence (AI) for asset intensive organisations, https://assets.kpmg.com/content/dam/kpmg/au/pdf/2023/navigating-the-digital-frontier-asset-intensive-organisations.pdf. Last accessed 2025/04/10
89. Annunziata F Fintech: tra infrastrutture tecnologiche e intelligenza artificiale… cit., p. 265

Geotaxation and Cyberspace Conceptual Paradigms for the Reappearing Digital Taxpayer

Piergiorgio Valente

Abstract Geotaxation focuses on international tax relations and their development under the influence of geographical factors. This includes considerations such as collaborations among states and the international impact of national or local tax policies. Today, geotaxation must expand to new spaces where human activity is evolving, such as cyberspace. Unlike traditional geo-tax subjects, cyber-reality challenges established social structures, norms, and international relations, including those in the tax domain. This article explores the implications of cyberspace and modern geotaxation for the state and the potential development of the international tax landscape.

Keywords Geotaxation · Metadata · Metaverse · Cyberspace · Global fiscal sovereignty · Collective intelligence · Digital world · Artificial Intelligence (AI) · Global finance · Geopolitics · International tax scenario

1 Geotaxation in the Information Age

1.1 Introduction

Geotaxation refers to a multidimensional legal and fiscal framework that integrates geographic, economic, and digital parameters into the design and application of tax systems. It addresses the allocation of taxing rights by reconciling traditional notions

---

This contribution builds upon previous works by the Author, published by various publishing houses in recent years. It is intended solely for academic and non-commercial purposes. No copyright infringement is intended.

P. Valente (✉)
Valente Associati GEB Partners/Crowe Valente, Milan, Italy
e-mail: p.valente@gebnetwork.it

Link Campus University, Rome, Italy

CFE Tax Advisers Europe, Brussels, Belgium

A. Borroni et al. (eds.), *Metaverses: Reshaping Law, Economy, and Society*, PoliTO Springer Series, https://doi.org/10.1007/978-3-032-08664-8_6

of territoriality with the transnational nature of modern economic activities, including those enabled by digitalization, dematerialization, and global value chains.

Geotaxation is emerging as a critical geopolitical issue at the intersection of fiscal sovereignty [1], global governance, and the territorial competition for tax revenues. In an increasingly interconnected and dematerialized world economy, taxation is no longer merely a technical matter of domestic policy, but a strategic instrument through which states assert influence, protect their economic base, and negotiate their position within the global order.

As global capital, data, and value chains transcend borders, traditional principles of source and residence taxation are being challenged by new geopolitical realities. The ability of states to define, defend, or expand their taxing rights has become both a matter of national interest and international contention. This struggle is manifest in debates over digital taxation, the reallocation of taxing rights under multilateral agreements (such as the OECD's Two-Pillar solution), and the race to regulate crypto-assets and platform economies.

Geotaxation thus reveals the tensions between unilateralism and multilateralism, between developed and developing countries, and between physical sovereignty and digital presence. It calls for a reconceptualization of tax law as not only a legal and economic tool, but also a vehicle of geopolitical agency.

1.2 Geotaxation as a Geopolitical Theme

In essence, geopolitics [2] focuses on the impact of geography on international political relations and vice versa. The principal actor in international political relations is the state. And the effective power to tax is historically a constitutive element of the state.

Tax revenue is, in fact, one of the fundamental resources for the modern state to perform its role, i.e. to ensure security of the people in its territory and to deliver its policy objectives [3]. Tax policy can, therefore, be considered of key relevance for national sovereignty [4]. In this light, geotaxation can be understood as the study of the interactions between geography and the international tax framework. It focuses on international tax relations and how they evolve under the influence of geographical factors. This includes examining collaborations among states at various levels (such as those fostered by the League of Nations, the OECD, the BEPS Inclusive Framework, the United Nations, the IMF, and the European Union) and assessing their impact on both national and international tax policy.

It also considers the implications of national or local tax policies for other states and subsequently for the international tax framework. A representative example of high impact national tax policy can be seen in so-called harmful tax practices [5].

Beyond the above, long-existing factors influencing tax policies, however, modern geotaxation is required to expand to new spaces, completely untracked. It is known as cyberspace [6], and it is the product of new communication technologies and their interaction in a virtual network.

Cyberspace marks the unlimited extension of the place where human activity can evolve, the multiplication of time, of the faces of the single user, through profiling, and subsequently the multiplication of relations and transactions.

Intrinsically different from all other geotax subjects, cyber-reality is challenging established social structures and norms, international relations, including taxation, as a whole. It is preaching a weaker role for the state, raising questions as to the authority (1) to fill the vacuum for spaces that until now were regulated thereby and (2) to produce appropriate rules to regulate the new space, cyberspace.

This work will explore the changes that cyberspace and modern geotaxation imply for the state, as traditionally the main actor in the international tax scenario and the respective potential development of such scenario. To this end, it includes an analysis of the traditional role of modern states and its evolution due to the influence of geotaxation factors, such as state cooperation and tax havens.

The last sections consider the future, outlining a direction towards an optimal international tax framework in the current context. Finally, the author concludes that the role of the state must be reinvented to respond to the cyber-reality: compromises are required and flexibility to adapt to continuous transition.

1.3 State: Main Actor of the International Tax Scenario

For international law, each state is a sovereign entity (*superiorem non recognoscens*), a subject of own, independent and exclusive rights. States' sovereignty is divided into (1) domestic sovereignty and (2) international sovereignty or independence.

On the one hand, domestic sovereignty means the formal organization of political authority within the state and the ability of public authorities to exercise effective control within the borders of their own polity [7]. The state exists if there is (1) some formal structure (2) exercising effectively power (monopoly over the means of force) (3) within a certain territory/space. The individuals residing within such a territory or space are subject to the effective control and authority of the state, operating within a framework of legal and administrative subordination [8]. Overall, all individuals are subject to the control of some state, depending on their location.

On the other hand, international sovereignty or independence is considered to result from 'mutual recognition between territorial entities that have formal juridical independence'. Relations among sovereign states are coordinate—no state is subordinate to any other state. Coordination means pluralism and decentralization. There is no global state; in the international arena there are only peers.

A core aspect of sovereignty—domestic and international—is that the state exercises its power over a specific, own territory and the individuals settled there. Each state excludes other states from interference with its domestic affairs evolving in such territory and is excluded from interfering with such affairs of other states in their own territory (mutual non-interference). As in the international context all states are equal, international public law provides the rules for states' interaction.

The territory identifies the limits of sovereignty [9]. This is the case also with fiscal sovereignty. A state's taxing power is thus limited by the same territory. Under generally acceptable international tax rules, in principle, taxing power may be exercised if there is a connection of the taxpayer with the territory (residence-based taxation) or if the taxable income is located within such territory (source-based taxation).

Taxing power is then a constitutive element of the state and a key indicator of its effective existence, i.e. of the state's ability to enforce its commands and to exclude other states from interfering with its territory [10]. In this context, international tax rules prescribe the interaction of states where their taxing power conflicts or overlaps with that of their peers. To regulate such situations, states have established an expanded bilateral tax treaty network. Any limitation to the state's taxing power implies a compromise of the state's sovereignty, a restriction of the state's dominion in (or in connection with) its territory, a loss of effective power [11].

The central question becomes: who stands to gain the power relinquished by traditional state structures, and how (if at all) should states respond to this shifting landscape [12]?

1.4 Tax Competition and Globalization

States' sovereignty—domestic and international—has historically been the key to the development of the international tax framework. Today, there are approximately two hundred sovereign states, which coexist on equal terms in the international tax arena. As there is no global fiscal sovereignty, international taxation is the product of the interactions of the various fiscal sovereignties.

The rules of this interaction have recently been changed, however, due to globalization and new technologies, marking an unprecedented mobility for taxpayers and taxable bases [13]. Such mobility effectively grants taxpayers a new margin of discretion in determining their jurisdiction of taxation, thereby introducing a fresh dimension of competition among taxing states [14].

Globalization has thus introduced a powerful element of competition among states, not only for capital flows, but also for residents, talent, and taxable bases [15].

While exercising their taxing power, states must carefully consider the tax frameworks (or "tax offers") of other jurisdictions if they wish to safeguard their own revenue base [16]. In other words, there is a new limit to fiscal sovereignty, which must be exercised in a manner that ensures competitiveness of the domestic tax framework.

Tax competition can be positive, as it can lead to optimization of the national and international tax framework and improve resource allocation, both in and outside the state [17]. Tax competition may, nevertheless, be considered harmful where it drives international taxation to the bottom [18]. This is the case where tax competition is unconstrained, leading to convergence of national tax rules towards (1) ever-lower nominal corporate tax rates, (2) ever-lower taxes on mobile revenue components, such as capital and interest and (3) broader taxable bases, shifting the burden to least mobile taxpayers, such as SMEs and individuals.

Harmful tax competition has stimulated international efforts to be curbed in a collaborative manner, for example in the context of the OECD [19] and the EU [20].

The globalization, coupled with the rapid expansion of the digital economy, has profoundly challenged the traditional notion of the nation-state and its fiscal systems. Globalization has eroded traditional territorial boundaries, enabling capital, goods, services, and even labor to move with unprecedented ease across jurisdictions [21]. In the next Section, we will show how the digital economy amplifies these challenges by decoupling economic activities from physical presence.

2 Tax Competition Versus Tax Coordination

2.1 Actors in the International Tax Scenario

The decline of states' fiscal sovereignty and their traditional dominance in the international tax landscape has coincided with the rise of new actors asserting influence over the rule-making process. These emerging stakeholders (ranging from supranational institutions and international organizations to multinational enterprises and digital platforms) are increasingly shaping global tax norms, often operating beyond the direct control or oversight of individual states [22].

As a result, the international tax arena has evolved into a more complex and pluralistic space, where authority is distributed among a broader set of participants. This shift challenges the classical state-centric model and raises important questions about legitimacy, accountability, and the future architecture of global tax governance.

International and supranational initiatives have been undertaken for the pursuit of goals shared among various states, especially where unilateral actions:

- cannot lead to the achievement of such goals or can do so less effectively than interstate cooperation; or
- can be harmful for the pursuit of such goals, for example where a lack of global perspective compromises the results of unilateral efforts.

Interstate cooperation is often driven by international institutions established to this end. Such institutions enjoy so much power as is conceded to them by the cooperating states or Member States. In other words, states remain sovereign and they choose to concede part of their sovereignty. International institutions, such as the OECD, must be distinguished from supranational ones, such as the EU. Supranational institutions are conceded the power to take decisions binding the states and the means to enforce such decisions, in case of non-compliance. This is not the case with the OECD, where decisions are made through ad hoc agreements among member countries [23].

In all cases, there is some concession of power by the states, signalling that certain interests need to be promoted collectively. This is illustrated at the OECD level, in the case of the OECD Model Tax Convention (OECD Model) and the commentary thereon, as well as the Transfer Pricing Guidelines (OECD Guidelines). Both instruments are not binding for OECD members. Nevertheless, today, they constitute standard points of reference for bilateral treaty negotiations, for treaty interpretation at the domestic level, for the domestic application of national tax law and bilateral treaties in relation to taxable income with cross-border elements. Furthermore, these standards have been adopted by a wide range of tax jurisdictions, irrespective of their membership in the OECD [24].

The Base Erosion and Profit Shifting Project is another illustrative example. Launched in 2013 under the aegis of the OECD on the basis of an Action Plan developed by the G20 together with the OECD member countries, it quickly transformed into an international cooperation initiative against tax avoidance and evasion, embraced by more than hundred jurisdictions (the BEPS Inclusive Framework) [25]. The result was a series of commonly accepted recommendations to be implemented voluntarily at the single state and tax treaty level in order to ensure coordination and to remedy mismatches and loopholes in the international tax framework. States' effective adoption of the recommended measures, including through signature of the so-called Multilateral Instrument [26] indicates a further compromise of their fiscal sovereignty—even more so taking into account the stipulation of mechanisms for the monitoring of states' compliance, for example peer review in relation to BEPS minimum standards [27].

At the same time, it indicates once more (similarly to the OECD Model and the OECD Guidelines) that under certain conditions, soft law can be an instrument more effective than hard law. In the case of European institutions, the compromise of Member States' sovereignty is even stronger, as is the commitment of the Member States to cooperate in the context of a union of shared economic and political goals and values. Thus, European institutions (including the European Commission, the EU Court of Justice and the European Central Bank) take decisions binding for Member States and enforceable, in the areas of their competence. However, the area of taxation constitutes one of the most significant pending question marks over European cooperation. Decisions on tax matters require, in principle, unanimous consent by Member States [28]. This requirement renders the decision-making process particularly difficult for such matters, explaining the lack of coordination of the tax laws of Member States, with few exceptions for VAT and excise duties. The role of European institutions is limited to monitoring and supervision of Member State compliance with EU law principles, for example equal treatment of taxpayers [29].

From the above, it becomes clear that international and supranational institutions have emerged as key players in the international tax arena, exerting significant influence over the development and evolution of the global tax framework. Their growing role reflects a shift away from purely state-driven policymaking toward a more multilateral and institutionalized model of tax governance, where consensus-building, technical expertise, and normative guidance increasingly shape national tax policies and cross-border coordination efforts. In addition, their influence both reflects and reinforces the retreat of states' fiscal sovereignty [30].

2.2 Transnational Corporations

Another critical actor in the international tax arena is the transnational corporation. In essence, a transnational corporation is a multinational corporation the management of which is largely centralized at a global level. Multinationals are distinguished for performing their business activities in several jurisdictions through a local presence, relatively independent in each jurisdiction. Transnational corporations perform their activities in several jurisdictions through local legal entities coordinated by a single management [31].

What emerges is a unified economic entity (an integrated organism) that conducts a single entrepreneurial activity across multiple jurisdictions through various local instruments, all under the direction of a central leadership [32]. Local entities cannot function effectively if separated from the organism. Thus, transnational corporations transcend in terms of economy national frontiers, irrespective of, but still complying with, legal requirements. To this end, such corporations (1) exploit economies of scale deriving from international integration (global dimension), (2) adjust their offer to the local market and consumer needs (local dimension) and (3) use the know-how from local markets to optimize their offer and resource allocation (learning dimension) [33].

It is argued that transnational corporations are 'political institutions having political relations with civil society [34] i.e. that their activity has real political impact, including in terms of taxation. The tax-related impact is twofold, in terms of:

- distribution of tax revenue among tax jurisdictions; and
- national tax law and subsequently fiscal sovereignty, as states seek to provide attractive frameworks.

To begin with distribution, the international tax framework includes widely accepted rules for the allocation of the business profits of transnational corporations among the various jurisdictions, i.e. Article 9 of the OECD Model and the OECD Transfer Pricing Guidelines. These rules embrace the arm's length principle, providing that transactions between related enterprises are to be priced as if they had taken place between independent enterprises, i.e. under free competition conditions. According to the OECD Guidelines, however, even so, transfer pricing is no exact science [35].

Judgment is required and thus there is margin for discretion [36]. As several different outcomes might be equally acceptable, transnational corporations can always manipulate transfer prices to a certain extent [37]. As transnational corporations have some discretion regarding the allocation of their taxable profits amongst states, national tax regimes gain relevance for their decision. In other words, to the extent that such corporations are able to manipulate the distribution of their profits, they may be expected to opt for a structure of their business activities that permits the most favourable taxation (among other considerations). And as states wish to attract investment and tax revenue, they may be expected to construe tax regimes addressing the aforementioned need of transnational corporations and to pursue enforcement on the same consideration. It follows that states' fiscal sovereignty is constrained by the preferences of transnational corporations, and such corporations have their own role to play in the new tax scenario, independently from the will of states.

2.3 *Global Minimum Tax and the Evolution of Fiscal Sovereignty*

The erosion of fiscal sovereignty has been accelerated by recent efforts to harmonize international tax frameworks. A landmark development in this regard is the 2021 agreement under the OECD/G20 Inclusive Framework on Base Erosion and Profit Shifting (BEPS), which introduced a Global Minimum Tax Rate for multinational enterprises (MNEs). The agreement sets a minimum effective tax rate of 15% on profits of MNEs with global revenues exceeding €750 million, aiming to address the longstanding issue of stateless income and to curb the negative effects of harmful tax competition among states [38].

Historically, the phenomenon of tax competition—where states reduce corporate tax rates to attract foreign investment—has been seen as a strategy to stimulate economic growth. However, it has led to a significant erosion of tax bases, shifting the fiscal burden disproportionately onto less mobile taxpayers, such as small and medium-sized enterprises (SMEs) and individuals. The Global Minimum Tax seeks to reverse this trend, ensuring a fairer distribution of tax revenues while reducing the incentives for profit shifting to low-tax jurisdictions [39].

While the introduction of a minimum global tax represents a significant step forward, its implementation is not without challenges. The success of the measure depends on uniform application across jurisdictions, which requires robust monitoring mechanisms and international coordination. Divergences in interpretation or enforcement could undermine its effectiveness, leading to the persistence of loopholes. Furthermore, there is a concern that some developing countries might perceive the measure as an encroachment on their fiscal sovereignty. For these nations, offering competitive tax regimes has often been a crucial strategy for attracting foreign direct investment and fostering economic development [40].

In parallel with the global minimum tax, the OECD's Pillar One initiative addresses the allocation of taxing rights, focusing on redistributing a portion of profits of large digital enterprises to market jurisdictions. This shift reflects the growing recognition that value creation in the digital economy often transcends physical presence, challenging traditional notions of source-based and residence-based taxation. While this represents a step toward greater fairness in the allocation of tax revenues, it also requires states to recalibrate their fiscal policies in light of these new paradigms [41].

The adoption of a global minimum tax exemplifies the tension between national sovereignty and the need for international collaboration. For many states, fiscal sovereignty has historically been a cornerstone of their autonomy, enabling them to tailor tax policies to their unique economic contexts. However, the interconnected nature of the global economy has blurred the boundaries of sovereignty, necessitating a shift toward cooperative frameworks. Instruments like the OECD's Multilateral Instrument (MLI) provide a template for addressing mismatches and closing gaps in bilateral tax treaties, but their voluntary nature underscores the importance of trust and consensus among participating states [42].

The global minimum tax marks a turning point in the international tax landscape, signalling a move toward a more coordinated and equitable system. However, its success will hinge on several factors: the willingness of states to cede limited aspects of their sovereignty, the establishment of clear and enforceable rules, and the ability to adapt to the ever-evolving digital economy [43]. As states navigate these challenges, they must also address the broader implications of globalization and technological advancement, ensuring that their fiscal policies remain resilient and responsive to a rapidly changing world.

Ultimately, the global minimum tax offers a framework for reconciling the competing demands of sovereignty, fairness, and efficiency in international taxation. By fostering greater cooperation and alignment, it has the potential to strengthen the global tax system and provide states with the resources needed to address shared challenges, from economic inequality to climate change. However, achieving these goals will require not only technical expertise but also political will and a commitment to collective action in an increasingly interdependent world.

2.4 *Janus in the Mirror*

From the above, it clearly arises that the state, in its Westphalian sense, is in severe crisis. State sovereignty has been injured by economic developments, while it has not managed to extend itself to the new spaces that have emerged in the Information Age. And where the state cannot effectively exercise authority, the socalled Pinocchio problem arises: "the strings that bound us to state and nation have snapped, and we are left adrift in a world of multiple authorities, allegiances, loyalties and identities" [44]. The next question is: are such multiple authorities adequate to undertake the role of guide? Is the state still needed?

There is fierce debate regarding whether—and to what extent—the society can function effectively without the involvement of the state, if it is possible to have governance without government [35]. Such governance would mean that an activity can be performed effectively without an institutional hierarchical organization, either due to an alternative equivalent organization or due to purely voluntary compliance. Nevertheless, experience has shown that this is not always the case. There is a need for some supervision by the state in order to incentivize players to continue to abide by certain standards. For example, although the risk of anarchy can be a strong incentive for compliance, its effectiveness is conditioned on the existence of clear and precise rules [45]. It is thus concluded that states are necessary to provide a so-called "shadow of hierarchy" [46], to warrant the continuous respect of minimum standards by setting the framework and the incentives for compliance [47].

Furthermore, the outline of the state's role in the Information Age (i.e. in physical space and in cyberspace) must be based on certain undeniable facts:

- territorial limits are no longer relevant for the determination of states' jurisdiction;
- in a borderless space, there may be jurisdiction of either all or of none;
- non-state actors, for example transnational corporations and internet giants, have a significant impact on international politics, including taxation;
- there is vast potential in new technologies and the private sector is already familiar therewith; and the coordinates of modern society are under continuous change that cannot be expected to stop anytime soon.

From the above, it follows, first and foremost, that states around the world need to establish strong cooperative relations and to reach agreement to lead together in the new era. Cyberspace can function only on international standards and such standards can be established only with worldwide consent, i.e. with conscious concession of important parts of sovereignty by states. Such cooperation could be put in place under the aegis of international bodies, established by states to have power thereon, independent therefrom. The alternative scenario seems to be the EU tax landscape, where the reluctance of Member States to concede sovereignty has led to double competition that harms the competitiveness of the single market. Furthermore, non-state actors need to be adequately engaged and incentivized to comply with the standards set. This could be achieved by effectively involving such non-state actors in the rule-making processes [48].

It is also of critical importance that states employ state of the art technological applications for the performance of their tasks. By way of an example, blockchain technologies warrant speed and precision that would enhance the potential of states to enforce their laws. And enforceability of the state's law is a precondition to the effective existence of the state itself. To the same end, states should closely follow technological and economic developments to ensure that the framework they set corresponds to reality. This is the only way to avoid having the Pinocchio problem arise again.

The idea of the modern state took shape in the 1600s, grounded on the concept of sovereignty—domestic and international. Sovereignty, including fiscal sovereignty, has been connected with a specific territory over which the state exercises its power and beyond which it must give precedence to other states. In the centuries since then, the coordinates of the space of human action have changed substantially. The planetary space is no longer a limit. Cyberspace has emerged to host an increasingly important part of human conduct, including highly valuable economic activities.

Yet cyberspace exists beyond state authority; in fact it exists beyond all states' authority. There is no territory, no limit, no clear and enforceable framework. In the tax area, this means no taxation or—in a best-case scenario—uncoordinated and difficult-to-enforce, manipulatable tax rules. The consequence is a lack of tax revenue for states, i.e. a lack of resources for governments to pursue their policies and perform the leading role assigned to them, their growing weaker and losing control. Such impact is enhanced by the rise of new actors or the expanded role of existing ones, influencing the tax scenario: international and supranational institutions, transnational corporations and tax havens.

In this regard, it has been argued that the human society could drive further into the Information Age without the state, and that the state could be replaced with better alternatives. However, there seems to be significant evidence to the contrary: states are still needed, although they need to transform from within to be able to perform a new role, a role adapted to the demands of the new world. The new coordinates cannot be ignored. States must coexist and cooperate with new powerful actors for the definition of future tax policies. At the same time, states must establish stable cooperative relations with one another at a broad level and in a structured manner, if they are to impose their tax policies in cyberspace (and subsequently in physical space as well). Recent initiatives provide evidence that states are indeed moving in the above direction. The BEPS Inclusive Framework connotes the fruitful cooperation of more than one hundred tax jurisdictions for the improvement of the international tax framework. But this is not always the case. A counterexample is the EU, where Member States are still reluctant to compromise their fiscal sovereignty, thus prolonging the lack of coordination of tax laws in the single market and despite the risk for broad tax arbitrage at the expense of their taxable bases. There is, nonetheless, no margin for such reluctance. In the new space, the economy is developing at the speed of light and the law needs to follow without delay, if recurrence of the BEPS problem is to be avoided. States need to take action to provide a clear framework and to ensure its enforcement. Such action should be taken in a coordinated manner and with a cooperative spirit. The structure and modus operandi of transnational corporations should be the leading example for states' cooperation with the purpose of barring all exits to stateless income.

In addition, the rule-making process, in the tax area but not only, should be inclusive. All key stakeholders (e.g. transnational corporations, civil society, individual taxpayers, tax advisers and NGOs) should be invited to contribute in order for the new rules to be practical, building on the input of all interested parties. On this premise, tax compliance could be expected to be enhanced, as taxpayers would be more aware of tax law and would better understand the reason for their tax obligations. Moreover, they could be motivated to comply with such laws, in order to preserve the privilege to take part in the rule-making process.

The flexibility of tax rules is of key relevance. As mentioned, soft law instruments have been proven largely operational where different cultures intermingle, and varying standards need to be compromised. Equally, for the same reason, they are easier to develop and to adjust to new standards.

Finally, new technologies should be exploited by governments, if they are to govern an ever more high-tech private sector. And their development should be promoted.

Change is the only constant; everything flows [49]. It has always been like this. The difference, today, is that change is an integral part of daily life—not just life as a whole—so that it must be embedded in the human mentality, social culture, government policies and law, including taxation.

The continuous transition unavoidably recalls Janus [50], the two-faced Roman god, signifying the passage towards a new start. Gazing both forwards and backwards, Janus incorporates the threshold between stability (the past) and fluidity (the future). The essence of Janus lies in the awareness that each system arises from disorder in the absence of form, from the uncertainty that succeeds the collapse of an existing form. Its two-faced image stands between beginning and end, between entrance and exit. Janus should inspire the new legal framework (which should be a global one) and every new legal framework going forward.

3 Cyber Tax Havens

3.1 Tax Havens

In addition to globalization and the digital economy, tax havens have been another key driver of change and development in international taxation [51]. The pressure that tax havens generate for other states has stimulated a strong international reaction, for example under the aegis of the OECD and the EU, towards the worldwide adoption of countermeasures that have significantly changed the tax landscape.

Tax havens have been defined by the OECD as fiscally sovereign jurisdictions that employ tax and other 'incentives to attract activities in the financial and other sectors [52]. In a 1998 report on harmful tax competition, the OECD distinguished four key features of tax havens:

- no or low-income taxation (usually limited to income with cross-border elements);
- lack of effective exchange of information with other jurisdictions (secrecy jurisdictions);
- lack of transparency in the operation of legislative, legal or administrative provisions; and
- lack of substantial activity requirements, which encourages investment of passive income flows (e.g. interest and royalties), as well as the booking of article profits.

The above features render such jurisdictions particularly attractive for the investment of mobile income flows, which are then diverted from the jurisdiction where the value was created (original taxing jurisdiction). Lack of transparency and exchange of information serves the protection of the arguable investment from tax audits. The result is extraction of taxable income from the place of creation, leading to base erosion and compromise of the respective state's fiscal sovereignty. In other words, the original tax jurisdiction is pressed to introduce measures:

- to remedy the lack of tax revenue (e.g. by reducing investment) or by increasing taxes on static income flows (e.g. real estate or personal income tax); and
- to enhance its tax competitiveness, for example by applying lower tax rates to mobile income flows, in order to disincentivize their transfer away therefrom.

For these reasons, tax havens are deemed to be the protagonists of the tax race to the bottom through harmful tax competition, while facilitating criminal activity, such as money laundering, fraud and evasion. Employing favorable tax law (without substance) to drain other states' resources, they oblige such other states to react through corresponding legislation; they de facto indicate the direction of other states' laws and the development of international taxation in general. In addition, their activity has caused the adoption of a series of norms at the national and bilateral treaty level aimed at curbing their effects. Such norms (1) impose substantial activity requirements, to ensure that income is taxed at the place of value creation and (2) enhance transparency and cooperation amongst jurisdictions, ensuring that tax authorities have access to the necessary information to assess transactions with cross-border elements [53].

3.2 Cyberspace

If the above is of fundamental relevance for geotaxation, to understand the evolution of the international tax framework, their impact is still modest, compared to the new entry: cyberspace. Its very definition has proven challenging: there are currently at least twenty-eight definitions, none of which is widely accepted [54]. Nonetheless, there seems to be convergence on cyberspace outlined as 'a global and dynamic domain characterized by the combined use of electrons and electromagnetic spectrum, whose purpose is to create, store, modify, exchange, share and extract, use, eliminate information and disrupt physical resources' [55].

Briefly, constituent elements of cyberspace are: (1) tangible telecommunications devices, (2) computer systems and software, (3) network between computer systems (intranet), (4) network of networks (internet), (5) users, (6) intermediaries and (7) data [56].

What is not part of cyberspace—and its most distinctive element—is a supervisory authority monitoring in a centralized manner the activities performed therein. There is no institution and no clear hierarchy for the whole-of-cyberspace and thus no ultimate accountability. The function of cyberspace is instilled by the continuous flow of information among users. The considerable increase in the value of intangible assets in relation to that of tangibles in the context of the digital economy is, in essence, due to the potential of cyberspace. In addition, as all takes place only virtually, the same user can apply multiple, different appearances (profiles) or choose complete anonymity. These characteristics of cyberspace explain the unprecedented challenge it poses to sovereignty, including fiscal sovereignty.

The problem is that cyberspace seems to be a domain 'with no system of law enforceable' [57]. In fact, the (potential) veil of anonymity renders the application of any rule highly burdensome, if not impossible. And even if anonymity were not the problem, that would be the lack of rules for a major part of the activities online, in cyberspace, that vary significantly from activities in the physical world and thus cannot fall under existing frameworks. A straightforward example is the lack of consistent rules for the valuation of intangibles [58]. The lack of proper rules for the taxation of digital business models is considered a main cause of the stateless income phenomenon and is the target of a series of legislative initiatives, at the national, international and supranational levels [59].

A prerequisite to enforcement is the establishment of laws. Yet, the law-making process in cyberspace needs to first address some burning tricky questions:

- In the absence of both a central cyber authority and a global state, which institution—if any—or which state/states has/have jurisdiction and competence to write the rules for cyberspace?
- If rules are to be enacted, who will fall under these rules? All users? Or only some users? How should the term 'users' be defined?
- And taking into account that cyberspace is global and unlimited, connoting an unprecedented interaction of cultures and values and systems of governance, what rules would be appropriate?

Second, until today, sovereignty has been recognized to states in connection with a specific territory. Territorial borders indicated the persons (and consequently the activities) within the jurisdiction and power of each state. There are no such borders in cyberspace; the same activity takes place by the same persons everywhere (and nowhere). In the physical world, physical presence will be in either State A or in State B and on this premise, there are rules that define in which of the two it should be identified and to what extent. In cyberspace, there is physical absence: there can be no physical presence in either State A or in State B (even if there can be major, even physical, impact in both of them). There is a jurisdictional conflict that cannot be resolved while remaining loyal to the current notion of sovereignty. Cyberspace introduces an aterritorial space, questioning the very foundations of state sovereignty. Under existing rules, the absence of any physical presence in any state means the absence of any state's competence, i.e. no power, no law, no tax. And absence is expanding, such that it is becoming the rule rather than the exception.

Third, there are specific new technologies that are increasingly challenging (even) specific roles that the state used to play within its territory. By way of an example, blockchain technology has enabled the function of peer-to-peer electronic payments, with the use of cryptocurrencies, such as the bitcoin [60].

In other words, it has enabled payments without financial intermediaries [61]. Similar is the case of other blockchain applications, such as land registration. Maintenance of the registry used to be a public function of the state, which thus had ultimate control over all transactions involving land. There are already states, however, experimenting with land titling based on blockchain, which will allow the transfer of real estate with recourse to the information in the blocks, i.e. without recourse to a public service [62].

4 Metadata and Tax Blind Spots

4.1 The Data Economy: On Evaluation and Taxation

What seemed like science fiction a few years ago is now science fact. This is referred to as the "Digital Revolution". And this is evident in the next generation of high technology, which is no longer developing intelligent computers and cyborgs, but is seeking to surpass human beings by creating genuine artificial intelligence (AI) [63].

The Digital Revolution has given rise to new ways of doing business. Taxation and law, in general, cannot remain unaffected. New rules are vital to regulate the new realities and to ensure smooth coexistence in new environments. With regard to international taxation, a significant worldwide effort has been, and is being, made to deal with the extent and nature of the implications of the aforementioned scenario and to provide appropriate responses.

In 2013, the OECD/G20 Base Erosion and Profit Shifting (BEPS) initiative [64] was launched. Action 1 of the OECD/G20 BEPS Action Plan was devoted to digital economy [65], which appeared to indicate the importance of the Digital Revolution. The OECD has also initiated a completely new project, "Going Digital" [66], with a view to addressing the whole spectrum of the consequences of digitalization. The OECD is keeping the tax debate going by making revolutionary suggestions for taxation in the future, which could, for example, even be imposed on robots [67].

As Digital Revolution primarily relates to new ways of doing business, i.e. the way of realizing taxable profit, the omens are not favourable.

The Final Report on Action 1 [68] lists a series of already widespread business models that exploit digital technology [69]. A prominent example of this is participative networked platforms [70]. These platforms facilitate online social interaction and the exchange of information among users, thereby establishing a framework for the creation, dissemination, and sharing of content within the network [71].

The value of the platforms lies with their users and is directly correlated to the user-network. However, the profit is principally derived from advertising the sale of products and services ancillary to the content exchanged among the users and/or donations. As such, the taxable profit is indirectly related to the value creation chain [72], the latter being largely driven by the intervention of network effects [73].

Another business model to be considered in the context of digitalization is sharing and/or collaborative economy [74]. In this case, the platform functions as an online marketplace, with the special characteristic that ownership is substituted with use and/or access. The owners of goods who are willing to license their use, possibly together with ancillary services, meet the potential short-term licensees of their goods and/or services online.

As a result, the platform gives rise to a three-partite relationship that can be exploited for profit or not. The three parties are: (1) a service provider, and/or proprietor of the relevant goods, who could be professional or an occasional peer; (2) a user of the goods and/or services; and (3) an intermediary, i.e. the platform operator.

The idea is not in itself new.

What is new is its relevance in terms of the income realized, i.e. digitalization has maximized the potential of sharing economic models [75]. Once again, a major part of the value created by the platform relates to the network of the platform. The more the users of the platform, the more efficient and reliable the platform, the more new users it attracts, the more value its users can create. For instance, reviews and comments on the goods and services available on the platform provide a reliable point of reference for other users, thereby improving the efficiency and attractiveness of the platform is immune in relation to the main risks inherent in the transactions performed. These risks are borne by the service providers themselves, who remain in possession of the goods.

The digital marketplace has also inspired a number of supportive online services. Online payment systems, online advertising and online consultation are an important part of the digital economy. These can resemble their physical versions, i.e. cash and/or credit card payment by way of electronic (e)-wallets, adverts, marketing and so on. There is, however, an important distinction that must be noted. Online services can be provided from anywhere in the world to anywhere else in the world, regardless of the location of assets, personnel and recipients [76]. The same applies to cloud computing and application stores [77]. These are free from location and space restrictions and exist everywhere and nowhere. But what is most challenging is their characterization for the purpose of allocating taxing rights. These stand in a grey area between the provision of services and the supply of goods [78].

The metaverse presents unique challenges for taxation. It is a three-dimensional digital space where users interact through avatars, purchase virtual goods, and participate in decentralized digital economies. These environments often use cryptocurrencies and non-fungible tokens (NFTs) as transaction tools, introducing new paradigms to fiscal frameworks [79]. Such evolving modes of conducting business in the Information Age lead to a number of fundamental conclusions, chief among them being the increasing predominance of intangibles over tangibles [80]. The transfer of ownership is being changed into a licence to use, thereby effectively blurring the distinction between services and goods. Putting people and/or users in contact is becoming more valuable than directly satisfying people's needs.

The collection and analysis of data can give rise to significant value creation, thereby facilitating the optimization of business [81]. Manufacture is being freed from facilities, organization and qualified personnel. It can function, at least, equally well anywhere in the world as long as the design is correct. In the digital economy, value increasingly resides in the idea rather than in the physical object, thereby allowing the place of value creation to be virtually anywhere (provided, of course, that a screen and a connection exist) [82]. In such circumstances, it becomes increasingly evident that a tax system rooted in the notion of physical presence is struggling (often in vain) to capture value in an economy that has transcended physical constraints and now thrives on the virtual and intangible [83].

Summarizing the points raised so far, key challenges of cyberspace include:

(a) tax residency: determining tax residency when users and transactions exist in decentralized, borderless digital spaces [84];
(b) valuation of digital assets: assessing the fiscal value of NFTs and cryptocurrencies, which are subject to volatile markets [85];
(c) compliance and enforcement [86].

It is increasingly arguable that the metaverse disrupts traditional tax concepts such as residence and source of income [87]. Some authors [88] highlight the risk of the metaverse becoming a new offshore tax haven, necessitating multilateral agreements to prevent double taxation or the non-taxation of metaverse-generated income [89].

4.2 AI and Metaverse

The use of Artificial Intelligence (AI), by means of machine learning algorithms for making tax decisions, forms the basis of a new relationship between tax authorities and taxpayers. On one hand, for Tax Administration, analyzing networks of connections is especially useful for identifying complex schemes, sometimes of a circular nature (mainly transnational) and suspicious transactions (seemingly inconsistent or overtly suspicious) involving entities from different countries that might otherwise be beyond reach without the use of AI.

On the other hand, for companies, studying networks of business connections (or business linkages) focuses on examining interactions among individuals and companies in their economic and financial transactions (as well as interactions among individuals and between companies/organizations), to gain a deeper and more detailed understanding of the intersubjective relational framework. Specifically, this type of analysis provides a coherent and organized view of relationships between people and companies, aimed at assessing investment opportunities, risk management, the reliability of financial intermediaries, and the credibility of institutional counterparts in various contexts (governance, sustainability, etc.). AI develops in connection with the metaverse—a fully immersive virtual world, more expansive than cyberspace, growing parallel to the real world—in the endless calculation of infinite numbers, bits, and information racing through the network. In the metaverse, all material boundaries are eliminated, and with technology, the permeability of material borders, the porosity of physical boundaries, and the multiplicity of places are determined. This gives AI the possibility to traverse meta-physical universes in search of new data, information, and discoveries. In such realm, interstitial spaces are created, without state dominion, positioned between real spaces (territories) defined within national borders, forming a "suspended" system—oblique (as it is stateless) and ubiquitous (as it is deterritorialized). In this open space, tax bases are generated without a territory—diasporic and (seemingly) free from tax jurisdiction. Simultaneously, a sort of marginalization of profits occurs, generated (through technology) in an extra-territorial scope beyond the reach of any single state.

These spaces, expanded by technology, experiment with a notion of transnationality without territory—hybrid and asymmetrical. It is characterized by expanded territories in spaces that, while involving traditional territories, transcend them in the "here" and "now," in an absent sovereignty. In contrast to concepts of limits and boundaries and inclusion/exclusion [90], essential in defining traditional social and political categories, the territorial aspect blurs, and the legal order transforms.

The world map loses the vibrant color of individual states, fades, and shrinks under the impact of the turbulent movement of AI, the internet, and social networks, which, like an electronic whirlwind, become a new digital Big Bang for a world transformed into unprecedented forms. The market's uniform, colorless hue takes over, and the tone of finance, devoid of shades, prevails; with the nullification of territoriality as the organizing principle of economic activities, the nation-state framework loses its centrality in global-scale economic activities [91].

In the forthcoming political, economic, and social order, new initiatives, proposals, and actions will emerge to foster taxpayers' trust in tax authorities, making compliance with tax regulations easier and less burdensome. To this end, AI could (and should) be used to streamline the tax administration's bureaucracy in selecting taxpayers subject to audits, optimizing time and costs, thereby reducing errors and omissions. Correspondingly, standards for transparency, privacy protocols (on taxpayers' personal and financial data), cooperative compliance, and information exchange based on bilateral and multilateral agreements will need to be defined. This outlines the framework for a new technology-driven relationship between tax authorities and taxpayers, requiring enlightened minds and continuous efforts in understanding and regulating to remain fair and ethical on both national and international levels [92].

4.3 Collective Intelligence and Digital World

The collective intelligence (CI) [93] suspended in the network is the product of artificial brains and human minds. The interdependence between the two defines the relationship between structure and process. These elements, bonded together in cyberspace, co-evolve in a digital ecosystem where the artificial, created by humans, renders individual intelligence virtual [94]. Individual intelligence merges with others of its kind, is enriched by them, and is enhanced through this interaction. In contact with other intelligences, it is neither depleted nor diminished through use, nor is it lost in fusion with others. If it can be described as "consumed," it is a consumption that is fruitful and generative of a new rationality, enriched and increasingly intelligent by "consuming" others' thoughts. From the interaction of various individual intelligences emerges a shared heritage, a true collective intelligence that elevates cyberspace to a living infosphere, a digital laboratory of human thoughts, desires, memories, and traditions accumulated over centuries [95].

With the advent of AI, the digital world becomes (even more) abstract, immaterial, and fragmented into its constituent components, dissolved in fluid connections that form and dissolve in an instant. Legislators are tasked with bravely navigating the chaos of this abstraction with interpretative tools of the law, to shape the contours of this intensified exchange of people, goods, data, and financial flows. As a representation of reality, law is abstract as well, like the nature of the object it seeks to regulate. And so, when reality becomes imperceptible in the manifestations of money, creativity, digital exchange, and in the labyrinth of AI calculations, law finds itself working upon itself.

Today, we are called to forcefully define the legal contours of abstraction in AI, in new sophisticated technologies, and robotics. This process is essential for the development of new law (intellectual property, legal liability, privacy and data protection, ethics, and transparency), to ensure coherence and justice in legal decisions, and to define the foundational criteria for AI-related taxation (classification of AI-linked assets, recognition of revenues and costs at the national level, tax credits and incentives, transfer pricing, royalties, and international transaction taxation).

In the necessary transition to a foundational phase for a new legal order and a new form of taxation, the dual nature of economy and law must be clarified in emancipatory terms, to overcome the predatory aspects of globalized economics. This duality signifies the capacity to align with the historical momentum shaped by technological evolution in both the natural sciences and human consciousness. Yet, at the core must always remain work—anchored in Humanity—liberated from the false simulacra of legal fictions, distorted economic reflections, and political hypocrisies. Only through this grounding can we cultivate a form of collective intelligence attuned to the highest faculties of the individual and their innate call for fulfillment [96].

References

1. Grounded in principles of fiscal sovereignty, equity, and efficiency, geotaxation proposes a reconfiguration of nexus rules, profit attribution standards, and jurisdictional boundaries. It aims to ensure that tax liabilities are fairly distributed across states in accordance with the actual geographic dispersion of value creation—whether physical, virtual, or hybrid in nature. See Valente P (2019) Geotaxation and the digital: Janus in the mirror, Intertax 47(4)
2. Geopolitics has been defined as "great power competition over access to strategic locations and natural resources" by Overland I (2015) Future Petroleum Geopolitics: Consequences of Climate Policy and Uncoventional Oil and Gas, in Handbook of Clean Energy Systems, Wiley. This definition underscores the core premise that geography (particularly in terms of territorial positioning, resource distribution, and access routes) plays a decisive role in shaping the power dynamics among states. In this context, geopolitical strategy is driven not only by military and diplomatic objectives, but also by economic imperatives, such as securing energy supplies, critical raw materials, trade corridors, and digital infrastructure. Historically, geopolitical tensions have often emerged where strategic control over land, sea, or resource-rich regions has been contested by rival powers. Today, such competition extends into new domains, including cyberspace and the global financial system, where the control of data flows, technological standards, and fiscal sovereignty has become equally central to geopolitical influence

3. Hearson M (2016) Bargaining away the tax base: The North-South Politics of Treaty Diffusion, Thesis for the Department of International Relations of the London School of Economics and Political Science for the degree of Doctor of Philosophy
4. Rosenbloom D (2009) Where's the pony? Reflections on the making of international tax policy. Canadian Tax J 57(3)
5. These typically refer to regimes that offer preferential tax treatment (such as abnormally low or zero effective tax rates, lack of transparency, or weak enforcement mechanisms) to attract mobile capital and profits without requiring substantial economic activity or presence within their jurisdiction. Such practices can incentivize aggressive tax planning, erode the tax base of other countries, and exacerbate global inequalities in tax revenue collection. This has prompted coordinated responses at the multilateral level, most notably through the OECD's work on Base Erosion and Profit Shifting (BEPS), and its subsequent Inclusive Framework. Through these initiatives, the international community has sought to identify, monitor, and curtail harmful tax practices, promoting a more transparent, cooperative, and substance-based approach to cross-border taxation. In this sense, the international tax framework is continuously shaped not only by consensus, but also by the tensions and externalities generated by unilateral national tax choices
6. Cyberspace is not a physical place, but a dynamic and layered construct where digital interactions occur (from financial transactions and data storage to communication and algorithmic decision-making). It is the realm where identities, assets, and power are increasingly exercised, shifting the geopolitical and legal landscape toward non-territorial logics of control and regulation (it includes everything from emails and websites to financial systems, cloud platforms, and online media). Cyberspace is primarily functional and informational, serving as the backbone of global digital connectivity.The metaverse, by contrast, is a more immersive, embodied, and interactive extension of cyberspace. It denotes a persistent, three-dimensional virtual environment in which users are represented by avatars and can interact in real-time across social, economic, and creative contexts. The metaverse blends virtual reality (VR), augmented reality (AR), and blockchain technologies to simulate physical presence, property, and identity in a shared digital world. While still evolving, it aspires to become a parallel digital society, with its own rules of governance, economics, and social dynamics
7. Krasner S (1999) Sovereignty: Organized hypocrisy. Princeton University Press
8. This relationship implies more than mere physical presence; it encompasses the individual's submission to the regulatory, fiscal, and judicial powers of the state, as exercised through its institutions. Such subordination manifests in various forms, including the obligation to comply with national laws, pay taxes, adhere to public policy directives, and accept the jurisdiction of domestic courts. It reflects the fundamental principle of territorial sovereignty, whereby the state not only exerts power over the geographic domain but also maintains exclusive competence over persons and activities within its borders. In the context of taxation, this control forms the basis for fiscal jurisdiction, justifying the state's right to impose taxes on individuals and entities located within its territory. It is this effective control (rather than mere physical boundaries) that determines the scope of the state's legal reach, reinforcing its legitimacy in matters of governance, enforcement, and fiscal policy
9. Kelsen H (2000) General theory of law and state 213. Etas, Milano
10. The state has also been defined as 'an organization with a comparative advantage in violence, extending over a geographic area whose boundaries are determined by its power to tax constituents' while it has been attributed 'two mutually reinforcing monopolies', i.e. military force to ensure payment of taxes and taxation that ensures maintenance of military force. See North D (1981) Structure and Change in Economic History, WW Norton & Co; Elias N (1939) The Civilizing process: Sociogenetic and psychogenetic investigations. Blackwell Publishing
11. In the words of Rosenbloom "there can be no other legal subject—no courts, no legislature, no functioning democracy—without a tax system. And conversely, as the tax system of a country fractures and loses legitimacy, so does representational government" (supra)

12. This inquiry is particularly urgent in an era marked by decentralization, technological disruption, and the rise of transnational actors (both corporate and digital) that increasingly influence areas once reserved for sovereign governance. States may find themselves compelled to redefine their roles, reassert regulatory authority, or even reinvent their institutional frameworks in order to remain relevant and protective of the public good. Whether through adaptive policy, international collaboration, or strategic retreat, the manner and extent of their response will shape the future balance of power in a world where borders are increasingly abstract, yet authority remains essential
13. Valente P (2017) Taxless corporate income: Balance against white income, grey rules and black holes, European Taxation 57(7)
14. In an environment where individuals and corporations can relocate capital, intellectual property, or even residency with relative ease, the ability to choose between tax regimes places pressure on states to adapt their fiscal policies. This phenomenon shifts part of the tax-setting power from the state to the taxpayer, compelling governments to engage not only in traditional tax administration, but also in strategic positioning to remain attractive in a globalized fiscal landscape. It underscores the growing interdependence between tax sovereignty and international mobility, where states must balance competitiveness with fairness, sustainability, and compliance.From this perspective, globalization seems to transform the international tax arena into a market where states act as providers of taxing frameworks that can be more or less attractive for taxpayers. See Strange S (1996) The Retreat of the State: The Diffusion of Power in the World Economy, Cambridge University Press
15. Dagan T (2018) International tax policy, Bar Ilan University Faculty of Law Research Paper No. 1805, Cambridge University Press. As economic activity becomes increasingly mobile and borderless, jurisdictions are compelled to offer more attractive fiscal, regulatory, and social environments to retain or attract high-value taxpayers and corporate investment. This competition can foster innovation and efficiency, but it may also trigger a downward spiral in tax rates, regulatory standards, and social protections (a phenomenon often referred to as a “race to the bottom”)
16. In this context, states must navigate the tension between maintaining national tax sovereignty and participating in a more harmonized global tax framework. The challenge lies in designing tax policies that remain competitive without undermining revenue sustainability, equity, or the broader social contract
17. In a globalized economy characterized by the high mobility of capital, labor, and digital services, domestic tax policy cannot be formulated in isolation. Competitive tax rates, incentive regimes, and compliance burdens in other states may significantly influence the behaviour of multinational enterprises and high-net-worth individuals, potentially diverting taxable activity abroad. As a result, states are increasingly compelled to align fiscal sovereignty with international realities, balancing the need for competitiveness with the imperative of maintaining stable, equitable, and sustainable public revenues
18. CFE Tax Advisers Europe (2018) Opinion Statement FC 10/2018 on the European Commission Platform for Tax Good Governance Discussion Questionnaire on Tax Competition and Competitiveness, http://taxadviserseurope.org/wp-content/uploads/2018/12/CFE-Opinion-Statement-Tax-Competitiveness-Final.pdf; Tiebout C (1956) A pure theory of local expenditures. J Politi Econ 64(5). The University of Chicago Press
19. Devereux M, Lockwood B, Redoano M (2003) Is There a race to the bottom in corporate taxes? an overview of recent research. https://warwick.ac.uk/; Coehlo A, Tavares-Lehmann A, Lehmann F, The influence of the tax burden in attracting foreign direct investment. https://www.researchgate.net/publication
20. OECD (1998) Committee on fiscal affairs, harmful tax competition: An emerging global issue; OECD, Centre for Tax Policy & Admin, The OECD’s Project on Harmful Tax Practices: 2006 Update on Progress in Member Countries.
21. Council Conclusions of the ECOFIN Council Meeting on 1 Dec. 1997 Concerning Taxation Policy

22. This increased fluidity has frequently outpaced the ability of states to effectively tax economic activities that were once clearly confined within their borders. As a result, the foundational principles of tax jurisdiction (based on physical presence, residence, and source) have been increasingly challenged by the transnational nature of modern economic activity. This disconnect has created both practical and conceptual difficulties for national tax administrations, particularly in addressing base erosion, profit shifting, and the digitalization of the economy. In this context, states are compelled to reconsider their tax frameworks, not only to preserve revenue, but also to ensure fairness, coherence, and international coordination in an evolving global tax landscape
23. Their growing influence reflects a shift in the locus of authority within international tax governance, where non-state actors now play a central role in setting standards, driving compliance, and influencing policy debates that traditionally fell within the exclusive domain of sovereign governments
24. In this framework, states are bound only to the extent that they voluntarily agree to and implement the proposed standards. The OECD thus operates on a consensus-driven model, lacking binding supranational authority, which allows member states to retain a significant degree of fiscal sovereignty while participating in the development of coordinated international tax policies
25. This reflects a form of voluntary constraint on individual states' fiscal sovereignty, undertaken in favor of adhering to commonly accepted international norms. By aligning their domestic tax frameworks with global standards (such as those developed under the OECD/G20 Inclusive Framework) states aim to enhance cooperation, reduce harmful tax competition, and strengthen the legitimacy and stability of the international tax system. This convergence illustrates a growing recognition that coordinated action is essential in addressing cross-border tax challenges in an increasingly globalized and digitalized economy
26. OECD (2017) Background brief: Inclusive Framework on BEPS
27. OECD (2016) Multilateral convention to implement tax treaty related measures to prevent base erosion and profit shifting. See also Valente P (2017) BEPS Action 15: Release of multilateral instrument, Intertax 45(3)
28. Thus, BEPS evidences that shared risks can be an opportunity for cooperation of sovereigns through the compromise of their sovereignty. See Morris W (2018) International Tax Reform: What will the future be? Tax J., https://www.taxjournal.com/articles/international-tax-reform-what-will-future-be-26092018
29. However, on 15 Jan. 2019, the European Commission proposed the progressive transition to a qualified majority decision-making system for certain areas of EU tax policy. Cf. European Commission (2019), Commission Launches Debate on a Gradual Transition to More Efficient and Democratic Decision-Making in EU Tax Policy. http://europa.eu
30. There is some compromise of the fiscal sovereignty of Member States, but still poor from an EU law perspective. The result is a single market with free movement rights between tax jurisdictions with different tax systems, inspiring (1) broad arbitrage by taxpayers and (2) double competition among Member States in the EU and between the EU and the rest of the world. The resulting loss of tax revenue can be seen as part of the price paid for Member States' continued reluctance to compromise their fiscal sovereignty. By prioritizing national autonomy over deeper fiscal integration or binding coordination, states risk undermining collective efforts to combat tax base erosion, profit shifting, and harmful tax competition—ultimately limiting the effectiveness of international tax cooperation
31. This gradual withdrawal is driven by the imperatives of the new world order, characterized by economic globalization, digitalization, and the growing interdependence of national tax systems. While states generally appear to acknowledge the need for coordinated responses and shared standards, those that resist such cooperation must often accept the consequences (namely, a reduction in their ability to tax mobile bases, such as digital services, capital, and intellectual property). This trade-off illustrates the evolving tension between the desire to preserve national autonomy and the practical necessity of engaging in multilateral frameworks to safeguard revenue, ensure fairness, and maintain legitimacy in international taxation

32. With operations, assets, and income streams spanning multiple jurisdictions, these entities possess the capacity to strategically allocate profits and structure their affairs in ways that often exploit mismatches between national tax systems. Their scale, complexity, and economic influence have made them central to discussions on base erosion and profit shifting (BEPS), and they are frequently at the heart of calls for greater tax transparency, fairness, and global coordination.Transnational corporations not only challenge the enforcement capacity of individual states, but also shape the evolution of international tax norms through their interactions with policymakers, lobbying activities, and adaptive tax planning strategies. As such, they are both subjects of regulation and participants in shaping the global tax order
33. Despite its legal fragmentation into separate subsidiaries or entities, this structure operates functionally as one enterprise, pursuing common objectives and coordinated strategies. This organizational reality challenges traditional tax and legal frameworks that treat entities as independent based on formal incorporation, rather than recognizing the underlying economic unity and control. It reinforces the need for tax systems to look beyond legal form and assess the substance and coherence of global business operations when allocating taxing rights
34. Bartlett CA, Goshal S (1998) Managing across borders: The transnational solution. Harvard Business School Press
35. Strange, supra n. 14
36. OECD (2017) Transfer pricing guidelines for multinational enterprises and tax administrations
37. CFE Tax Advisers Europe, Professional Judgment in Tax Planning, 2021: "Attitudes and practices with regard to tax planning have changed very considerably in the recent past. Regulators, businesses, advisers and academics all acknowledge that many tax-planning arrangements which were common 25 years or so ago would generally be considered abusive by today's standards. This is to say that they involved manipulation and artificiality in the design of transactions, structures and arrangements beyond the substantive operations of the businesses and the lives of the individuals concerned, which were nevertheless considered acceptable as long as 'the letter of the law' was respected. It is important to stress that individuals and businesses decided on their tax strategies, in the final instance, on the basis of their risk appetite—and they still do so today. But 25 years ago there was a very different perspective on what is abusive and therefore high risk—or, put simply in another way, what is acceptable. The question 'If it is legal, is it acceptable?' did not present itself as it does today". This discussion paper is founded on CFE's commitment to high professional standards in tax advice and seeks to promote ethical professional judgment across all tax advisers in Europe. While tax advisers play a valuable role in the proper functioning of tax systems, this role can be undermined by the promotion of abusive tax arrangements within legal parameters. CFE has issued this paper to stimulate discussion on how to tackle this problem among all who have an interest in how our tax systems function in Europe, not just tax specialists. The question ("If it is legal, is it acceptable?") comes down to whether there is manipulation and artificiality in tax planning, for example, where planning arrangements are designed without any genuine underlying economic purpose other than achieving the tax saving
38. Alternatives to the arm's length principle have been discussed from time to time; although the formulary apportionment method has been supported by a series of jurisdictions, the arm's length principle is still the focus of most jurisdictions' consent. OECD, supra n. 64, paras 1.16 et seq
39. OECD (2021) Statement on a two-pillar solution to address the tax challenges arising from the digitalisation of the economy
40. IISD (2021) Inclusive framework agreement on the global minimum tax
41. OECD (2021) Global anti-base erosion model rules (Pillar two)
42. Simontacchi S, Scandone F S, Lorenzi U (2022) Tax aspects of the metaverse: Is a Pillar Three coming?, International Transfer Pricing Journal, vol 29, no 5, pp 301–308. This paper explored a possible "Pillar Three" within the OECD's BEPS framework, addressing the unique attributes of digital assets
43. OECD (2021) Inclusive Framework on BEPS: Multilateral instrument and implementation
44. OECD (2021) Global minimum tax and implementation challenges

45. Grande E, Pauly L (2005) Complex sovereignty: reconstituting political authority in the twenty-first century, University of Toronto Press; Hall R & Biersteker T (2002) The emergence of private authority in global governance. Cambridge University Press
46. Goldstein J, Martin L (2000) Legalization, trade liberalization and domestic politics: A cautionary note, international organization, vol 54, no 3. The MIT Press, pp 603–632
47. Having accepted that the presence of the state is still needed, it is relevant to determine the role that the state should be expected to play in present circumstances, considering that its traditional role has been heavily questioned. In this respect, what is certain is that the state exists if it is effective and to the extent that it is effective—in other words: to the extent that it can make laws that are enforceable
48. Boerzel T, Risse T (2010) Governance without a state: Can it work? Wiley
49. Giving the floor to a broad range of stakeholders, it may be expected that future rules will be more practical and flexible, more business-like. In addition, the more stakeholders understand the purpose behind a rule, the more they can be expected to comply therewith
50. Platon, Kratilos, para. 402a—saying attributed to Heraclitus of Ephesus, a pre-Socratic Greek philosopher
51. Valente P, Bagetto L (2017) Geofiscalità, Il Dilemma di Giano tra Cifra Tellurica e Continente Digitale, Eurilink, Roma. Valente P (2002) Il Continente Digitale, il Sole-24 Ore, Milano. Valente P (2001) L'Impresa Invisibile, il Sole-24 Ore, Milano, 2001
52. Valente P (1998) Cyber (Tax) planning—from an Italian perspective, Tax Plan Int Rev:17
53. OECD (1998) Harmful tax competition, supra n. 19
54. OECD (2015) Base erosion and profit shifting. http://www.oecd.org/tax/beps/
55. Krammer F, Starr S, Wentz L (2009) Cyberpower and national security: Policy recommendations for a strategic framework, in Cyberpower and National Security, National Defense University
56. Mayer M, Martino L, Mazurier P, Tzvetkova G (2014) How would you define cyberspace?, experimental online laboratory, PhD in Politics, Human Rights and Sustainability, Scuola Superiore Sant'Anna. https://www.academia.edu/
57. Ibid.
58. Mayer M, de Scalzi N, Martino L, Chiarugi I (2013) International politics in the digital age: power diffusion or power concentration? From the XXVIIth SISP Conference, University of Florence, International Relations Section
59. OECD (2015) Aligning transfer pricing outcomes with value creation—actions 8–10. Final Reports, OECD/G20 Base Erosion and Profit Shifting Project
60. Valente P (2018) Time to establish a modern, fair and efficient taxation standard for the digital economy, Intertax 47(4)
61. Nakamoto S (2008) Bitcoin: A peer-to-peer electronic cash system. https://bitcoin.org/bitcoin.pdf
62. Valente P (2018) Bitcoin and virtual currencies are real: Are regulators still virtual?, Intertax 46(6)
63. World Bank Group (2017) Natarajan H, Krause S & Gradstein H), Distributed Ledger Technology (DLT) and Blockchain, FinTech n. 1, http://documents.worldbank.org/ curated/en/177911513714062215/pdf/122140-WP-PUBLIC-Distributed-Ledger-Technology-and-Blockchain-Fintech-Notes.pdf. The vacuum of state or other institutional jurisdiction and authority in cyberspace has left room for action to non-state actors. The dominant position that such actors have gained is evidenced in the often highly impressive impact of cyber-campaigns - for example the case of the Zapatistas in Mexico (McLaughlin W S (2003) The Use of the Internet for Political Action by Non-State Dissident Actors in the Middle East, 8 First Monday 11) or in the partnerships of states with private industry for cybersecurity purposes (Gallagher P (2013) The Partnership Between NIST and the Private Sector: Improving Cybersecurity, Testimony, https://www.nist.gov/speech-testimony). See Whyte C (2015) On the Future of Order in the Cyberspace, Strategic Studies Quarterly, Air University, Q9(2)
64. The effects of the Digital Revolution are innumerable and multifaceted. Dispute resolution, education, employment, healthcare, human rights and transport are just a few of the areas that

are affected. Robots now permeate every aspect and sphere of human life, both personal and professional. Thinking machines and humanized robots are becoming the new doctors, the new drivers, the new judges and the new professors, as well as the new chefs, the new secretaries and the new vendors, and this is only the beginning. Modern relationships now more and more involve humans and quasi-humans

65. The OECD/G20 BEPS initiative was launched in 2013 with OECD, Addressing Base Erosion and Profit Shifting (OECD 2013), International Organizations' Documentation IBFD. This report sought to identify the circumstances in which domestic and international tax rules were not aligned with modern business practices and procedures and, therefore, facilitated tax avoidance. The OECD/G20 BEPS initiative evolved into 15 Actions dealing with the primary challenges identified. These were set out in OECD, Action Plan on Base Erosion and Profit Shifting (OECD 2013), International Organizations' Documentation IBFD. In October 2015, the Final Reports on the 15 Actions were published. In June 2017, the core of the OECD/G20 BEPS initiative was completed with the signature of the Multilateral Convention to Implement Tax Treaty Related Measures to Prevent Base Erosion and Profit Shifting (24 Nov. 2016), Treaties IBFD [hereinafter: the "MLI"]. The MLI is intended to amend existing treaty networks to bring these into line with the outcomes of the OECD/G20 BEPS initiative
66. The digital economy is defined by the OECD as "the result of a transformative process brought by information and communication technology (ICT), which has made technologies cheaper, more powerful, and widely standardized, improving business processes and bolstering innovation across all sectors of the economy". See OECD, Action 1 Final Report 2015 - Addressing the Tax Challenges of the Digital Economy Executive summary, p. 11 (OECD 2015), International Organizations' Documentation IBFD, also available at www.oecd.org/tax/addressing-the-tax-challenges-of-the-digital-economy-action-1-2015-finalreport-9789264241046-en.htm
67. See www.oecd.org/going-digital
68. Oberson X (2017) How taxing robots could help bridge future revenue gaps. www.oecd.org/
69. OECD (2015) Action 1 final report
70. The outlines of a law of abstract things - such as representations of money (finance), of ideation (copyrights), and of telecommunications (e-commerce, digital content, information and advertising) - trace the connection between representation and the tangible sphere of productive activity, the growth of social wealth, and, more broadly, the political existence of communities. In this light, the political identity of the age of globalization is reconstructed, countering the narrative that depicts it solely as the imposition of mass debt by voracious financial oligarchies
71. Participative networked platforms are usually divided into the following four categories: (1) social networking, for example, Facebook; (2) online games, for example, Online Chess; (3) participative community platforms, for example, Wikipedia; and (4) Internet publishing and broadcasting platforms, for example, YouTube. See OECD (2007) Participative Web and User-Created Content: Web 2.0, Wikis and Social Networking, www.oecd.org and The Economic and Social Role of Internet Intermediaries (2010). www.oecd.org
72. The content transmitted through these platforms is generated and owned by the users themselves, who retain full intellectual property rights over their contributions. The platforms act primarily as intermediaries, providing the technological infrastructure for hosting, sharing, and distributing the content, but do not acquire ownership or proprietary rights over the material.This distinction reinforces the principle that platforms, while facilitating communication and engagement, are not content creators and therefore bear limited responsibility for the substance of user-generated content (subject, however, to applicable terms of service, community guidelines, and regulatory obligations concerning content moderation, copyright enforcement, and data protection)
73. The term "value chain" is defined as a set of activities that an enterprise operating in a specific industry performs to deliver a valuable product or service for the market. See Porter M (1985) Competitive Advantage: Creating and Sustaining Superior Performance, Simon and Schuster, New York

74. According to the OECD, network effects relate to the fact that the decisions of the users may have a direct effect on the benefits received by other users. See OECD (2015), Action 1 Final Report
75. The European Commission defines the sharing and/or collaborative economy as "encompassing business models where activities are facilitated by collaborative platforms that create an open marketplace for the temporary usage of products and services often provided by private individuals". See European Commission, Communication from the Commission to the European Parliament, the Council, the European Social and Economic Committee and the Committee of the Regions, A European Agenda for Collaborative Economy (COM (2016) 356 final) (June 2016), http://ec.europa.eu. Typical examples of collaborative economy business models are Airbnb and BlaBlaCar
76. Shaheen S, Sperling D, Wagner C (1999) A Short History of Car-Sharing in the 90s, 5 J. World Transport Policy & Prac 3. http://innovativemobility.org
77. Typical value chain analysis to determine the place and process of value creation entails the weighing the functions performed, together with the assets used, the risks assumed and the people involved
78. According to the OECD, "application stores typically take the form of central retail platforms, accessible through the consumer's device, through which the consumer can browse, view information and reviews, purchase and automatically download and install the application on his device". See OECD (2015) Action 1 Final Report
79. Beyond the service sector, production is experiencing an even more profound revolution. It is becoming quasi home-made. By way of 3D or layered printing, consumers can purchase licences, whether or not online, in respect of designs from manufacturers and build the desired product at home or, in the worst case, at a nearby specialized printing centre. The process uses the same concept as printing. The only difference is that materials are not limited to paper and several layers may be needed to arrive at the 3D result.With regard to such premises, the OECD insists that this involves the transformation of production and sales into licensing. This would imply materially different tax treatment. See OECD (2015) Action 1 Final Report, para 271
80. Economic activities within the metaverse include the buying and selling of virtual goods, services, and digital real estate. NFTs and cryptocurrencies represent a new form of wealth, often unregulated by traditional tax laws. The absence of physical boundaries further complicates the allocation of taxing rights. See Visca S (2022) Life in the metaverse: let there be tax? Tax adviser, p 30
81. In today's digital economy, value creation is often driven less by physical assets and more by intangible elements such as intellectual property, data, algorithms, brand equity, and user engagement.This shift challenges traditional economic and legal frameworks, particularly in areas such as taxation, valuation, and ownership, which were historically designed around the exchange of tangible goods and physical presence. As a result, policymakers and scholars are now compelled to rethink how value is measured, regulated, and taxed in an economy where intangibles dominate the landscape. See Valente P (2019) The Data Economy: On Evaluation and Taxation, European Taxation, p 251
82. If the potential of data to create value is currently increasingly acknowledged, the situation was different only a few years ago when the collection of data was deemed to be low value adding activity. See European Commission, Report of the EU High Level Expert Group on Taxation of the Digital Economy, May 2014 and Olbert M (2016) International Taxation in the Digital Economy: Challenge Accepted?, Mannheim
83. This shift marks a profound departure from traditional economic models, where value was typically tied to the production, exchange, and ownership of tangible goods within defined geographic boundaries.Now, creativity, code, user data, and network effects generate immense economic worth, often independently of any fixed location or physical infrastructure. This decentralization of value challenges existing legal, economic, and fiscal frameworks, which must now contend with an economy that is simultaneously everywhere and nowhere
84. This digital economy celebrates borderless transactions, dematerialized assets, and decentralized operations, challenging the very foundations upon which traditional tax frameworks were

built.As economic activity becomes progressively detached from geography, relying instead on data flows, digital interfaces, and user interactions, the mismatch between conventional tax rules and the realities of value creation becomes more pronounced. This calls for a fundamental rethinking of how and where economic presence is defined (and how it should be taxed)

85. Carman PD (2023) Pinning down a cloud: source rules in the metaverse, Tax Notes International, vol 109, no 6, pp 727–735, here the author emphasizes the complexities of source rules in the metaverse. He notes that the lack of a physical reference point for transactions complicates traditional frameworks and suggests alternative criteria based on server locations, user IP addresses, or the effective use of digital assets
86. Young RK (2024) Taxing the Metaverse, Georgetown Law Journal, Cardozo Legal Studies Research Paper No.724. This paper offers novel arguments justifying metaverse taxation. Because economic activity within the metaverse satisfies both the Haig-Simons and Glenshaw Glass definitions of income, its exclusion from taxation could create a virtual tax haven. Furthermore, this emerging technology provides policymakers with an opportunity to modernize the tax system. The author argues that the metaverse's capacity to record all digital activity and track individual wealth presents a unique opportunity for governments to tax income immediately upon receipt. This would allow tax authorities to overcome the traditional realization requirement, which often incentivizes deferral. An immediate taxation approach (such as a mark-to-market system) could prove more efficient and equitable, provided it can address the inherent challenges of valuation and liquidity
87. By ensuring compliance with tax regulations in an environment where anonymity and decentralization prevail. See Franssens G, Van West J P (2022) Taxing the metaverse: the EU perspective, Tax Notes International, vol 105, no 13, pp 1535–1537, where the authors analysed the applicability of European tax law, suggesting that VAT and digital services taxes could be adapted for virtual transactions. However, they identify significant challenges, including tracking transactions and identifying economic operators responsible for tax compliance
88. In essence, as economic activity becomes increasingly digitized and decentralized, individuals and businesses are able to operate within virtual environments that transcend national borders. This development challenges the ability of tax authorities to apply traditional principles based on physical presence or geographic location. The concept of "residence" becomes blurred when avatars, digital assets, and blockchain identities supplant physical footprints. Likewise, identifying the "source" of income grows more complex when value is created, exchanged, and monetized within virtual ecosystems that lack clear territorial connections. As the metaverse continues to evolve, existing tax systems will need to adapt to these emerging realities (or risk becoming obsolete)
89. Owens JP, Oliveira Costa N (2022) The tax treatment of the metaverse economy and the potential for a new offshore tax haven. Tax Notes Int 107(5):537–546
90. This potential for tax avoidance or arbitrage arises from the metaverse's inherently borderless nature, where users, assets, and transactions exist in a virtual space disconnected from traditional notions of physical presence or jurisdiction.Such developments may require urgent multilateral coordination to establish clear rules for taxing metaverse-generated income and to prevent scenarios of double taxation (or, conversely, non-taxation) due to mismatches in national tax systems. Without international consensus, the metaverse could exacerbate existing challenges in global tax governance and undermine the fairness and sustainability of domestic tax bases
91. Valente P, Bagetto L (2017) Geofiscalità, Eurilink, Roma, p 100
92. On one side, the global economy operates in markets without national barriers, inextricably interdependent, lacking geographical anchoring, and detached from the real economy; on the other, the nation-state finds itself permeable to the free flow of dematerialized wealth. In a double reversal of power dynamics, the state repurposes its structures in relation to the market, and politics becomes subordinated to the economy, becoming an instrument of supranational and transnational interests that transcend the logic of peoples and communities
93. Such a mission translates into the pursuit of common guidelines for AI usage through synergy among administrative authorities, international organizations (UN, OECD, EU), businesses, and tax professionals, in a spirit of global collaboration

94. Lévy P (1999) Il virtuale, Cortina, Milano, trad. it., p 115
95. Lévy P (1997) Il virtuale, Cortina, Milano, trad. it., p 109
96. The self-sustaining nature of collective intelligence enhances it through a circular interplay and mutual relationship between past and future
97. An intelligence capable of embracing the entirety of human knowledge and the boundless expanse of the web, in service of the common good of all the world's peoples

Outer Space Metaverse as Key Space Traffic Management Enabler

Piercarlo Rossi and Giulia Roasio

Abstract The metaverse concept has recently gained prominence across various sectors, including its potential application to outer space operations. This innovative technological paradigm offers transformative possibilities for outer space—a domain that remains largely deregulated and governed by outdated international treaties, posing peculiar challenges. Furthermore, the absence of robust data transparency and sharing frameworks poses significant barriers to realising a metaverse tailored to this context. Addressing these challenges could unlock the metaverse's potential to support the development of legal norms for Space Traffic Management, shared collision avoidance procedures, and the improved regulation of insurance, liability, and damage compensation mechanisms. Building on advancements pioneered for commercial metaverse applications. This chapter examines the potential of the metaverse as a critical enabler for integrating technological innovation with regulatory advancements in outer space governance.

Keywords Space traffic management · Collision avoidance procedures · Metaverse · Legal design · Space law

1 Introduction

Nearly seventy years have passed since the launch of the first satellites, during which space activities have expanded significantly in diversity, purpose, and global participation. From Earth's orbital environment, exploration has extended further into deep space, the Moon, other planets, and beyond. The terrestrial orbital environment, where space exploration began, is a finite resource increasingly exploited by states, international intergovernmental organisations, and non-governmental entities.

P. Rossi (✉) · G. Roasio
University of Turin, Turin, Italy
e-mail: piercarlo.rossi@unito.it

G. Roasio
e-mail: giulia.roasio@unito.it

A. Borroni et al. (eds.), *Metaverses: Reshaping Law, Economy, and Society*, PoliTO Springer Series, https://doi.org/10.1007/978-3-032-08664-8_7

The Earth's orbital environment comprises several layers, each offering unique characteristics and applications critical to modern satellite operations [1]. Closest to the planet lies Low Earth Orbit (LEO), located a few hundred kilometres above the surface. This region is heavily utilised for Earth observation, scientific research, and an increasing number of communication satellites and constellations that provide high-speed internet services.

Beyond LEO, Medium Earth Orbit (MEO) spans altitudes ranging from approximately 2,000 to 35,000 kms. It is predominantly utilised for satellite navigation systems, including GPS and Galileo, providing precise positioning and timing services across the globe.

The Geostationary Orbit (GEO) is about 36,000 kms above the equator, allowing satellites to synchronise with Earth's rotation and maintain a fixed position relative to the surface. This makes GEO suitable for telecommunications, television broadcasting, and meteorological monitoring.

Additional specialised orbits, such as highly eccentric elliptical trajectories and polar orbits, address specific operational needs. Polar orbits, in particular, allow satellites to monitor the entire planet traversing over the Earth's poles. This makes these orbits ideal for global environmental monitoring and reconnaissance missions.

Collectively, these orbital regions facilitate a broad spectrum of services, including communication, navigation, scientific exploration and environmental surveillance. They are indispensable to daily life, underpinning a rapidly growing space economy [2]. However, as the number of satellites and space missions increases, so do the challenges related to orbital congestion and space debris proliferation. The growing complexity of space operations, the emergence of prominent constellations, and the rising risks of collision and interference with the functioning of space objects impact the long-term sustainability of space activities [3].

Recognising the growing risks associated with orbital congestion and debris proliferation, the United Nations Committee on the Peaceful Uses of Outer Space (COPUOS) has emphasised the importance of sustainable space activities. Established in 1959 by the UN General Assembly [4], COPUOS defines sustainability in space as maintaining space activities indefinitely, ensuring equitable access to benefits while preserving the orbital environment for future generations [5].

Key elements for maintaining sustainability include Space Situational Awareness (SSA), Space Traffic Management (STM), and space debris mitigation. These interconnected components are vital to ensuring safe and sustainable operations in Earth's orbital regions [6], but effective regulation poses considerable challenges given the complexity of these issues.

This chapter focuses on the urgent issue of STM regulation and highlights how advanced technologies, particularly the metaverse, can offer innovative solutions to address this challenge.

2 Specialty and Critical Issues of the Outer Space Sector

Orbital congestion and debris proliferation, particularly in LEO, has reached concerning levels [7], posing a substantial threat to the sustainability of space activities and to people, property, public health, and the terrestrial environment [8].

They also complicate space safety. Orbital congestion increases the complexity of space operations and can compromise ongoing missions. Without effective STM measures and fully reliable SSA systems, the likelihood of accidents and the degradation of orbital conditions could rise dramatically.

From an economic perspective, this situation substantially burdens space companies. The expansion of satellite constellations in LEO adds thousands of satellites to an already crowded, debris-filled environment. While losing a few satellites within a constellation might not jeopardise the entire mission or its commercial value, deploying prominent constellations intensifies collision risks and potential interference. This, in turn, heightens the liability risks for states, as outlined in Article 7 of the 1967 Outer Space Treaty (OST) [9] and the 1972 Liability Convention [10].

Debris proliferation and orbital congestion also challenge orbital services development, as managing collision risks complicates SSA and STM system advancements [11]. If left unaddressed, the growing satellites and debris volume problem could yield severe economic, social, and political consequences, threatening future space exploration.

The probability of a significant accident generating debris, such as the 2009 collision between the commercial satellite Iridium-33 and the non-functional Russian satellite Cosmos-2251, which produced thousands of trackable debris fragments and innumerable smaller, untraceable ones [12], remains high. The so-called 'Kessler Syndrome' describes the worst-case scenario [13], a theory suggesting that a critical mass of Earth orbital debris will further fragment in successive collisions, triggering a chain reaction and creating a non-navigable debris belt surrounding Earth. This phenomenon would render any space activity practically impossible, with severe social, economic, and political ramifications [14]. This scenario is becoming increasingly plausible as the development of a high-fidelity three-dimensional simulation model of LEO predicts that even without future launches, the growth rate of collision debris would exceed its natural decay rate within a few decades [15] [16]. Protecting space and the Earth's orbital environment is a global and cross-cutting priority demanding coordinated international action to standardise and regulate the sector.

To ensure the long-term safety and sustainability of space activities, substantial and urgent actions are required to address the Earth's orbital debris problem in three key areas: collision avoidance, prevention of new debris generation, and removal of existing and potential future debris [17]. A reliable STM regulatory framework would significantly contribute to the first two priorities and would at least support the development of Active Debris Removal (ADR) solutions.

STM is a critical concept within the governance of space activities, and even in the absence of a concrete internationally accepted definition of STM, a widely referenced one is provided by the International Academy of Astronautics, which describes it as "the set of technical and regulatory provisions for promoting safe access into outer space, operations in outer space, and return from outer space to Earth free from physical or radio-frequency interference" [18]. This definition highlights its role in preserving the principles established by the 1967 Outer Space Treaty, ensuring equitable and non-interfering access to space for all nations.

STM supports debris mitigation efforts by enabling the management of satellite trajectories to avoid collisions and limit the creation of new debris. Effective STM relies heavily on Space Situational Awareness (SSA) systems to track and predict the movement of objects in orbit. It seeks to address several operational challenges, such as orbital congestion, ensuring collision avoidance and safe operations; debris mitigation, implementing protocols for end-of-life disposal, active debris removal (ADR), and ensuring compliance with international standards; and coordination mechanisms, facilitating pre-emptive measures against potential collisions.

The international space law framework, in terms of binding legal instruments, currently comprises the five United Nations treaties on outer space: the 'Outer Space Treaty' (1967), the 'Rescue Agreement' (1968), the 'Liability Convention' (1972), the 'Registration Convention' (1976), and the 'Moon Agreement' (1984), which are binding at least for all signatory states. However, none of these treaties contain detailed STM regulations nor a reference to further instruments. Fortunately, there is broad consensus among private actors and states regarding the importance of space sustainability. This commitment is evident in the strong support given to the recent COPUOS Guidelines for Space Long-term Sustainability (LTS guidelines) [19], which were finally adopted by the United Nations in 2019 and to the previous guidelines on space debris mitigation [20], drafted by IADC, subsequently adopted by COPUOS and, upon the latter's proposal, by the UN General Assembly in 2007 [21].

Despite significant international efforts regarding space debris mitigation and LTS guidelines, these remain soft law instruments; therefore, their practical application to all stakeholders depends on states' compliance on a voluntary base through the implementation of domestic legislation. Furthermore, these guidelines lack specificity and do not include a detailed regulatory framework and operational procedures to avoid collision, ensure safe STM and implement an effective and transparent SSA system. Implementation at the national level is inadequate to ensure uniform practices, and additional and detailed regulatory levels are required, encompassing technical guidance, design, and operational standards, all of which should be embedded within binding regulations for effective, consistent enforcement.

Given the increasing complexity of space activities, there is growing recognition of the need for an international STM system, potentially modelled on traffic management systems in the aviation and maritime sectors. However, establishing such a framework requires overcoming political and technical barriers, including data sharing and enforcement mechanisms.

STM represents a cornerstone of modern space governance, aiming to balance the growing demand for space utilisation with the imperative of long-term sustainability.

Integrating advanced technologies, international cooperation, and robust regulatory frameworks will determine its success in addressing contemporary and future challenges in the orbital environment.

The metaverse and digital twin technologies are emerging as potential tools to enhance STM capabilities by enabling real-time simulations and predictive modelling. Such technologies could revolutionise how space traffic is monitored and managed, offering scalable solutions for sustainable space operations.

3 The Role of Metaverse in the Space Traffic Management Regulation

The emergence of the metaverse presents transformative opportunities for STM. The metaverse can enable advanced modelling, analysis, and coordination of satellite trajectories by leveraging immersive virtual environments, digital twins, and real-time simulations. It provides an innovative framework for monitoring, predicting, and mitigating potential collisions, optimising orbital pathways, and supporting sustainable space operations. As technological frontiers expand, integrating metaverse-driven tools into STM systems will ensure activities' long-term viability and safety within Earth's orbital regions and beyond.

The metaverse offers a virtualised, interactive platform where operators, regulators, and stakeholders can visualise and manage space traffic in real-time. Like autonomous intersection management (AIM) systems developed for terrestrial traffic [22], STM can benefit from digital twins and simulations that dynamically predict satellite movements and evaluate collision risks.

The metaverse in STM can facilitate dynamic collision avoidance modelling, distributed data-sharing frameworks and virtual negotiation mechanisms. AI algorithms can model collision risks and propose avoidance manoeuvres using real-time data, while blockchain technologies can enable transparent data sharing. Operators would thus be able to simulate and negotiate manoeuvre priorities in virtual environments, establishing protocols for equitable resource use.

Beyond technical innovations, the metaverse can assist in refining STM regulatory frameworks by providing a testing ground for legal scenarios. Simulated environments enable stakeholders to test policies on liability, damage compensation, and compliance with international treaties, offering insights into the practical application of regulatory measures.

Simulating worst-case scenarios, including chain-reaction collisions, the metaverse could inform policymakers about effective collision prevention and active debris removal strategies, supporting the development of enforceable standards.

STM simulation in the metaverse could help allocate orbital slots and manoeuvre rights. Such mechanisms incentivise compliance with regulatory standards while streamlining congestion management.

By harnessing the metaverse's capabilities, STM can transition from fragmented, manual processes to automated, data-driven workflows. This evolution ensures long-term sustainability and security for space operations, promoting equitable access to orbital resources and protecting the space environment for future generations.

The metaverse, generally defined as a persistent, interconnected, and interactive virtual environment, offers a compelling framework to address the complex challenges of STM. Beyond being a mere gaming platform, the metaverse, in this context, represents a sophisticated digital twin of Earth's orbital environment, integrated with real-time data and advanced simulation capabilities.

Key applications of the metaverse for STM may include the Real-time Space Situational Awareness (SSA) and Visualization that may allow a shared orbital picture. All authorized stakeholders (satellite operators, regulators, researchers) could access a single, high-fidelity visualization of all tracked objects, their trajectories, and associated uncertainties.

Rather than abstract data tables, operators could experience the orbital environment in 3D, enabling intuitive understanding of complex conjunctions, debris clouds, and satellite constellations. This could utilize augmented reality (AR) or virtual reality (VR) interfaces for deeper immersion.

Another promising application is connected to the AI-Powered Anomaly Detection. Integrated artificial intelligence could continuously monitor the virtual environment, identifying abnormal behavior, potential collision courses, or unauthorized maneuvers, flagging them for human review.

From an institutional and systemic design perspective, the metaverse may provide a platform for seamless, real-time collaboration among geographically dispersed stakeholders in order to implement virtual control rooms. Operators from different entities could converge in virtual control rooms to discuss conjunction warnings, negotiate maneuver parameters, and coordinate actions in a shared, immersive space.

In cases of conflicting maneuver plans or suspected interference, the metaverse could serve as a neutral virtual ground for direct communication and negotiation, potentially mediated by an international regulatory body's avatars.

The immersive nature of the metaverse offers an unparalleled training ground for future space operators and regulators able to set realistic training simulations. New personnel can train in highly realistic scenarios involving conjunction warnings, debris impacts, and emergency procedures without risking actual assets. The metaverse could host globally accessible training modules, fostering a shared understanding of STM best practices and regulatory requirements across the industry.

4 Conclusions

This chapter posits that leveraging metaverse technologies is essential for addressing contemporary STM challenges. Integrating advanced simulations, digital governance tools, and collaborative regulatory frameworks will mitigate current risks and shape the future of sustainable space operations.

Implementing a metaverse-based STM solution is not without significant challenges.

A first challenge is related to data fidelity and integration. Ensuring real-time, high-fidelity data streams from diverse SSA sensors (ground-based radars, telescopes, space-based sensors) into the metaverse is crucial. Data reconciliation and fusion from disparate sources will be complex. Moreover, a globally adopted STM metaverse would require robust interoperability standards for data formats, communication protocols, and virtual environment interactions, demanding unprecedented international cooperation.

Given the critical nature of STM, a second challenge concerns that a metaverse platform must be highly secure against cyber threats, unauthorized access, and data manipulation. Redundancy and resilience mechanisms are paramount. Ensuring equitable access to the metaverse platform for all spacefaring nations and commercial entities, regardless of their technological capabilities or economic status, is vital to prevent digital divides in space.

Finally, there are questions regarding data privacy, algorithmic bias in automated decision-making, and the potential for a "metaverse jurisdiction" need careful consideration. While automation and AI can enhance STM, maintaining human oversight and ultimate decision-making authority is essential, especially for critical maneuvers.

The rapid growth of space activities necessitates an urgent re-evaluation and reinforcement of Space Traffic Management regulation. The current fragmented and reactive approaches are insufficient to guarantee the long-term sustainability and safety of Earth's orbital environments. The metaverse, with its capabilities for real-time visualization, immersive collaboration, advanced simulation, and transparent operational logging, presents a groundbreaking opportunity to transcend these limitations.

References

1. ESA Website. https://www.esa.int/Enabling_Support/Space_Transportation/Types_of_orbits, (every online reference was last accessed on 08/01/2025)
2. De Oliveira Bittencourt Neto O (2023) Revisiting the Delimitation of Outer Space in Light of the Long-Term Sustainability of Space Activities. Air & Space Law 48(S.I.):93–111.
3. Committee on the Peaceful Uses of Outer Space (2021) Guidelines for the Long-term Sustainability of Outer Space Activities, Vienna, p 1. https://www.unoosa.org/res/oosadoc/data/documents/2021/stspace/stspace79_0_html/st_space79E.pdf
4. UN Doc. RES 1472 (XIV) adopted on 12 Dec. 1959.
5. LTS Guidelines, *supra* n. 3, p. 2.
6. Johnson K (2020) Key governance issues in space. Center for Strategic and International Studies. https://www.csis.org/analysis/key-governance-issues-space
7. ESA's Space Environment Report (2023). https://www.esa.int/Space_Safety/ESA_s_Space_Environment_Report_2023.
8. LTS Guidelines. *supra* n. 3, pp 1–2
9. Treaty on Principles Governing the Activities of States in the Exploration and Use of Outer Space, including the Moon and Other Celestial Bodies (1967), adopted in 1966 by the UN General Assembly with Resolution 2222 (XXI)

10. Convention on International Liability for Damage Caused by Space Objects (1972), adopted in 1971 by the UN General Assembly with Resolution 2777 (XXVI)
11. Masson-Zwaan T, Zhao Y (2023) Towards an international regime for space traffic management. Air & Space Law 48(S.I.):75–92
12. Weeden B (2010) Iridium-Cosmos collision fact shee. Secure World Foundation. https://swfound.org/media/6575/swf_iridium_cosmos_collision_fact_sheet_updated_2012.pdf
13. Kessler DJ, Cour-Palais BG (1978) Collision frequency of artificial satellites: The creation of a debris belt. J Geophys Res 83:2637–2646
14. Migaud MR (2020) Protecting earth's orbital environment: policy tools for combating space debris. Space Policy 52:1–9
15. Bradley AM, Wein LM (2009) Space debris: Assessing risk and responsibility. Adv Space Res 43:1372–1390
16. Liou J, Johnson NL (2008) Instability of the present LEO satellite populations. Adv Space Res 41(7):1046–1053
17. Maclay TD, McKnight DS (2021) Space environment management: Framing the objective and setting priorities for controlling orbital debris risk. J. Space Saf. Eng. 8:93–97
18. Contant-Jorgenson C, Lala P, Schrogl K (2006) The IAA Cosmic Study on space traffic management. Space Policy 22:283–288
19. LTS Guidelines, *supra* n. 3.
20. United Nations Office of Outer Space Affairs (2010) Space debris mitigation guidelines of the committee on the peaceful uses of outer space, Vienna. https://www.unoosa.org/pdf/publications/st_space_49E.pdf
21. UN Doc. A/RES/62/217, adopted on 22 Dec. 2007
22. Li B, Cao D, Tang S, Zhang T, Dong H, Wang Y, Wang F (2023) Sharing traffic priorities via cyber–physical–social intelligence: A lane-free autonomous intersection management method in metaverse. IEEE Trans. Syst., Man, Cybern.: Syst. 53(4):2025–2036

Implications of Artificial Intelligence for the Labour Market

Elena Signorini

Abstract Technological innovation has inevitably transformed the management of labor relations. Assuming that generative artificial intelligence (AI) is more likely to integrate into existing jobs rather than eliminate them, automation significantly impacts work performance, changing its intensity and mode due to new AI-driven organizational models. Additionally, AI alters traditional employer powers in organization, management, and control, introducing new risks for workers. These risks have significant implications for data processing, monitoring, and evaluation methods, potentially affecting worker privacy and job security. Considering these premises, the article aims to explore the evolving landscape of labor relations in the age of AI and highlights the need for policies that balance innovation with worker protections.

Keyword Artificial Intelligence · Labour market · Transitions

1 Introduction and Delimitation of the Context for Reflection

The growing use of artificial intelligence (AI) raises further questions that require us to go beyond the well-established perspectives on digitization, society in general, and the labour market.

Although AI has existed since 1956 [1], the acceleration of its development and deployment has been observed mainly since 2010. It has now gained significance at an institutional level, revealing its complexity [2]. At the European level, on 13 June 2024, Regulation (EU) 2024/1689 of the European Parliament and of the Council was adopted to establish harmonized rules in this field [3].

This phenomenon differs from the past: while earlier technologies required complex and specialized technical knowledge, today AI can be described as speaking

E. Signorini (✉)
Università Degli Studi Di Bergamo, Bergamo, Italia
e-mail: elena.signorini@unibg.it

A. Borroni et al. (eds.), *Metaverses: Reshaping Law, Economy, and Society*, PoliTO Springer Series, https://doi.org/10.1007/978-3-032-08664-8_8

a "common language". It enables anyone, regardless of their academic background, to use these technologies to create and produce objects [4].

The effects of this democratization of technology—first through computers and now directly accessible to individuals—are destined to permeate most aspects of life and human activities [5]. However, they require a specific regulatory framework to mitigate the pervasiveness of the phenomenon and its impact on daily life. This need for protection led the European Union to adopt a general regulation based on a risk-based approach. The aim is to develop flexible and adaptable tools that strike a balance between supportive measures and safeguards, ultimately protecting human dignity and individual autonomy [6].

According to this regulation, an AI system is defined as "a machine-based system that is designed to operate with varying levels of autonomy and that may exhibit adaptiveness after deployment, and that, for explicit or implicit objectives, infers, from the input it receives, how to generate outputs such as predictions, content, recommendations, or decisions that can influence physical or virtual environments" [7].

The document also emphasizes the ambivalence of AI systems. While they are powerful tools capable of delivering significant benefits, they can also, depending on the circumstances, pose risks and harm public interests or fundamental rights protected under EU law [8].

The impact on environments (commonly referred to as inferential capacity), whether physical or virtual, and the potential benefits and risks require thoughtful reflection, particularly from the perspective of labor law [9]. This includes considerations of the new jobs and skills that will be needed, but above all, the potential effects on the contractual relationship between worker and employer, starting from the profiling phase. Labor law, respecting its "protective inspiration", is positioned as a necessary instrument to ensure the protection of workers in the current scenario [10].

The Regulation itself specifies: "in the context of employment and protection of workers, this Regulation should therefore not affect Union law on social policy and national labour law, in compliance with Union law, concerning employment and working conditions, including health and safety at work and the relationship between employers and workers" [11].

We are entering a new era characterized by novel environments where relationships [12] develop—such as the metaverse and cyberspace. In these domains, challenges [13] coexist in an immaterial yet realistic world, almost like "a perfectly transparent three-dimensional chessboard extending to infinity" [14], also marked by consensually shared digital experiences and hallucinations by those who inhabit and interact within these spaces [15].

This is a silent and creeping digital revolution that, on the one hand, brings considerable uncertainty and, on the other, unlocks new potential. This potential demands the establishment of a robust and cohesive digital regulatory framework capable of managing the positive impact of generative AI—not only in the world of work but also across other domains. Indeed, the phenomenon affects and engages multiple areas of law [16].

The AI Act was conceived as a set of rules adopted by the European Parliament's Committee on Legal Affairs and Citizens' Rights, designed to establish a comprehensive legal framework within the Euro-Unitary System [17], intended to be uniform and consistent, and capable of governing and addressing the effects of the adoption and use of AI systems on the European market. At the same time, it plays a complementary role to existing Union law on employment and worker protection [18].

This moment feels reminiscent of the past, when technical evolution enriched everyday life with mechanical and electrical objects, creating intermediaries between the human body and the surrounding environment. However, in this case, it is AI that is being placed at the service of individuals to promote and ensure a high level of protection for health, safety, and fundamental rights, as enshrined in the Charter of Fundamental Rights of the European Union.

This anthropocentric approach is explicitly referenced in Recitals 1 and 6, as well as Article 1, paragraph 1 of Regulation (EU) 2024/1689. It underscores the commitment to place human well-being at the center of AI development and deployment [19].

2 Generative AI and Jobs: Policies to Manage the Transition

The digital transformation is expected to have far-reaching effects on various aspects of the working relationship, impacting multiple fronts. These include business organization and work [20] (encompassing types, skills, training, and the digital divide), contracts (addressing time, place, duration, and workload), employment conditions (such as conciliation, work-life balance, teleworking, and smart working), and enterprise organizational arrangements (in terms of management, workload monitoring, remote working, and worker control) [21].

While the transformation offers positive efficiency effects, it also raises ethical concerns related to the risks posed by the widespread adoption of AI. These risks touch on workers' rights, particularly regarding the management of employment relationships, health, and privacy issues.

The use of AI systems in the workplace has become increasingly prevalent. In some cases, this adoption even "starts from the bottom", with workers independently using AI systems to carry out their tasks. It is estimated that 75% of knowledge workers are already utilizing AI in their activities [22].

A recent study revealed that the demand for roles and skills related to digital technologies experienced significant growth during the first six months of 2024. Overall, the demand increased by 150%, with roles associated with generative AI witnessing an impressive growth of 246% [23]. In this context, it is important to note the distinction between tasks that are non-automatable and those where nuances emerge, allowing human activity and artificial intelligence to either coexist or play exclusive roles, depending on the situation. What is essential here is to emphasize that human control, particularly when dealing with high-risk artificial intelligence systems, remains irreplaceable and cannot be circumvented.

According to the International Labour Organisation (ILO), AI systems are less likely to eliminate jobs entirely; instead, they are more likely to complement them by altering their quality and the level of worker autonomy. The study highlights that the outcomes of this transition are not predetermined but rather depend on human decisions made during the initial stages of this process [24].

Regarding the interconnection between AI, algorithmic systems, and work performed through digital platforms (as addressed by Directive No. 2831 of 2024), this topic falls outside the scope of today's discussion. It is mentioned here only to underscore how this interconnection raises questions about the classification of employment relationships, which is currently experiencing a phase of "confusion" [25].

The ongoing debate on this issue reveals contrasting perspectives. Some argue for the fragmentation and evolution of the traditional notion of structured employment relationships [26]. One of the most compelling theories in this regard proposes a "matrix technique" approach, organized around values and principles rather than the traditional logic of fact-species-effects [27]. This approach conflicts with a segment of legal doctrine that emphasizes how courts have historically based their decisions on the recurrence of specific elements and indicators within the execution of the employment relationship [28]. Such a scenario, which places facts and indicators in a position of diminished importance [29], risks rendering them incapable of providing judges with meaningful tools to resolve disputes using reasonable, predictable, and non-discretionary criteria. It is equally difficult to argue that artificial intelligence, regardless of its advanced capabilities, could effectively replace human intelligence in the complex task of weighing and balancing various indicators present in specific cases. This process, essential for determining their relevance to one abstract legal framework or another, requires a nuanced and context-sensitive approach that remains beyond the reach of AI systems [29].

3 The Implications of AI in the Exercise of Employer Powers

The ability of AI to influence physical and virtual environments is vast, with consequent impacts on the structure of the working relationship. According to the Regulation, all those who use an AI system under their authority, whether they are natural or legal persons, fall into the category of c.d. deployer (except where such systems are used in the course of a personal non-professional activity), specifying that such uses may also involve other persons [30]. The Regulation takes a risk-based approach to develop a proportionate and effective set of rules, but specifies that "in any case, no decision producing an adverse legal effect on a person should be taken based solely on the output of the remote biometric identification system" [31]. The intention is to keep the employer's powers within the limits established, and to urge greater control on the part of the employer regarding the negative consequences that improper use of such systems may entail [32]. For this purpose, a system of general prohibitions has been developed such as those related to the use of biometric categorization systems [33]; those related to the use of AI systems aimed at identifying emotions [34]. It is also worth noting the prohibition of assessing and judging individuals not on their actual behaviour but on behaviour foreseen by IA based solely on profiling [35].

The elaborate regulatory system aims to mitigate the risks associated with the use of AI, which can, in fact, amplify an employer's powers and, in some cases, even operate beyond the employer's control [36]. One significant effect of technological control in the workplace has been the transformation of the trust-based relationship that traditionally characterizes employment relationships [37]. This shift is largely a consequence of the pervasive use of remote surveillance [38], whose ubiquity has altered the nature of work [39] to such an extent that workers often feel as though they are under constant supervision [40] by dystopian monitoring techniques [41]. Without proper regulation of video surveillance, workers' rights risk being eroded in what some might describe as a "cyber epidemic" [42].

The daily routines of workplaces are now subject to constant observation through tools such as webcams, access badges, GPS tracking, and systems that monitor movements and emails [43]. This reality necessitates a delicate balance between the production and organizational needs of the company and the protection of the worker's dignity and privacy.

Central to this balance is the focus on transparency and proportionality in monitoring practices, as highlighted in Article 4, paragraph 3 [44]. This is particularly relevant when considering the evidentiary implications of information obtained through such surveillance tools, which must be used responsibly to avoid infringing on workers' rights.

In addition to when outlined among the other risks potentially there are those related to the violation of privacy and data processing areas [45] both national and Community legislator has intervened on several occasions to try to govern the collection and processing of the huge amount of personal data and metadata that users put on the net. Issues relating to identification, categorization of the worker's biometric data, or acquisition of information through devices that workers use (iris), or postural changes (showing emotional responses to certain stimuli) or acquired through the individual's reaction times. The Regulation devotes a large number of recitals and articles to the link between AI and biometrics, highlighting the sensitivity of the issue that requires that processing and collection must take place in compliance with the requirements and regulatory framework of the GDPR to avoid the risk of compromise of these [46]. The fear is that there may be uncontrolled uses of leaked or stolen biometric data giving rise to deepfake models in mass [47].

The Regulation also highlights the additional implications of the use of AI systems with respect to occupational risk prevention for workers' health and safety[48]. To this end, it underlines the need to limit IA systems identified as high risk that have a significant adverse impact on the health, safety and fundamental rights of individuals and workers. The Regulation specifically dedicates several recitals to health and safety at work. For instance, Recital No. 9 addresses work platforms, stating that the Regulation should be without prejudice to the provisions for improving working conditions through digital platforms, as outlined in the Directive of the European Parliament and the Council on the improvement of working conditions through digital platforms. It also refers more generally to the protection of individuals or products. This provision serves as a declarative statement that outlines the ultimate aim of the regulation, which is to enhance the protection of workers. It should be understood within the framework of worker protection established by Directive 89/391, which mandates, as a general provision, that employers are responsible for ensuring the safety and health of workers in all aspects related to their well-being at work [49].

Technological innovation also necessitates the establishment of knowledge pathways that involve active worker participation in decision-making processes. This aspect is addressed by Article 88 of the GDPR and Article 23 of the Regulation on forms of teleworking, issued by the Ministry of Enterprises and Made in Italy (MIMIT) on March 10, 2023 (prot. no. 6964) [50]. The first paragraph of the regulation specifically addresses "health and safety at work and risks associated with the use of technological devices" [51]. This is particularly relevant in the context of cybersickness [52], a phenomenon related to digital work environments that manifests in various ways, including perceptual and spatial disorientation, which can affect men and women differently [53].

4 Algorithms and Security

Taking up what has already been discussed in the previous paragraphs and still on the basis of the EU Regulation, it should be reported as all the IA systems used "*in* employment, workers management and access to self-employment, in particular for the recruitment and selection of persons, for making decisions affecting terms of the work-related relationship, promotion and termination of work-related contractual relationships, for allocating tasks on the basis of individual behaviour, personal traits or characteristics and for monitoring or evaluation of persons in work-related contractual relationships" should be classified as high risk by virtue of their impact on workers. The implications in this area are numerous, with one of the most notable being the risk of discrimination, which has already emerged in various aspects of digitization. Among these concerns, we can highlight issues such as algorithmic discrimination, particularly in platform-based work, as well as potential threats to data protection and privacy security [54].

According to Annex III of the Regulation, high-risk AI systems also encompass those used for employment, workforce management, and access to self-employment (Sector No. 4, Annex III). This category specifically includes AI systems for staff selection and recruitment, as well as decision-making regarding employment conditions, promotions, terminations, task assignments, or performance evaluations [55]. Another significant risk in this context is the phenomenon of "hallucinations". This occurs because AI, unlike humans, lacks a biological brain and operates through algorithms and learning models. Hallucinations can result from how machines process and interpret data, leading to a distorted perception of reality. When AI fails to provide an adequate response to a query, the algorithm may generate skewed or inaccurate results, creating alternate, and sometimes problematic, versions of reality [56].

These phenomena fall into two distinct categories: (a) overfitting, which occurs when AI over-adapts to the specific training data, leading to a situation where the model, based on the observed sample, struggles to generalize to situations that differ from the training data. This happens because, with an excessive number of parameters relative to the number of observations, AI may lose its ability to accurately reflect the reality of the data. The second category is (b) overgeneration, which arises when AI overgeneralizes, creating incorrect or overly simplistic connections between input data [57]. The Regulation's decision to classify AI systems by risk levels stems from the goal of enhancing health and safety protections while safeguarding fundamental rights. This approach aims to address the identified risks within AI systems—whether unacceptable, high, limited, minimum, or zero [58]—by defining the obligations placed on all participants in the AI value chain, including suppliers and deployers [59].

5 Concluding Remarks

The above highlights that rapid technological change continues to raise legal challenges and necessitates that labor law find new approaches to shaping a modern world of work, where sustainability, subsidiarity, and resilience serve as guiding principles for relationships between individuals [60]. AI tools have reinforced the need for legislation that can adapt to the fast-paced evolution of technology while ensuring fundamental legal certainty. The issue of safe, reliable, and ethical AI remains a crucial concern [61]. AI ethics refers to "a set of values, principles, and techniques that employ widely accepted standards of right and wrong to guide moral conduct in the development and use of AI technologies" [62].

The philosophy that guided the European legislator in crafting the Regulation consists on the identification and application of ethical guidelines, entrusted to experts (the High-Level Expert Group on AI—HLEG), to ensure the reliability and ethical compatibility of AI [63]. These guidelines, first outlined in 2019, identify seven non-binding ethical principles, including: human intervention and supervision; technical robustness and security; privacy (confidentiality) and data governance; transparency; diversity, non-discrimination, and equity; social and environmental well-being; and responsibility [64].

The current situation underscores the importance of a multidisciplinary approach to AI management, requiring an ongoing update of skills not only in the implementation of these systems but also in their management, from both technical and legal perspectives.

The world is undergoing a process of digital transition. This also applies to the world of work, and it is becoming increasingly clear how important it is for people and AI to continue cooperating. It will be essential to keep the operational dimension sought out: in particular, we must avoid overlapping and substitution between the protagonists of the new relationships that are being woven into this new relational context aiming at the creation of an ethical system of mutual relations designed to concretize the point of balance between man and technology. AI is increasingly the beating heart of companies: it is crucial to ensure that the results of using such systems will be accepted uncritically and passively or if they will be subject to human assessment (ethical and legal) [65]. The final decision and its possible impact on the occupational population involved must be assessed [66].

The draft bill No. 1146 of 2024, titled "Provisions and Delegation to the Government on Artificial Intelligence" is driven by these concerns [67]. Aware of the need to respond to the evolving demands of a rapidly changing world, the legislator has crafted a meta-law [68] composed of meta-norms. These norms are based on the understanding that the physical and virtual worlds are set to coexist simultaneously. The bill includes a section specifically addressing labor issues, with the goal of ensuring that digital technologies can be leveraged for social good. It aims to provide rights-based solutions for managing risks and protecting workers from unintended or harmful consequences that may arise from the use of AI.

The bill includes provisions directly related to labor law (articles 10, "Provisions on the use of artificial intelligence in labor matters" [69]; articles 11 [70] and 12, "Provisions concerning intellectual professions" [71] of Bill No. 1146/2024). It also includes a series of provisions that, while not strictly labor-related, will influence the conditions and methods of work. These provisions regulate the integration of AI systems in specific sectors and activities, emphasizing the role of AI in decision-making processes (articles 7, paragraph 5; article 14, paragraphs 1 and 2; article 12, paragraph 1; article 13, paragraph 2). The focus is on ensuring transparency and protecting workers from discrimination, creating a legal framework that safeguards workers' rights both in the market and the workplace [72].

In this process, legal experts must also observe how the new digital production regulations will influence the transformation of businesses, which are increasingly characterized by digital production units. These units will require reflection on how the regulations for traditional production units—inhabited not by robots or digital systems but by individuals—can be extended [73]. The advent of AI has sparked a silent revolution across production sectors, which, in response to new demands, are undergoing a silent transformation [74]. This change requires focused attention to ensure that the benefits of AI are channelled toward enhancing protections for workers' safety and health, while minimizing the risks associated with AI practices under the concept of algorithmic management [75]. These issues require both institutional research and reflection.

Despite the efforts of European legislation and the regulatory path taken by our government (including Decree No. 60 of 2024 on active policies), it is crucial that legislative actions remain forward-looking. They must aim to "regulate new legal categories, seizing the opportunities provided by the development of new technologies, while safeguarding those who may be affected by a regression in protections or by risks related to the respect of fundamental rights" [76].

References

1. "Artificial intelligence coined at Dartmouth", in www.dartmouth.edu
2. Ranieri M "Intelligenza artificiale (IA) in educazione", en www.unitn.it; Russell, S, Norvig, P (2010) Artificial intelligence: A modern approach, Prentice Hall
3. Regulation (EU) 2024/1689 of the European Parliament and of the Council of 13 June 2024 laying down harmonised rules on artificial intelligence and amending Regulations (EC) No 300/2008, (EU) No 167/2013, (UE) n, 168/2013, (UE) 2018/858, (UE) 2018/1139 e (UE) 2019/2144 and Directives 2014/90/EU, (EU) 2016/797 and (EU) 2020/1828 (Artificial Intelligence Regulation), https://eur-lex.europa.eu/legal-content/IT/TXT/PDF/?uri=OJ:L_202401689
4. La tecnologia parla la nostra lingua, https://innovacionparaunmundomasseguro.elmundo.es/la-inteligencia-artificial-puede-hacer-del-mundo-un-lugar-mas-seguro-pero-como.html
5. Greenstein S (2021) Preserving the ruole of law in the era of artificial intelligence, in AI and law
6. Casonato C, Marchetti B (2021) Prime osservazioni sulla proposta di regolamento dell'Unione Europea in materia di intelligenza artificiale. BioLaw J 3:436, with extensive bibliography
7. Art. 3, Regulation cit
8. Recitals 4 and 5, Regulation cit
9. Roser M "AI timelines: What do experts in artificial intelligence expect for the future? Our World in Data", www.ourworldindata.org; Relazione 2023 sullo stato del decennio digitale, 2023, in https://digital-strategy.ec.europa.eu/it/library/2023-report-state-digital-decade; Special Eurobarometer 532, The Digital Decade, Report Field-work, 2023, testo consultabile sul sito https://europa.eu; Bussola per il digitale 2030: il modello europeo per il decennio digitale, testo consultabile sul sito della https://commission.europa.eu/; European Parliament resolution of 20 October 2020 with recommendations to the Commission on the framework for ethical aspects of artificial intelligence, robotics and related technologies (2020/2012(INL))
10. Ciucciovino S (2024) La disciplina nazionale sulla utilizzazione della intelligenza artificiale nel rapporto di lavoro, in Lavoro Diritti Europa, n. 1
11. Recital 9, Regulation 2024/1689
12. Rodriguez P (2022) Le métavers, un monde aux frontières du réel, in La révolution métaverse. Le défi de la nouvelle frontiére d'Internet, (dir. F. Rodriguez) Paris, Dunod, « Hors collection », pp 11–31
13. Dwivedi YK, Hughes L, Baabdullah AM, Ribeiro-Navarrete S, Giannakis M, Al-Debei MM, Dennehy D, Metri B, Buhalisj D, Cheung CMK, Conboy K, Doyle R, Dubey R, Dutot V, Felix R, Goyal DP, Gustafsson A, Hinsch C, Jebabli I, Janssen M, Kim YG, Kim J, Koos S, Kreps D, Kshetri N, Kumar V, Ooi KB, Papagiannidis S, Pappas IO, Polyviou A, Park SM, Pandey N, Queiroz MM, Raman R, Rauschnabel PA, Shirish A, Sigala M, Spanaki K, Wei-Han Tan G, Kumar Tiwari M, Viglia G, Fosso Wamba S (2022) Metaverso oltre l'hype: prospettive multidisciplinari su sfide emergenti, opportunità e agenda per la ricerca, la pratica e la politica. Int J Inf Manag 66:102542. https://www.sciencedirect.com/science/article/pii/S0268401222000767
14. Gibson W (2012) La notte che bruciammo Chrome in Distrust that Particular Flavor
15. We talk about cyberspace as an immaterial but very real world, a hallucination lived by consensus every day by billions of legal operators and a virtual environment accessible through computers, in Neuromante di W. Gibson, 1984
16. The contexts concerned range from the industrial sector, to that of art, Bayneix I (2023), L'artification du luxe par les NFT, in Droit de la propriété intellectuelle et du numérique, Ed.

Dalloz, n.7, 2, il settore musicale, Enser N (2022), Après les domaines du luxe, de l'art et du jeu vidéo, les NFT semblent avoir conquis un nouveau domaine: celui de la musique, in Ed. Dalloz IP/IT, Intelligence artificielle, https://www.dalloz-actualite.fr/flash/sacem-cap-sur-nft; a quello finanziario e del turismo Loukili G (2023), Le metaverse: leurre ou lueur pour le tourisme ?, in Juristourisme, n. 256, 2ss. These are contexts populated by phenomena such as NFTs (the phrase NFT (short for non-fungible token) has been made word of the year by Collins Dictionary. It says use of the abbreviation rose by more than 11,000% in 2021. NFTs are certificates to say that you own something digital - so original versions of viral videos, memes or tweets can be sold as if they were art), la blockchain, Cryptocurrencies and the metaverse

17. Di Cerbo V (2024) Digitalizzazione e rapporto di lavoro. L'AI Act nel quadro normativo a tutela del lavoratore, in Lavoro Diritti Europa, 4, p. 2 ss. in www.lavorodirittieuropa.it
18. Recital n. 9. Cefaliello A, Kullmann M (2022), Offering false security: how the draft artificial intelligence act undermines fundamental workers rights, European Labour Law Journal, 13 (4), 542–562, https://doi.org/10.1177/20319525221114474 where the 2021 proposal is referred to: one of the objectives of the proposal is to ensure consistency with existing Union legislation applicable to areas where high-risk AI systems are already used or likely to be used in the near future, which includes the EU social acquis
19. Regulation cit. https://eur-lex.europa.eu/legal-content/IT/TXT/PDF/?uri=OJ:L_202401689
20. P. Tosi (2024) Qualche riflessione su intelligenza artificiale e diritto del lavoro, in Lavoro Diritti Europa, 1. www.lavorodirittieuropa.it
21. In this contribution it will not be possible to go into all the questions related to the installation of art. 4 of the Workers' Statute, already widely subject of the attention of the doctrine, and for which they see: Bellavista A (2023), Sorveglianza elettronica, protezione dei dati personali e tutela dei lavoratori, in LDE, 1, p. 1; Bellavista A (2022), Controlli tecnologici e privacy del lavoratore, in A. Bellavista, R. Santucci (a cura di), Tecnologie digitali, poteri datoriali e diritti dei lavoratori, Giappichelli, p. 99, ss; Coelho Moreira T (2016), The Electronic Control of the Employer in Portugal, in LLI, 2, p. 1; Colosimo C (2018), La moderna declinazione del potere di controllo, in P. Curzio, L. Di Paola, R. Romei (a cura di), Diritti e doveri nel rapporto di lavoro, Giuffrè, vol. II, p.75 ss.; Donini A, Novella M, Vallauri M. L (2022), Prime riflessioni sul lavoro nel metaverso, in LLI, 2022, 8, 2; Ingrao A (2018), Il controllo a distanza sui lavoratori e la nuova disciplina privacy: una lettura integrata, Cacucci; Ingrao A (2019), Il braccialetto elettronico tra privacy e sicurezza del lavoratore, in DRI, 3, 895; Ingrao A (2021), Il controllo a distanza sulla prestazione dei ciclofattorini tra Scoober App e GPS, in LLI, 7, 1, 165; Ingrao A (2021), Il potere di controllo a distanza sull'attività lavorativa e la nuova disciplina della privacy nella sfida delle nuove tecnologie, in Nuove tecnologie e diritto del lavoro. Un'analisi comparata degli ordinamenti italiano, spagnolo e brasiliano, a cura di G. Ludovico - F. Fita Ortega - T. Christina Nahas, Milano University Press, 111; Marazza M (2016), Dei poteri (del datore di lavoro), dei controlli (a distanza) e del trattamento dei dati (del lavoratore), WP CSDLE, It., n. 300; Marazza M (2019), Social Media e Relazioni industriali. Repertorio di questioni, in LLI, vol. 5, n. 2, I ss.; Maresca A (2017), Controlli tecnologici e tutele del lavoratore nel nuovo art. 4 St. lav., in Controlli a distanza e tutela dei dati personali del lavoratore, a cura di P. TULLINI, Giappichelli, 1; Marazza M (2017), I controlli a distanza del lavoratore di natura "difensiva", in P. Tullini (a cura di), Controlli a distanza, Giappichelli, p. 27; Novella M (2021), Poteri del datore di lavoro nell'impresa digitale: fenomenologia e limiti, in LD, 3–4, 451; Maio V (2022), I controlli difensivi e la tutela del patrimonio aziendale, in C. Pisani, G. Proia, A. Topo, Privacy e lavoro. La circolazione dei dati personali e i controlli nel rapporto di lavoro, Milano, Giuffrè, p. 416; Piccinini I, Isceri M (2021), Questioni attuali sul potere di vigilanza e controllo del datore, in LDE, n. 4; Pisani C, Proia G, Topo A (2022), Privacy e lavoro. La circolazione dei dati personali e i controlli nel rapporto di lavoro, Giappichelli; A. Pizzoferrato (2023), I limiti al potere di controllo datoriale nell'era digitale, in LDE, 1, p. 2; Pizzonia D, Poletti D (2022), Gli intermediari dei dati, in EJPLT, 1; ROMAGNOLI U (1979), Sub art. 8, in G. Ghezzi -. F. Mancini - L. Montuschi -. U. Romagnoli, Statuto dei diritti dei lavoratori, Bologna, Zanichelli; Schiavoni E. C (2022), Gli obblighi informativi in caso di sistemi decisionali o di monitoraggio automatizzati (art. 4, co. 1, lett. a, d.lgs. n. 104/2022) (art. 1, co. 1, lett. s, e art. 1-bis, d.lgs.

n. 152/1997, inserito dall'art. 4, co. 1, lett. b, d.lgs. n. 104/2022), in Trasparenza e attività di cura nei contratti di lavoro. Commentario ai decreti legislativi n. 104 e n. 105 del 2022, a cura di D. Garofalo–M. Tiraboschi–V. Filì–A. Trojsi, Adapt University Press, 211; Sitzia A, Ramos Quintana M. I (2019), Sorveglianza difensiva "occulta" sui luoghi di lavoro e dignità nella prospettiva della Grande Camera della Corte EDU: la Sentenza López, in LDE, 3; Sitzia A (2016), Il controllo (del datore di lavoro) sull'attività dei lavoratori: il nuovo articolo 4 st. lav. e il consenso (del lavoratore), in LLI, n. 1, 83 ss.; Tebano L (2021) La digitalizzazione del lavoro tra intelligenza artificiale e gestione algoritmica, in Ianus, 24, 43

22. Cavallaro A (2024) L'impatto della Gen. AI sul futuro del lavoro: come cambiano i ruoli, in www.agendadigitale.eu
23. Casadei C (2024) Intelligenza artificiale generativa, domanda in crescita del 246%, in Il Sole 24 Ore, in www.ilsole24ore.com
24. International Labour Organization (2023) Generative AI and jobs: A global analysis of potential effects on job quantity and quality, in www.ilo.org
25. Tosi P Qualche riflessione su intelligenza artificiale e diritto del lavoro, cit. p. 3
26. The origin of these theories was Cass. 24 gennaio 2020, n. 1663, in https://www.lavorodirittieuropa.it/images/Cassazione_Foodora-.pdf
27. Brollo M (2024) Le dimensioni spazio temporali dei lavori, relazione alle Giornate di studio AIDLASS di Campobasso del 25–26 maggio 2023, in Le dimensioni spazio temporali dei lavori, Atti delle giornate di studio Aidlass, La Tribuna, pp 13–150; Perulli A, Treu T (2022) In tutte le sue forme e applicazioni- Per un nuovo statuto del lavoro, Torino
28. Tosi P Qualche riflessione su intelligenza artificiale e diritto del lavoro, cit. p. 3
29. Tosi P, cit
30. Recital 13, Regulation cit.; Di Cerbo V, Digitalizzazione e rapporto di lavoro. L'AI Act nel quadro normativo a tutela del lavoratore, cit. p. 8
31. Recital 35, CIT Regulation
32. Failla L (2023) Intelligenza artificiale e lavoro. Una sfida giuridica, ma anche etica, in www.ipsoa.it; Tullio A (2023) L'intelligenza artificiale cambia i modelli organizzativi aziendali. Attenzione ai limiti, in www.ipsoa.it; Peruzzi M (2023) Intelligenza artificiale e lavoro. Uno studio su poteri datoriali e tecniche di tutela, Giappichelli; Zappalà L (2021) Informatizzazione dei processi decisionali e diritto del lavoro: algoritmi, poteri datoriali e responsabilità del prestatore nell'era dell'intelligenza artificiale, in Biblioteca 20 maggio, n. 2
33. Recitals 30 and 35 as well as art. 3, points 40) and 34), Regulation cited.., All. III
34. Recital 44 and Art. 3, point 39, Regulation cit
35. Recital no. 42; on the subject Ciucciovino S (2024, Risorse umane e intelligenza artificiale alla luce del regolamento (UE) 2024/1689 tra norme legali, etica e codici di condotta, in Diritto delle relazioni industriali, 3, p. 576
36. Ciucciovino S, La disciplina nazionale sulla utilizzazione della intelligenza artificiale nel rapporto di lavoro, cit.; Peruzzi M (2024) IA e obblighi datoriali di tutela del lavoratore: necessità e declinazioni dell'approccio risk based, Federalismi, n. 30
37. European Commission, White Paper on Artificial Intelligence—A European approach to excellence and trust (COM(2020) 65 final). https://eur-lex.europa.eu
38. Maresca A, Controlli tecnologici e tutele del lavoratore nel nuovo art. 4 St. lav., cit. p. 4
39. Signorini E (2023) Lavoro e tecnologia: connubio tra opportunità e rischi, Federalismi, con ampia bibliografia cui si rimanda, in www.federalismi.it; Aloisi A, Gramano E (2019), Artificial intelligence is watching you at work: Digital surveillance, employee monitoring, and regulatory issues in the EU context, 41(1) Comp Lab L & Pol'y J:95–121
40. Parkes H (2023) Dystopian worker surveillance techniques likely to disproportionately affect young people says IPPR (Institute for public policy research). www.ippr.org
41. Examples of devices that in practice also allow for the timely recording and recording of work activity: through the automatic saving of the history, the employer can acquire a huge amount of information about the ways and times of work but also of non-work. Remote control: Ministero del Lavoro, nessuna liberalizzazione; norma in linea con le indicazioni del Garante della Privacy, Nota n. 20, 18 giugno 2015, Ministero del Lavoro e delle Politiche Sociali, in https://www.lavoro.gov.it

42. Della Giustina C, De Gioiacarabellese P (2023) Le scelte tragiche durante l'emergenza sanitaria: quando le risorse sono scarse, chi curare? Una riflessione comparatistica, in Stato, Chiese e pluralismo confessionale, n. 2, 82, not. 59 «The risk is that we could move from a state of health emergency to a state of cybernetic or information emergency, which could determine, as an ultimate consequence, the disappearance of mankind»
43. Sitzia A (2016) Il controllo del datore di lavoro su Internet e posta elettronica: quale riservatezza sul luogo di lavoro? NGCC 6(1):901
44. Maresca A (2017) Controlli tecnologici e tutele del lavoratore nel nuovo art. 4 St. lav., cit. p 4
45. Maresca A (2017), Controlli tecnologici e tutele del lavoratore nel nuovo art. 4 St. lav., in Controlli a distanza e tutela dei dati personali del lavoratore, a cura di P. Tullini, Giappichelli, 1 ss; Cappai M (2024) Intelligenza artificiale e protezione dei dati personali nel d.d.l. n. 1146: quale governance nazionale? Federalismi, n. 30
46. Castigli M (2022) Metaverso: le autenticazioni biometriche sono il punto critico. https://www.cybersecurity360.it/news/
47. Trend Micro (2022) Ricerca leaked today, Exploited for life. How social media Biometric Patterns Affect Your Future. https://www.trendmicro.com/vinfo/it/security/news/internet-of-things/leaked-today
48. G. Ludovico (2021) Nuove tecnologie e tutela della salute del lavoratore, in Nuove tecnologie e diritto del lavoro, a cura di G. Ludovico-F. Fita Ortega-T.C. Nahas, Milano University Press, 79 ss.; Pelusi LM (2020) Nuove competenze per la prevenzione dei rischi nella IV rivoluzione industriale. Da nuovi rischi, nuove professionalità, in Working Paper. Salus Adapt, 3; Piglialarmi G (2020) Il 5G e i nuovi ambienti di vita e di lavoro. Quale sfida per il diritto?, in Working Paper. Salus Adapt, 3; Tiraboschi M (2020) Esposizione a campi elettromagnetici prodotti da telefoni cellulari, malattia professionale a eziologia multifattoriale, tutele del lavoro, in Dir. rel. ind., 2, 555 ss.; European Agency for Safety and Health at Work (2018), Documento finale dell'European Risk Observatory, Foresight on new and emerging occupational safety and health risks associated with digitalisation by 2025, https://osha.europa.eu/en/publications/foresight-new-and-emerging-occupational-safety-and-health-risks-associated
49. Employment Buddy (2022) The Metaverse: Employment Law implications, in Employmentbuddy, https://www.employmentbuddy.com/hr-articles/the-metaverse-employment-law-implications/; Treu T (2024) CRSD, direttiva sui lavoratori delle piattaforme e valutazione dei rischi, Federalismi, n. 30
50. https://olympus.uniurb.it/index.php?option=com_content&view=article&id
51. Timellini C (2023) Verso una Fabbrica Intelligente: come l'AI invita a ripensare la tutela della salute e della sicurezza dei lavoratori, in Variazioni su Temi di Diritto del Lavoro, 4 832 ss
52. Baroni S (2023) Che cos'è la cybersickness (e perché discrimina le donnne?). https://www.lasvolta.it/6734/che-cose-la-cybersickness-e-perche-discrimina-le-don-ne#:~:text=Si%20tratta%20di%20un%20disturbo,oculare%20e%20difficolt%C3%A0%20di%20concentrazione
53. Di Carluccio C (2024) Quando a stressarsi è l'avatar, tecnologie immersive e nuove sfide per la tutela della salute e della sicurezza sul lavoro in Riv. Giuridica Ambiente Diritto, 1, https://www.ambientediritto.it/dottrina/quando-a-stressarsi-e-lavatar-tecnologie-immersive-e-nuove-sfide-per-la-tutela-della-salute-e-della-sicurezza-sul-lavoro/
54. Recital 57, Regulation 2024/1689
55. Annex III, Regulation 2024/1968
56. Aloisi A, De Stefano V (2020) Il tuo capo è un algoritmo, Laterza
57. Gallese Nobile C (2023) Legal aspects of the use artificial intelligence in telemedicine. J Digit Technol Law 1(2):314–336
58. https://www.europarl.europa.eu/topics/it/article/20230601STO93804/normativa-sull-ia-la-prima-regolamentazione-sull-intelligenza-artificiale
59. Tebano L (2023) Poteri datoriali e dati biometrici nel contesto dell'AI Act, in Fe-deralismi, pp 204–207
60. Cingolani R, Andresciani D (2020) Robots, macchine intelligenti e sistemi autonomi: analisi della situazione e delle prospettive. In: Alpa G (a cura di) Diritto e intelligenza artificiale, pp

23–56; Signorini E, Zambardino F (2024) La blockchain e la telemedicina: quali opzioni per la sicurezza e l'integrità dei dati sanitari, Federalismi, 27, p. 205 ss., in www.federalismi.it

61. European Council, Extraordinary meeting 1–2 October 2020, Conclusions
62. Leslie D (2019) Understanding artificial intelligence ethics and safety: A guide for the responsible design and implementation of AI systems in the public sector
63. Cazzaniga and others (2024) Gen-AI: Artificial intelligence and the future of work, IMF Staff Discussion Note SDN2024/001, International Monetary Fund, Washington, DC, p. 24 laddove precisa che "Women and highly educated workers are consistently more exposed to, but also more likely to benefit from, AI". https://www.imf.org/en/Publications/Staff-Discussion-Notes/Issues/2024/01/14/Gen-AI-Artificial-Intelligence-and-the-Future-of-Work-542379
64. European Commission (2019) Shaping Europe's digital future. Ethical guidelines for trustworthy AI. www.europa.eu
65. European Parliament resolution of 20 October 2020 with recommendations to the Commission concerning the framework for the ethical aspects of artificial intelligence, robotics and related technologies (2020/2012(INL)), https://www.europarl.europa.eu/doceo/document/TA-9-2020-0275_IT.html
66. To give a concrete example from the case of the Swedish company, if in the end the employees with the lowest IQ will be dismissed, or if this will be mediated with the selection criteria and legal parameters provided by the relevant regulations, as would be desirable
67. DDL n. 1146, 20 May 2024, whereas in art. 10, 11, 12 of addresses the relation-ship between AI and work, https://www.senato.it/leg/19/BGT/Schede/FascicoloSchedeDDL/ebook/58262.pdf; Dagnino E. (2024), Intelligenza artificiale, lavoro e diritto interno: note a margine del disegno di legge di iniziativa governativa, in Lavoro Diritti Europa, 4; Biasi M. (2024), Intelligenza artificiale e processo: verso un robot giudice per le controversie lavoristiche? In: M. Biasi (a cura di), Diritto del lavoro e intelligenza artificiale, Giuffré, pp. 737–759
68. Bensoussan A (2022) Métavers: bienvenue dans le monde hybride, in Solution Notaire Hebdo n° 19; Bensoussan A, Forster F et al (2022), Dossier IA et Métavers: Le réseau Lexing® vous informe. https://www.alain-bensoussan.com/avocat-metavers/
69. Art. 10, (Provisions on the use of artificial intelligence in the field of work 1. Artificial intelligence is used to improve working conditions, protect the psychophysical integrity of workers, increase the quality of work performance and people's productivity in accordance with European Union law. 2. The use of artificial intelligence in the workplace must be safe, reliable, transparent and cannot be carried out in contrast with human dignity or violate the confidentiality of personal data. The employer or principal is required to inform the worker of the use of artificial intelligence in the cases and in the manner referred to in Article 1-bis of Legislative Decree No. 152 of 26 May 1997. 3. Artificial intelligence in the organisation and management of the employment relationship ensures the observance of the inviolable rights of workers without discrimination on the basis of gender, age, ethnic origin, religious beliefs, sexual orientation, political opinions and personal, social and economic conditions, in accordance with European Union law
70. Art. 11. (Observatory on the adoption of artificial intelligence systems in the world of work) 1. In order to maximize the benefits and contain the risks deriving from the use of artificial intelligence systems in the workplace, the Observatory on the adoption of artificial intelligence systems in the world of work has been established at the Ministry of Labor and Social Policies, with the task of defining a strategy on the use of artificial intelligence in the workplace, monitoring the impact on the labor market, identify the work sectors most affected by the advent of artificial intelligence. The Observatory promotes training of workers and employers in artificial intelligence. 2. The Observatory is chaired by the Minister of Labour and Social Policies or by his representative. A decree of the Minister of Labour and Social Policies, to be adopted within ninety days of the date of entry into force of this law, shall establish the components, operating procedures, as well as the additional tasks and functions of the Observatory referred to in paragraph 1. The members of the Observatory are not entitled to remuneration, attendance fees, reimbursement of expenses or other emoluments however denominated. 3. The establishment and operation of the Observatory shall not entail new or increased burdens on public

finances and shall be ensured with the financial, human and instrumental resources available under current legislation

71. Art. 12. (Provisions on intellectual professions) 1. The use of artificial intelligence systems in the intellectual professions is permitted only to carry out instrumental and support activities for professional activities and with a prevalence of the intellectual work that is the subject of the work. 2. In order to ensure the relationship of trust between the professional and the client, the information relating to the artificial intelligence systems used by the professional shall be communicated to the recipient of the intellectual service in clear, simple and exhaustive language
72. Dagnino E, Intelligenza artificiale, lavoro e diritto interno: note a margine del disegno di legge di iniziativa governativa, cit. p.10; Biasi M (7 maggio 2024), Il lavoro nel disegno di legge italiano sull'AI: antipasto dell'AI Act, in Agenda Digitale, in www.htpps//www.agendadigitale.eu.
73. Faioli M (2021) Unità produttiva digitale. Perché riformare lo Statuto dei lavoratori, in Lavoro Diritti Europa, 3
74. Macrì P (2023) Metaverso, IA e Digital Twin. Il futuro dell'automazione industriale secondo Siemens. https://www.industriaitaliana.it/siemens-automazione-metaverso-digital-twin/
75. De Stefano V, Wouters M (2022) AI and digital tools in workplace management and evaluation. An assessment of the EU's legal framework, in European Parliamentary Research Service Study Panel for the Future of Science and Technology, 17; Marassi S (2024), Intelligenza artificiale e sicurezza sul lavoro, Lavoro Diritti Europa, 4, www.lavorodirittieuropa.it
76. Hearing of the Minister in the XI Commission (Public and Private Employment), Chamber of Deputies (26 October 2023), Fact-finding investigation on the relationship be-tween artificial intelligence and the world of work, with particular reference to the impacts that generative artificial intelligence can have on the labour market, https://www.lavoro.gov.it/documenti/audizione-ministro-calderone-intelligenza-artificiale-del-26-ottobre-2023

MirageXR: A Framework for Metaverse-Enabled Cross-Platform Integration of Physical Hardware and XR Environments

Rabee Aladdin Qiblawi, Layth Sliman, Ziad Osman, and Ali Massoud Haidar

Abstract The metaverse has evolved from a concept in science fiction into an interconnected network of immersive digital experiences. Yet it remains fragmented—especially in integrating virtual components with real hardware. Existing solutions often rely on disparate communication protocols and siloed workflows, making it difficult to establish a unified, scalable approach. To unlock the metaverse's full potential across education, industry, and entertainment, developers and researchers need a framework that streamlines synchronization between the physical and virtual realms. By eliminating barriers to interoperability and providing standardized, flexible integration pathways, such a framework can facilitate rapid prototyping, iterative development, and the deployment of metaverse applications that bridge digital vision and tangible reality. In this paper, we introduce MirageXR, a real-time framework designed to close the gap between the real and virtual worlds. Although it does not address every facet of the metaverse, MirageXR offers a pivotal solution to one of its central challenges: seamlessly connecting virtual 3D objects, particularly virtual electronics, with real hardware. By leveraging standardized communication protocols and intuitive design, MirageXR enables users to build unified, hands-on XR experiences that blend digital and physical environments with ease, paving the way for future advances in metaverse integration.

Keywords Augmented reality · Extended reality · ESP32 · Network sockets · Unity · Metaverse

R. A. Qiblawi (✉) · Z. Osman · A. M. Haidar
Department of Electrical and Computer Engineering, Beirut Arab University, Debbieh, Lebanon
e-mail: r.qiblawi@bau.edu.lb

L. Sliman
Efrei Paris, Pantheon Assas University, Paris, France

A. Borroni et al. (eds.), *Metaverses: Reshaping Law, Economy, and Society*, PoliTO Springer Series, https://doi.org/10.1007/978-3-032-08664-8_9

1 Introduction

The term Metaverse generally refers to a persistent, networked virtual world that incorporates real-time interaction, 3D environments, and social presence [1]. Although often traced back to Neal Stephenson's novel Snow Crash [2], the metaverse concept has steadily expanded through widespread internet connectivity, the rise of social gaming platforms, and advancements in immersive technologies such as Virtual Reality (VR), Augmented Reality (AR), and Mixed Reality (MR) [3]. Major tech companies and platforms now explore metaverse-like ecosystems in diverse ways: for instance, Meta Horizons uses VR headsets to host collaborative social spaces, while Roblox offers a user-generated gaming environment where millions create and share 3D worlds [4]. In each case, users are increasingly empowered to craft custom content, ranging from simple 3D objects to elaborate interactive experiences.

Beyond entertainment and social connectivity, Metaverse has begun to influence education, where Extended Reality (XR) platforms open immersive, hands-on learning opportunities. In these educational metaverses, learners can manipulate 3D representations of historical artifacts, scientific simulations, or engineering prototypes, often in real time and with peers in shared virtual spaces [5]. This type of immersive digital environment is particularly useful for teaching STEM-related fields, since students not only absorb theoretical knowledge but also engage with virtual labs that emulate real-world scenarios. By bridging physical and digital realms, educators can provide interactive lessons, offer instant feedback, and facilitate deeper conceptual understanding—all hallmarks of a metaverse-driven learning experience [6].

However, despite the potential of these emerging platforms, the metaverse's success depends on overcoming significant technical, ethical, and practical obstacles [7]. These challenges matter not only due to technical limitations, but also because ethical issues—such as privacy, consent, and surveillance—can erode user trust and hinder widespread adoption. Moreover, Metaverse still remains fragmented in terms of standards, protocols, and underlying infrastructure [8]. This fragmentation limits accessibility and inclusion, as inconsistent ecosystems may exclude users without specific devices or platforms, further deepening the digital divide. It also hinders collaboration across educational or industrial domains, where smooth interoperability is critical for cross-platform learning and development.

This fragmentation extends to hardware integration, where real-world devices—such as sensors, robotics kits, or embedded systems—are often relegated to proprietary or siloed interfaces that do not translate well across different XR ecosystems. As a result, educational and industrial innovators seeking to merge tangible electronics with immersive, multi-user virtual spaces frequently confront obstacles in interoperability, scalability, and real-time data exchange. These are not just technical barriers they limit innovation, discourage creative content creation by non-experts, and reduce the educational value of real-time feedback loops that are essential in both learning and decision-making.

Addressing this gap requires a framework that can seamlessly synchronize physical hardware with virtual simulations while emphasizing interoperability and user-friendly abstractions. By lowering technical barriers, such a framework could help developers create immersive experiences that blend physical electronics with their virtual counterparts in a bidirectional manner, thereby establishing a continuous feedback loop.

How This Chapter is Organized

This chapter is organized into several key sections that are built upon one another. It begins with an introduction that outlines the challenges of integrating physical hardware with XR environments, followed by a review of current approaches in AR, IoT, and XR in the State-of-the-Art section, which highlights the limitations of existing solutions. Next, the Proposed Framework section introduces the MirageXR framework by detailing its lifecycle, user interaction, and component design—with supporting UML diagrams—to demonstrate its efficacy in bridging the physical and virtual electronic components. The chapter concludes by exploring advanced applications such as immersive mixed reality, digital twins, and agentic AI integration, and offers insights into potential future research directions.

2 State of the Art

Recent improvements in augmented reality (AR), Internet of Things (IoT), and extended reality (XR) technologies are enabling deeper integration between physical hardware and immersive environments. As such integration becomes more common, especially within the evolving metaverse—developers are exploring new ways to bridge the virtual and physical. However, combining XR with electronics poses an inherent technical barrier, as it demands expertise across multiple domains, including embedded systems, networking, and 3D development. This complexity can be especially limited for developers with limited technical knowledge.

While various AR and IoT frameworks offer partial solutions, few provide standardized, extensible platforms that enable real-time, bidirectional interaction. The following section reviews notable work in this space and introduces MirageXR as a framework that addresses these challenges head-on.

Several recent proposals aim to integrate AR and IoT for practical applications. Aaditya Auti et al. [9] developed an IoT-based home automation system using an ESP32 microcontroller coupled with AR to offer remote control of appliances via a visual interface. Chakraborty et al. [10] took a similar approach, integrating AR for remote control and visualization to enhance interactivity in smart environments. Blanco-Novoa et al. [11] went further by proposing an open-source framework that bridges AR/MR devices (e.g., Microsoft HoloLens) with IoT systems using MQTT, HTTP, and Node-RED. This framework demonstrated real-time interactions (with latencies under 100 ms) in a smart power socket control use case—but its reliance on nonstandard libraries and limited documentation reduced its adaptability. Seungwoon

Lee et al. [12] advanced these ideas by outlining a convergence platform based on oneM2M standards to integrate mixed reality with IoT devices. However, being primarily conceptual, their solution still awaits a fully usable implementation.

To address these limitations, several additional frameworks have been proposed. For instance, MARIoT (2021): Developed by Kuyucu and İnce, MARIoT is a low-code framework that allows developers to create custom mobile AR interfaces for smart-home devices. It uses a Node-RED–based backend and an MQTT broker to enable bidirectional communication while offering an intuitive drag-and-drop interface for rapid prototyping [13]. Although it simplifies IoT–AR integration, its reliance on marker-based AR and external servers can limit its deployment to simple smart-home scenarios.

MechARspace (2022): Zhu et al. introduced MechARspace as an immersive authoring system that connects IoT-enhanced physical objects (e.g., interactive toys) with AR content via plug-and-play sensor/actuator modules. The system allows users to define trigger–action rules so that interactions with physical objects are immediately reflected as virtual feedback [14]. Despite its ease of use in tangible environments, MechARspace's specialized hardware requirements limit its general applicability.

ROS Reality (2018): For industrial and robotics applications, ROS Reality (Rosen et al.) links ROS-enabled robots with XR headsets through a WebSocket-based ROS bridge. This setup supports real-time teleoperation and visual monitoring in immersive environments [15]. While the approach is highly extensible within robotic systems, it assumes familiarity with the ROS ecosystem and is less suited for rapid prototyping in educational settings.

Digital Twin with GraphQL (2020): Tu et al. proposed a digital twin framework for industrial machinery where real-time sensor data is synchronized with a virtual model hosted on a HoloLens application via a GraphQL API. This method leverages modern web technologies for low-latency, bidirectional data exchange but may require significant customization to adapt to diverse IoT devices [16].

The MetaGadget framework [17] converts and transmits user input from metaverse platforms into control signals for external IoT devices. The system's architecture relies on asynchronous HTTP POST events rather than persistent connections, enabling multi-user device control and integration without requiring advanced programming knowledge. Workshops demonstrated the viability of MetaGadget for educational and prototyping purposes, showcasing real-time control, user-friendly development, and cross-space interactions. However, compared to MirageXR, it offers limited real-time performance, lower extensibility, and a narrower scope (Table 1).

Despite the progress in AR-based electronics education and IoT-integrated augmented reality applications, existing solutions often lack a standardized, extensible framework that allows real-time, bidirectional interactions between physical hardware and virtual environments. Current tools predominantly focus on either visualization (e.g., AR-based educational tools) or remote control (e.g., IoT-based automation), without offering a seamless, synchronized interaction model

Table 1 Comparison of different framework related to integration of IOT and XR

Framework	Year	Description/approach	Pros/cons
ROS reality	2018	Connects ROS robots with XR headsets over WebSocket for real-time control	Extensible in robotics; needs ROS expertise
Digital Twin with GraphQL	2020	Industrial twin syncing sensors with HoloLens via a GraphQL API	Low latency; requires customization
MARIoT	2021	Low-code AR for smart homes using Node-RED/MQTT with drag-and-drop prototyping	Rapid prototyping; requires Node-RED expertise
MetaGadget framework	2025	Converts metaverse inputs to IoT control via async HTTP POST events	User-friendly; limited performance/scope
MirageXR (our framework)	2025	Allows bidirectional communication over web sockets using Arduino Code	Developer friendly requires Unity3D expertise to build prototypes

that supports dynamic experimentation and prototyping. The MirageXR framework addresses this gap by introducing a structured approach that unifies AR-enhanced electronics education with real-time hardware interaction. By integrating an Arduino-compatible library for ESP32 microcontrollers with a Unity-based AR interface, MirageXR enables developers and educators to create interactive, hybrid physical-virtual experiences.

3 The Proposed Framework

Implementing an AR/MR-IoT framework presents various challenges that are often overlooked but should be carefully considered during the design phase (see Fig. 1, which illustrates the system flow). One of the primary challenges is the heterogeneity of technologies required to develop a comprehensive system, as it involves making low-level architectural and software decisions across multiple domains of expertise. Our proposed framework, MirageXR, addresses this complexity by leveraging low-level C++ and low-level socket communication between C++ and Unity. The Mirage library enhances Arduino by managing virtual pins instead of physical ones, using JSON and socket communication. It maintains an Arduino-style interface while enabling seamless networked interactions. In Unity, a corresponding system synchronizes these virtual pins with ESP/Arduino, allowing real-time control of game objects and ensuring smooth integration between the digital and physical components of the AR/MR-IoT ecosystem. This approach not only reduces technical barriers but also

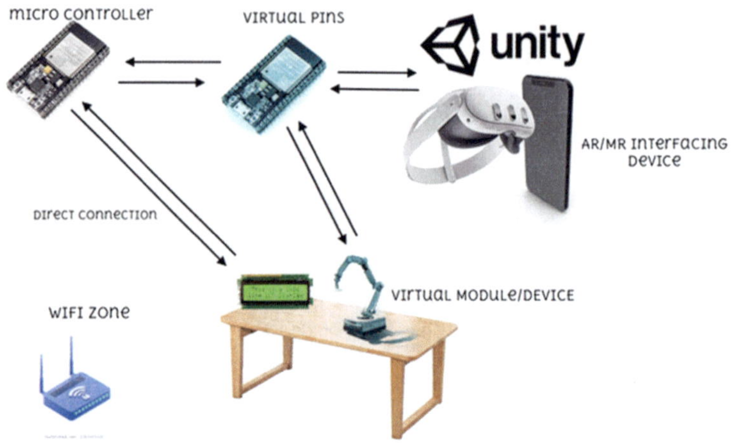

Fig. 1 Visual representation of MirageXR system

significantly enhances accessibility, supporting diverse applications in education and interactive entertainment within the evolving landscape of the metaverse.

3.1 The Life Cycle

To understand the system, it's important to examine its lifecycle (Fig. 2). The MirageXR lifecycle outlines the sequence of operations that enable seamless communication and synchronization between the physical hardware and the virtual XR environment. The core stages are as follows:

- **Pins Setup**: After a successful connection, the system initializes all virtual pins to a known state. The `initDigitalPins function` sets each pin to `INPUT` mode and assigns a default value of `LOW`. This approach mirrors the safe startup behavior of real microcontrollers, avoiding unintended outputs and ensuring a controlled launch state.
- **Handshake**: The first message from the ESP32 to the Unity runtime defines the number of virtual pins. While users can customize the pin count in Arduino code, the system defaults to 24 generic pins, which are typically sufficient for most applications.
- **Game Loop**: Since Unity is a game engine, the system's runtime behavior can be likened to a game loop. In this phase, the ESP32 continuously streams virtual pin values as the single source of truth. Any modifications initiated from Unity are sent as requests using predefined Enums, as explained in Sect. 3.3.3 on the Unity UML structure.

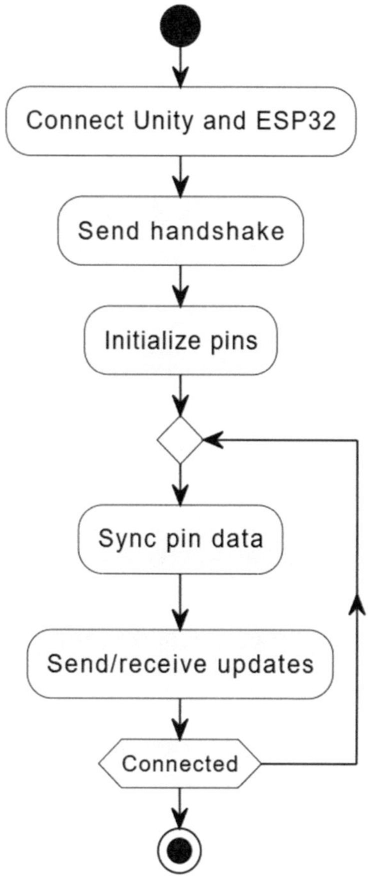

Fig. 2 Visual block diagram of the system's life cycle

3.2 User Story

In the MirageXR framework, the user story revolves around the developer's journey to seamlessly integrate physical hardware with virtual environments. Here's how the developer interacts with the system (Fig. 3).

Setup Communication

The developer configures the initial communication setup between the ESP32 device and the Unity environment using the Mirage Network Manager. This ensures that both physical and virtual components can exchange data in real-time.

Configure Virtual Pins

Through the Arduino library's Pin Management, the developer maps virtual pins to physical inputs and outputs. This configuration allows the developer to mirror real-world hardware interactions within the XR environment.

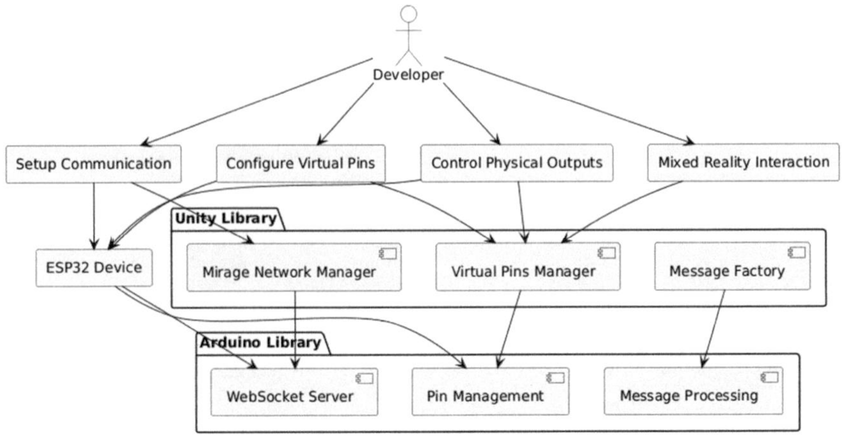

Fig. 3 User story diagram for the MirageXR framework

Control Physical Outputs

Utilizing the Virtual Pins Manager, the developer can manipulate virtual and physical outputs, controlling devices like sensors and actuators from within the Unity application. This step is critical for testing and deploying interactive hardware projects.

Mixed Reality Interaction

With the Message Factory, the developer integrates mixed reality interactions, enabling seamless communication between the real and virtual worlds. This includes manipulating virtual objects through real-world actions and vice versa, enhancing the immersive experience.

In essence, the developer's story in using MirageXR involves setting up, configuring, and integrating physical devices into dynamic XR environments, facilitating efficient development and testing of hybrid interactions.

3.3 Component Design

The component design of the MirageXR framework is shown in Fig. 4. It illustrates the interaction between the ESP32 Library and Unity Library, connected via the WebSocket Protocol for real-time communication.

Web Sockets handle the core communication between the ESP32 microcontroller and the Unity runtime, enabling fast, two-way data transfer for real-time interaction. To ensure everything stays in sync and avoids miscommunication, we implemented a handshake system at the start of the connection. This step verifies that both the ESP and Unity are ready to exchange data. We also use a Message Factory pattern (on

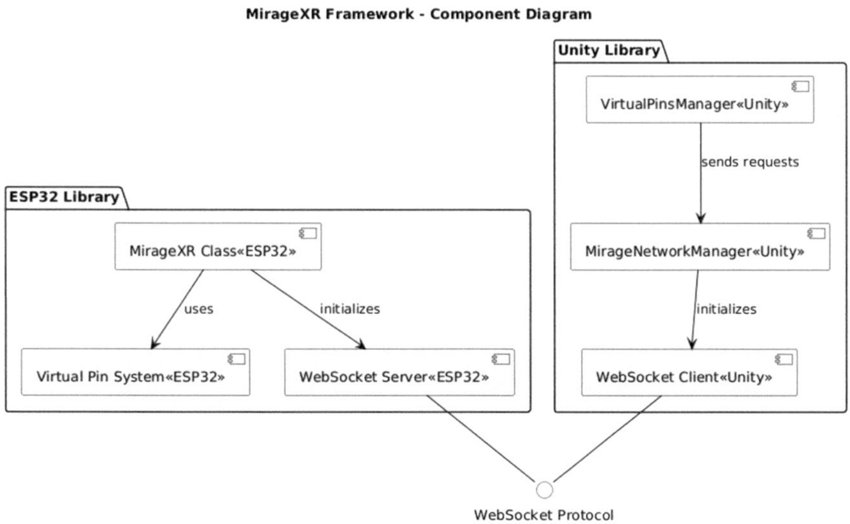

Fig. 4 MirageXR framework component design showing real-time interaction between the ESP32 and unity libraries via WebSocket

Unity Side) to manage incoming messages. It identifies the type of each message, performs different actions accordingly, and distinguishes between various incoming commands from the ESP to Unity.

3.4 UML Design

To clearly represent the architecture and data flow of MirageXR, we created detailed UML diagrams for both the Unity and Arduino (ESP32) libraries. These diagrams help visualize the internal class structure and how different components interact through object-oriented programming (OOP) principles.

On the Arduino side, the framework is written in C++ using encapsulation, modular design, and abstraction to manage virtual pins and network communication. On the Unity side, the system is implemented in C#, leveraging polymorphism and event-driven programming to create a responsive and extensible XR environment.

3.4.1 ESP32 Side (C++ Arduino Code Implementation)

The ESP32 library is built with a focus on simplicity and Arduino-like syntax, but it still leverages OOP concepts for structure and maintainability (Figure). Key components include (Fig. 5).

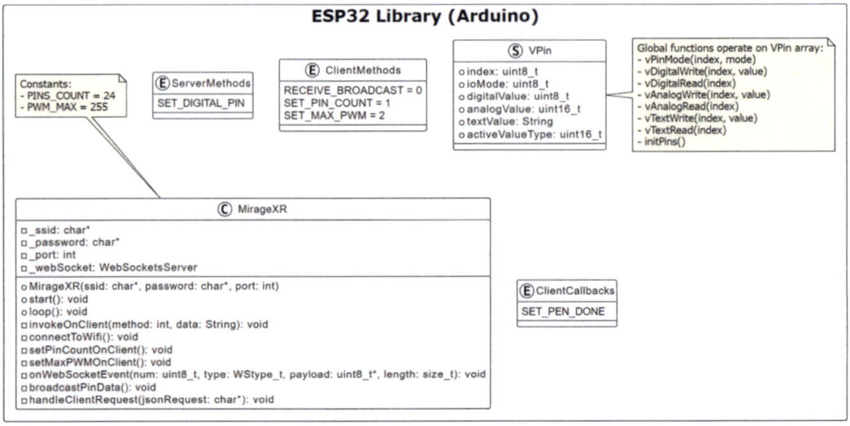

Fig. 5 UML diagram Arduino library

`MirageXR` **Class**: This is the main controller class. It encapsulates Wi-Fi credentials, WebSocket setup, and core methods like `start()`, `loop()`, `broadcastPinData()`, and `handleClientRequest()`.

`VPin` **Struct**: A custom data structure used to simulate virtual pins. Each instance stores digital, analog, and text values along with its mode (input/output). This abstraction makes it easy to work with a flexible pin system that is not tied to physical hardware.

Client and Server Methods Enums: These define structured constants for message types (e.g., `SET_DIGITAL_PIN`, `RECEIVE_BROADCAST`), improving clarity and type safety in network messaging.

Global Functions: Helper methods like `vDigitalWrite`, `vAnalogRead`, and `vTextWrite` operate on an internal `VPin` array. This mirrors standard Arduino function signatures but internally targets a virtual data layer.

This structure ensures code reusability, readability, and ease of extension—developers can add more virtual functionality (like PWM or I2C simulation) without altering the message-handling core.

3.4.2 Unity3D Side (C# Implementation)

Figure 6 shows the UML class diagram for the Unity-side of the MirageXR framework, built in C# using object-oriented principles like encapsulation, abstraction, inheritance, and modularity. These principles ensure the system remains scalable, maintainable, and well-suited for real-time XR interactions with physical hardware.

MirageNetworkManager: This class acts as the central hub for managing the network connection between the Unity runtime and the ESP32 microcontroller. It encapsulates the logic for initiating, maintaining, and terminating WebSocket connections. It includes fields like `_ip`, `webSock`, and `connected`, and exposes

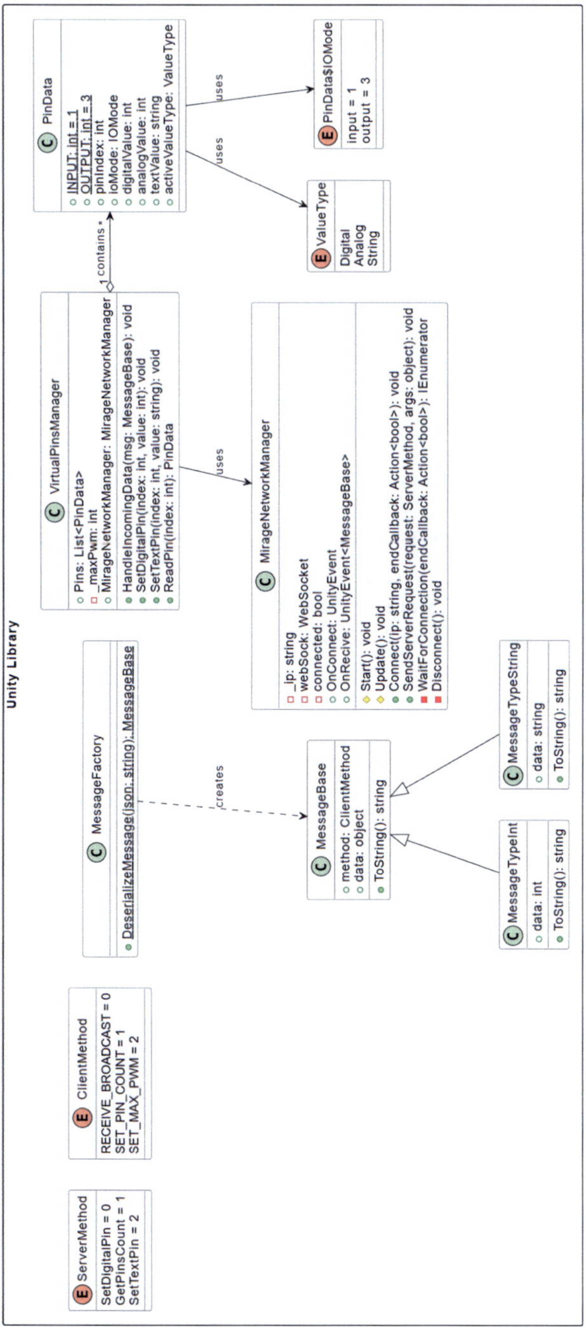

Fig. 6 UML diagram unity package

events such as `OnConnect` and `OnReceive` that are triggered when a connection is established, or data is received. In addition to handling connection state and data transmission, this class defines key enumerations for **server and client methods** (e.g., `SET_DIGITAL_PIN, RECEIVE_BROADCAST`). These enums are tightly synchronized with the ESP32-side equivalents written in C++, ensuring consistent message interpretation across both platforms. This mirrored design minimizes communication errors and simplifies debugging by standardizing the available actions and commands.

MessageFactory: This class is responsible for **deserializing raw JSON messages** into strongly typed C# objects. It follows the **Factory Design Pattern**, dynamically creating instances of message types based on their contents. This helps isolate the parsing logic from the rest of the system, improving code readability and maintainability. For example, an incoming message might contain a pin update, such as a change in digital or analog value, or a system-level command like setting the maximum number of pins. The `MessageFactory` interprets the JSON structure and returns an appropriate subclass of `MessageBase`, such as `MessageTypeInt` or `MessageTypeString`, which are then handled downstream by other components.

VirtualPinsManager: This high-level class acts as the **main interface for developers** to interact with virtual pins inside Unity. It maintains a list of `PinData` objects, each representing the state of a virtual pin (including mode, value, and type). It provides public methods like `SetDigitalPin()`, `ReadPin()`, and `SetTextPin()` that allow users to control pin behavior through code.

One of its key features is **editor integration**. The `Pins` list is exposed within the Unity Editor, allowing developers to **monitor pin states visually** during runtime without writing any additional code. This makes it particularly useful for rapid prototyping and debugging XR applications where virtual and physical data need to stay tightly aligned.

4 MirageXR in Practice

In the previous section, we explained how the framework was developed and described its lifecycle. In this section, we shift our focus to the developer's perspective—how MirageXR is used in practice. We'll walk through a simple experiment that demonstrates how to interface the framework with Unity objects, highlighting how physical hardware can interact with virtual components in real time.

4.1 Installing Framework Libraries

Installation on ESP32

To use MirageXR with your ESP32 board, you'll need to install the MirageXR Arduino library. The library is also hosted on Github.[1] To use the library download the project as a zip file then in the Arduino IDE add the library by heading to Sketch > Include Library > Add.ZIP Library.

Installing MirageXR in Unity

To start using MirageXR in Unity3D, you'll first need to install the framework via the Unity Package Manager. The Unity-side of the framework is bundled as a Unity package and hosted in a Git repository[2] on GitHub.

To install a Git-based package in Unity:

1. Open Unity and go to **Window > Package Manager**
2. Click the "+" button in the top-left corner
3. Select "**Add package from Git URL...**"
4. Enter the MirageXR Git repository URL
5. Click **Add** to complete the installation

4.2 Using MirageXR Within Unity

MirageXR can be used in different ways, although it mainly was created to work with XR systems, it can work on 2D flat screens, as the focal point of the system is to interface with virtual pins in a convenient manner, non the less XR and especially mixed reality would create the most compelling and engaging experience. In this section we'll explore how to use the framework within unity in general and in the next section we'll talk about using it with Meta's SDK.

4.2.1 Adding the Essential Prefab

After installing the required package, In the **Project window**, navigate to: Packages > miragexr-electrobits-core-unity > Runtime > MirageXR.ElectroBits > Prefabs. Then Locate the prefab named **ElectrobitsCore**. Drag and drop the **ElectrobitsCore** prefab into the.

By following this process, you will create an instance of the prefab that can be easily interfaced with the ESP device once its IP address is provided. The IP address can be manually entered in the Unity Editor or assigned programmatically, as it is

[1] https://github.com/rabeeqiblawi/mirage-xr-arduino.

[2] https://github.com/rabeeqiblawi/miragexr_electrobits_core_unity.

Fig. 7 IP address field in unity editor

exposed as a public field and passed to the Connect method within the Connection Manager script referenced earlier (Fig. 7).

4.2.2 Interfacing with Virtual Pins

While the Connection Manager operates in the background to manage network connections, the Virtual Pins Manager is responsible for interfacing with the virtual pins. It maintains a data structure that stores the real-time values of these pins.

These values can be visualized in the Unity Inspector, which is useful for real-time visual debugging. More importantly, they are accessible through code via a public array. This allows developers to use pin values to dynamically affect the metaverse—for example, to move objects, emit light, trigger animations, control audio, change textures, or initiate various interactive behaviors as shown in Fig. 8.

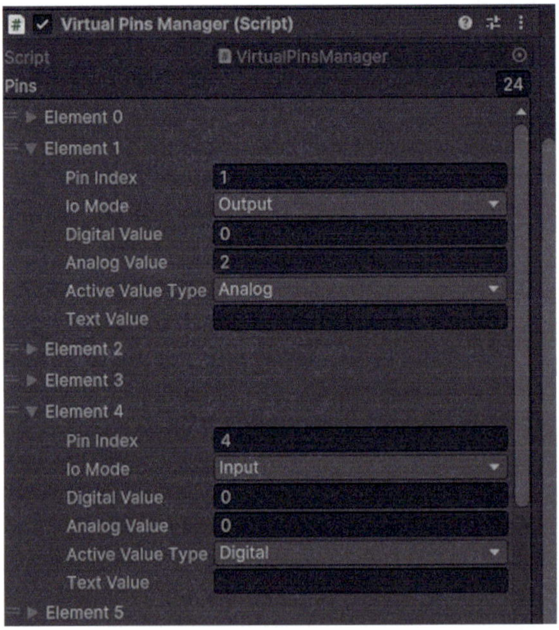

Fig. 8 Values visuazantion

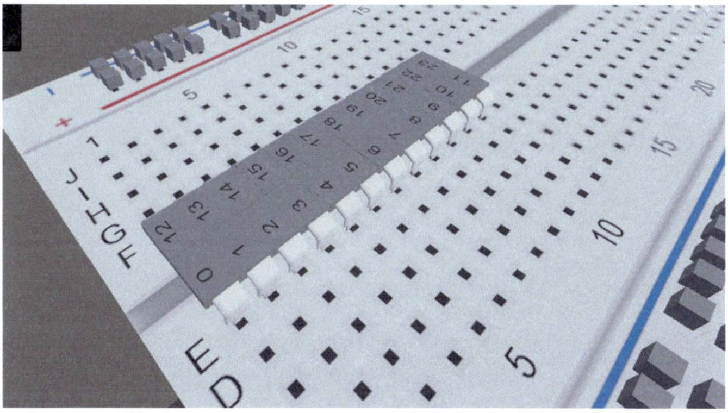

Fig. 9 Visualizing virtual pins in the Unity3D

4.2.3 Visualizing the Pins Within Unity's 3D Space

One of the useful things that can be done is the ability to visualize pins. This is done by creating an array of objects that represent the pins, and when connected to the ESP, only enabling the number of pins specified in the data structure. As illustrated in Fig. 9, this is visualized over a virtual breadboard and works on both flat screens and in XR (Metaverse) environments.

4.3 Using MirageXR with Meta' SDK

MirageXR can be used in different ways, although it mainly was created to work with XR systems, it actually can work on 2D flat screens, as the focal point of the system is to interface with virtual pins in a convenient manner, nonetheless XR and especially mixed reality would create the most compelling and engaging experience.

Meta Quest devices are widely regarded as mainstream options, particularly with the release of the Meta Quest 3 and 3S, which focus on mixed reality as of the time of writing. In contrast, Apple's Vision Pro is seen as a more advanced device, though its high price and limited availability present significant barriers to broader adoption in the mixed reality space. Therefore, we will focus on working with the Meta Quest SDK for Unity, as it is considered an accessible and beginner-friendly way to get started.

To install the Meta SDK, with unity being opened navigate to the Unity Asset Store[3] in the web browser, search for "Meta All-in-One SDK,[4]" and select the option to open it with Unity. Once installed, you will notice some changes in the Unity

[3] https://assetstore.unity.com/.

[4] https://assetstore.unity.com/packages/tools/integration/meta-xr-all-in-one-sdk-269657.

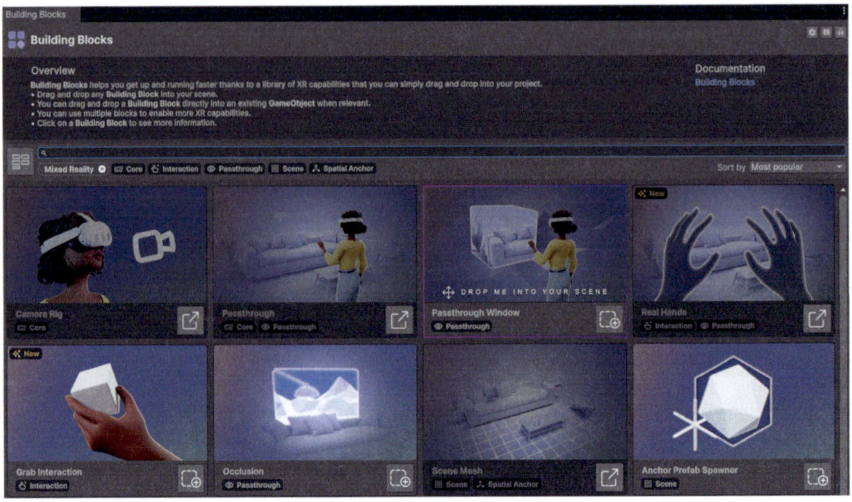

Fig. 10 Meta's SDK building blocks for mixed reality

Editor's user interface. The most relevant change is the addition of Meta's Building Blocks, a feature in the Meta SDK that helps make development easier.

Diving deeply into XR development is out of the scope of this paper, yet It's really recommended to check out the samples that come with Meta's SDK that cover most of the essential things related to start diving into XR development (Fig. 10).

With Meta's SDK inside Unity, it becomes easy to interact with virtual components such as virtual leds, screens, picking up and connecting wires, this requires using the interaction system that comes with the Meta's SDK, in practice all it requires is adding different "Interactable Components" to the Game Objects within Unity.

4.4 *Controlling a Virtual Robotic Arm (Experiment)*

In its prefect form Mirage Framework would be integrated with a simulator-like environment where non-unity users can use an app hosted on a device that supports Mixed Reality to connect virtual electronics to a real ESP, this requires building the wiring system and creating virtualized active and passive electronics and, in a Utopian scenario virtual battery would be added to the system. Yet for now, with the current state of the framework, Unity Developers can create their own custom experiences by interfacing with the system manually by accessing the data in the `Virtual PinsManager.cs.`

In this section we will explore the experiment of interfacing a physical joystick to control a virtual robotic arm (Fig. 11).

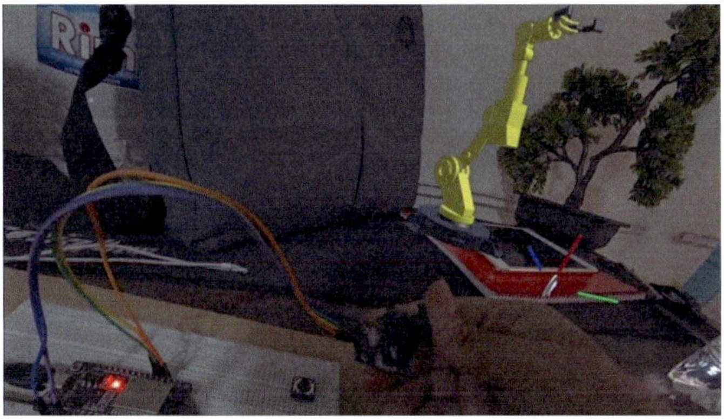

Fig. 11 Mixed reality experience of controlling a virtual robotic arm

4.4.1 ESP/Arduino IDE Side

Mirage library makes interfacing with the virtual pins a straightforward process lifting the networking part from the developer, all it requires is instantiating the MirageXR class and setting up the real pins then assigning virtual pins that should resemble the virtual pins in Unity's Side, Fig. 12 shows the initialization in the Arduino IDE.

'oom ...

mirage.ino mirage.h

```
#include <Arduino.h>
#include "mirage.h"

MirageXR mirage = MirageXR("Tenda_331C18", "ak739008", 1984);

const int potPin = 34;
const int xPin = 32;
const int yPin = 35;
const int robotArmInputX_virt = 1;
const int robotArmInputY_virt = 2;

void setup()
{
  Serial.begin(115200);
  mirage.start();
  pinMode(xPin, INPUT);
  pinMode(yPin, OUTPUT);
  vPinMode(robotArmInputX_virt, OUTPUT);
  vPinMode(robotArmInputY_virt, OUTPUT);
}
```

Fig. 12 Initializing MirageXR in Arduino IDE

```
void loop()
{
  mirage.loop();
  int xValue = analogRead(xPin);
  int yValue = analogRead(yPin);

  int threshold = 400;
  int center = 2048;

  int xMapped = 2; // Default to center value
  if (xValue > center + threshold) xMapped = 3;
  else if (xValue < center - threshold) xMapped = 1;

  int yMapped = 2; // Default to center value
  if (yValue > center + threshold) yMapped = 3;
  else if (yValue < center - threshold) yMapped = 1;

  vAnalogWrite(robotArmInputX_virt, xMapped);
  vAnalogWrite(robotArmInputY_virt, yMapped);
  delay(10);
}
```

Fig. 13 The control loop

For the loop shown in Fig. 13, the code itself isn't more complex—we're simply reading data from the physical joystick, mapping the values to a suitable range for more predictable behavior, and then passing the signal to the Unity runtime.

4.4.2 The Unity Code

On the Unity side, our focus will be on the interfacing component rather than walking through the entire project setup process. Setting up a new Unity project, importing packages, was mentioned before. For the virtual robotic arm, we selected a pre-built model available on the Unity Asset Store. This model was initially configured to be controlled using traditional keyboard inputs, which makes it a good starting point for modification (Fig. 14).

Figure 15 demonstrates taking the values from the `ViruaIPinsManager`, then using them to replace the values that were used to control the arm through standard keyboard input.

```
int xRaw = VirtualPinsManager.Pins[XInputIndex].analogValue;
int yRaw = VirtualPinsManager.Pins[YInputIndex].analogValue;

int x = 0;
if (xRaw == 1) x = -1;
else if (xRaw == 3) x = 1;

int y = 0;
if (yRaw == 1) y = -1;
else if (yRaw == 3) y = 1;
```

Fig. 14 Using the virtual pins as a control signal for the virtual arm

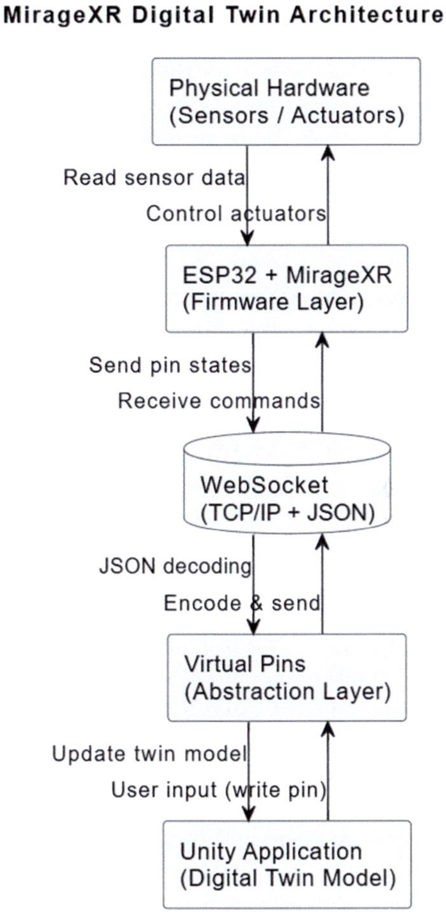

Fig. 15 MirageXR digital twin architecture and data flow

This approach highlights how the integration between the Arduino and Unity has been significantly simplified and optimized using the MirageXR library. By acting as a seamless communication bridge, MirageXR enables real-time data exchange and more dynamic interaction between physical and virtual components. While this example focuses on controlling a robotic arm, the underlying principle can easily be adapted to other Unity-based applications. For instance, the same setup could be tuned to control a virtual car, drone simulation, or any interactive 3D environment that benefits from external hardware input.

5 Beyond the Framework

The MirageXR framework described so far provides a solid foundation for integrating physical hardware with Unity in general. Beyond its current capabilities, there are expansive opportunities to extend MirageXR into broader XR applications within the metaverse. In this section, we explore how developers can leverage and extend MirageXR for advanced Mixed Reality (MR) applications, including collaborative multi-user environments and robotics integration. These explorations are forward-looking and theoretical, but they offer technical depth and practical insight into what could be built on top of MirageXR's core. By examining these scenarios, we aim to inspire developers and researchers to push **beyond the framework** and envision how MirageXR can power rich, interactive experiences in the evolving metaverse.

5.1 MirageXR in Context of Immersive Mixed Reality

MirageXR's design makes it an effective starting point for building MR applications. Developers can use the framework's Unity integration and Arduino-compatible library to rapidly prototype experiences that blend real and virtual elements. The process involves minimal setup: the hardware side (e.g., an ESP32 microcontroller) runs the MirageXR firmware to expose sensors and actuators as networked "virtual pins," while the Unity side uses a provided prefab and API to link those pins to virtual objects in an MR scene. This abstraction allows developers to treat physical components as part of the virtual scene, updating in real-time without writing low-level networking code.

In one of our Mixed Reality prototypes developed for the Meta Quest 3, we utilized the MirageXR framework to embed virtual pins directly within a 3D model of an electronic chip, integrating them seamlessly into a realistic circuit assembly that included a breadboard and connecting wires. This setup served as a powerful demonstration of the framework's ability to bridge virtual and physical components in real time. By visualizing and interacting with these pins in an immersive MR environment, we were able to validate the framework's utility in prototyping and educational scenarios. This prototype highlighted MirageXR's potential not only for technical experimentation but also for intuitive, spatial understanding of electronic systems, which is especially valuable in XR-based electronics training and visualization.

5.2 *Digital Twining with MirageXR*

The idea of digital twins has been gaining momentum across many industries to better monitor, simulate, and control physical systems. At its core, a digital twin is a virtual replica of a real-world device or setup—constantly updated with live data. This real-time connection allows developers, engineers, and students to keep track of performance, try out changes before applying them to actual hardware, and catch issues early. What makes a digital twin truly effective is its ability to stay closely synchronized with its physical counterpart in both directions.

When we introduce digital twins into Extended Reality (XR) environments, their value grows even further. Instead of just viewing data on a screen, users can interact with it in 3D space—seeing, manipulating, and learning from it in real time. And as XR applications become more socially and spatially connected, these experiences begin to align with what we think of as the metaverse—shared digital spaces where people, devices, and intelligent systems coexist. In this context, a digital twin isn't just a monitoring tool—it becomes an interactive participant in a broader, immersive ecosystem.

However, making this seamless connection between the real and the virtual remains a technical challenge. The key is ensuring data flows reliably and instantly between physical hardware and its virtual twin, keeping both environments in perfect sync. That connection is what transforms digital twins from static models into truly dynamic, real-time companions in XR.

Figure 7 illustrates this architecture, showing how data flows between physical hardware and its virtual counterpart through the MirageXR framework, enabling continuous, real-time alignment between the two.

Consider, for example, a virtual model of a microcontroller-based sensor system—such as a temperature-controlled motor setup. Using MirageXR, sensor data (e.g., current temperature) can be streamed from the physical device to the Unity application, where it dynamically updates UI elements or animates physical behavior on a virtual twin (e.g., changing motor speed, triggering visual alerts). Conversely, interactions with the virtual model—such as adjusting a virtual dial or toggling a virtual switch—can immediately update the corresponding physical outputs via MirageXR's socket-based command layer. This real-time loop of data enables the virtual model to evolve into a full-fledged digital twin.

5.3 *Enhancing the Framework with Agentic AI (Educational Use)*

As educational environments evolve toward more immersive and intelligent platforms, integrating Agentic AI into systems like MirageXR introduces a powerful new dimension in learning. Agentic AI refers to context-aware artificial intelligence agents equipped with speech recognition (STT), text-to-speech (TTS), and the ability to perform environment-specific actions. These capabilities allow learners to engage in natural, conversational interactions with the system while receiving intelligent, adaptive guidance, significantly improving the process of learning how to code microcontrollers and work with XR environments. Image Z illustrates the high-level interaction between the student, the AI system (via an LLM API wrapper), the MirageXR-enabled Unity application, and the physical ESP32 hardware. It emphasizes how voice input is processed, interpreted, and used to trigger hands-free, real-time interactions—enabling a more intuitive and accessible learning experience (Fig. 16).

5.3.1 AI as a Context-Aware Learning Assistant

In a traditional setup, students often rely on tutorials, documentation, or peer instruction to understand the logic behind connecting sensors and programming microcontrollers. With the integration of Agentic AI [18], MirageXR transforms into an interactive learning assistant capable of understanding the current state of the virtual environment and the hardware configuration.

As learning increasingly takes place in immersive digital spaces, including those envisioned as part of the metaverse, this kind of intelligent, context-aware support becomes even more impactful. Not only does it provide technical guidance, but

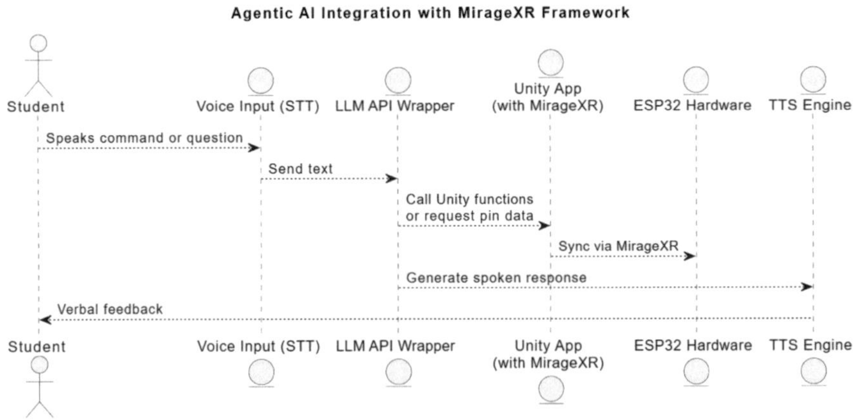

Fig. 16 Sequence diagram of agentic AI integration with MirageXR framework

it also introduces a layer of social interaction between the learner and an AI agent, simulating a conversational, collaborative experience. This aligns with a key pillar of the metaverse: fostering shared, interactive spaces where humans and digital entities engage meaningfully.

For example, when a student asks a voice question such as "How do I use this light sensor in Unity?", the AI system can interpret the query, reference the state of connected devices through the MirageXR pin layer, and respond using natural speech. This response might include explanations of pin assignments, usage recommendations, or next steps to configure the system within the Unity application.

Importantly, the AI is not acting in isolation—it is continuously aware of the runtime state of the system. If a student incorrectly configures a virtual pin or mislabels a mode, the AI can detect the inconsistency and offer corrective suggestions in real time. This contextual awareness, powered by MirageXR standardized data model, makes the learning process more intuitive, interactive, and socially engaging—key components of the emerging educational landscape in the metaverse.

5.3.2 Hands-Free Mobile Operation and Voice Control

In mobile and headset-based learning environments, where hands-free interaction is often necessary, Agentic AI enables seamless control through voice commands. Rather than navigating menus or using controllers, students can interact with the MirageXR-enabled Unity application by simply speaking instructions such as "*Start the simulation*," or "*Show analogue pin values*."

These voice commands are interpreted by the AI agent, which then triggers specific actions within the Unity scenes such as starting animations, toggling visual elements, or initiating virtual experiments. This approach is especially beneficial for learners who are engaged in physical tasks simultaneously, such as wiring circuits or operating real hardware, and cannot manually interact with the digital interface.

5.4 Expanding the Metaverse with MirageXR

One of the defining characteristics of the metaverse is its ability to support meaningful social interaction [19]—not just between users, but also between users and intelligent systems, all within a shared, immersive environment. This isn't just about avatars in virtual spaces; it's about collaborative presence—people working, learning, and experiencing things together in real time, often anchored to the same physical or virtual reference points.

MirageXR is uniquely positioned to support this kind of interaction. Many modern XR SDKs—such as Meta's SDK for Unity [20]—include support for shared spatial anchors, which allow multiple users to localize to the same point in physical space. This feature enables consistent placement of virtual content across devices. When combined with MirageXR's architecture, these shared anchors can serve as the foundation for collaborative learning experiences anchored to the real world.

A typical example might involve a teacher demonstrating a hands-on electronics lesson in mixed reality. The microcontroller—running the MirageXR firmware—acts as a lightweight server, broadcasting sensor data and pin states to nearby students connected on the same local network. Because the ESP32 (or similar microcontroller) is only responsible for serving pin state updates, and not handling spatial mapping or heavy networking, its computational load remains minimal. The Unity clients (used by students) handle rendering, spatial awareness, and shared anchoring, while MirageXR maintains real-time communication between the virtual and physical components.

In this scenario, students can stand side-by-side with their instructor, view the same virtual circuit diagram floating over a real breadboard, and observe live feedback as the teacher interacts with the physical hardware. The experience is not only technically synchronized—it feels social, immersive, and shared. This local, collaborative setup is a practical step toward enabling metaverse-aligned educational spaces, where interaction is not isolated to a screen, but integrated into the physical environment in a way that supports real-time, multi-user learning.

By leveraging MirageXR's lightweight client–server model in conjunction with XR platforms that support shared anchors, developers and educators can create scalable, device-friendly applications that bring the social promise of the metaverse into everyday classrooms and workshops.

6 Summary

This chapter introduced *MirageXR*, a cross-platform framework designed to bridge the gap between physical hardware and immersive XR environments. By abstracting hardware interactions into a unified system of virtual pins and facilitating real-time, bidirectional communication via TCP/IP sockets, MirageXR enables seamless integration of ESP32-based microcontrollers with Unity applications. The framework offers a modular and extensible architecture that supports low-latency synchronization, a standardized communication protocol, and familiar development paradigms modelled on Arduino. These features collectively reduce technical barriers, promote rapid prototyping, and enhance accessibility for both educational and industrial applications within the metaverse.

Beyond the core architecture, the chapter explored MirageXR's broader implications for immersive mixed reality experiences. The framework facilitates the creation of real-time digital twins, collaborative XR-based educational settings, and intelligent, voice-controlled learning assistants through integration with agentic AI systems. These extensions underscore MirageXR's potential not only as a technical solution for real-time hardware-XR integration, but also as a foundational platform for constructing shared, interactive metaverse environments. In doing so,

MirageXR contributes meaningfully to the evolving landscape of XR-enhanced learning, prototyping, and digital interconnectivity.

References

1. Dionisio JDN, Burns WG, Gilbert R (2013) 3D virtual worlds and the metaverse: current status and future possibilities. ACM Comput Surv 45(3):1–38
2. Stephenson N (1992) Snow crash. Bantam Books
3. Mystakidis S (2022) Metaverse. Encyclopedia (MDPI) 2(1):486–497
4. Invesco Ltd (2022) Metaverse: new horizons for the digital economy. White Paper, Invesco Ltd.
5. Kaddoura S, Al Husseiny F (2023) The rising trend of metaverse in education: challenges, opportunities, and ethical considerations. PeerJ Comput Sci
6. Tsirulnikov D, Suart C, Abdullah R, Vulcu F, Mullarkey CE (2023) Game on: I mersive virtual laboratory simulation improves student learning outcomes & motivation. FEBS Open Bio 13(3):396–407
7. Doughan Z, ALMubasher H, ElBizri M, Haidar A (2022) The metaverse: a virtual world in the palm of your hand. BAU J Sci Technol 4(1):1–8. Article 10
8. World Economic Forum (2023) Interoperability in the metaverse. World Economic Forum Report
9. Auti A et al (2022) IoT based home automation system using augmented reality. Int J Innov Res Technol (IJIRT) 8(9) (2022)
10. Chakraborty S et al (2023) Augmented reality and IoT based system for enhanced interactivity in smart environments
11. Blanco-Novoa U et al (2020) Creating the internet of augmented things: an open-source framework to make IoT devices and augmented/mixed reality systems talk to each other. Sensors (2020)
12. Lee S et al (2022) Novel architecture of OneM2M-based convergence platform for mixed reality and IoT. Comput Mater Continua 71(1):51–69
13. Kuyucu F, İnce M (2021) MARIoT: a low-code framework for mobile AR-based smart home prototyping. In: Proceedings of the 2021 international conference on IoT and smart environments
14. Zhu Y et al (2022) MechARspace: an immersive authoring system for tangible AR experiences with IoT-enhanced objects. IEEE Trans Visual Comput Graph
15. Rosen D et al (2018) ROS reality: bridging the gap between virtual teleoperation and physical robots via AR/VR interfaces. In: Proceedings of the 2018 IEEE/RSJ international conference on intelligent robots and systems
16. Tu X et al (2020) Digital twin integration in mixed reality: a GraphQL-based approach for real-time industrial IoT. J Ind Inf Integr
17. Kurai R, Yanagawa H, Hiroi Y, Hiraki T (2025) MetaGadget: an accessible framework for IoT integration into commercial metaverse platforms [arXiv preprint arXiv:2501.15931]
18. Pounds E (2021) What is agentic AI? NVIDIA blog
19. Hennig-Thurau T, Aliman DN, Herting AM, Cziehso GP, Linder M, Kübler RV (2022) Social interactions in the metaverse: framework, initial evidence, and research roadmap. J Acad Market Sci
20. Meta Horizon OS Developers (2025) Interaction SDK overview

From Virtual Realms to Tangible Reality: Privacy at the Nexus of Artificial Intelligence and the Metaverse

Paolo Tortorano

Abstract This article delves into the intricate and ever-evolving domain of personal data protection, with a particular focus on the profound challenges posed by the metaverse and artificial intelligence (AI). Commencing with an exposition of the foundational principles underpinning the discipline of privacy and data governance, it provides a conceptual framework for the reader. The narrative progresses to an examination of the mechanisms devised to safeguard privacy within the metaverse, an emergent digital realm replete with immersive and interconnected virtual experiences. The discourse further explores the transformative implications of AI, whose unparalleled capacity for data processing and analysis raises unprecedented questions concerning privacy. In this context, the article offers a reflective critique of strategies oriented toward data minimization, underscoring their vital role in upholding ethical and regulatory standards amidst these technological innovations. Concluding with initial observations, the paper advocates for adaptive and forward-thinking regulatory paradigms to navigate the complexities of data protection in an era of relentless digital advancement.

Keywords Personal data protection · Privacy · Artificial intelligence · Data minimization · Regulation

1 An Exordium to the Discipline of Personal Data Protection: Foundational Principles

In the contemporary era, characterized by an increasingly interconnected and interdependent global landscape, the forces of globalization and the relentless march of technological advancement have profoundly reshaped the processes governing the collection, management, and dissemination of personal information. This sweeping transformation has given rise to novel and increasingly pervasive mechanisms of

P. Tortorano (✉)
Università Degli Studi Della Campania "Luigi Vanvitelli", Naples, Italy
e-mail: paolo.tortorano@unicampania.it

A. Borroni et al. (eds.), *Metaverses: Reshaping Law, Economy, and Society*, PoliTO Springer Series, https://doi.org/10.1007/978-3-032-08664-8_10

surveillance, exercised not only by state apparatuses but also by private entities operating within an intricate and often opaque digital ecosystem. As a direct consequence of these developments, there has emerged a growing and urgent imperative for the implementation of robust protective measures designed to safeguard the rights, freedoms, and fundamental dignity of individuals, with particular attention devoted to the realm of personal privacy [1].

Within this shifting paradigm, the very essence of the right to privacy has undergone a substantive evolution, progressively moving away from its classical conceptualization as the "right to be let alone". Instead, the contemporary discourse surrounding privacy now places paramount emphasis on an individual's capacity to exert agency and control over the dissemination, access, and utilization of personal information. Furthermore, concerns related to privacy are increasingly examined through the broader lens of modern power structures, wherein informational infrastructure has ascended to the status of a fundamental and indispensable pillar underpinning contemporary governance and societal organization [2].

From a historical standpoint, the juridical underpinnings of concepts such as privacy and confidentiality can be traced back to the latter part of the nineteenth century. It was during this pivotal period that Samuel D. Warren and Louis D. Brandeis introduced the seminal notion of privacy into the legal domain through their groundbreaking 1890 publication, "The Right to Privacy" [3]. This pioneering scholarly work played an instrumental role in catalyzing the formal recognition of privacy as a distinct legal construct, laying the intellectual groundwork for its subsequent evolution and entrenchment within judicial and legislative frameworks [4].

Nevertheless, the true consolidation and authoritative recognition of the right to privacy did not occur instantaneously but rather emerged through a prolonged and intricate process, one that was profoundly shaped by the critical and often determinative role played by judicial interpretation. The judiciary, in the absence of substantial academic engagement on the subject following Warren and Brandeis' exposition, assumed an indispensable role in delineating, affirming, and elaborating upon privacy rights. Indeed, in the decades following the publication of their seminal article, scholarly discourse on privacy within the United States remained conspicuously sparse, if not altogether absent. This lacuna in academic engagement accentuated the role of the judiciary, as courts, through a succession of landmark rulings, elucidated fundamental tenets of American legal doctrine. More significantly, they engaged in a dynamic and evolutionary interpretation of the provisions enshrined in the Bill of Rights, thereby identifying and fortifying a constitutional foundation for the right to privacy [5].

The decade of the 1960s heralded a momentous shift in both societal consciousness and governmental structures, marking a critical juncture in the trajectory of privacy jurisprudence. This period was characterized by a fundamental transition from the archetypal liberal state model to a more pluralistic state apparatus governed by the principles of the rule of law [6]. The concomitant systemic changes served to heighten societal awareness and deepen public discourse surrounding the necessity of safeguarding the private sphere of individual existence from undue encroachment [7].

In the United States, the conceptualization of privacy occupies a markedly different position relative to its European counterpart. Unlike in Europe, where privacy is enshrined as a fundamental human right, the American legal tradition predominantly situates privacy within the domain of consumer rights. This positioning necessitates a delicate and often precarious balancing act between the imperative of individual protection and the commercial exigencies of business enterprises operating in an increasingly data-driven economy [8].

Conversely, within the legal and regulatory framework of the European Union, the evolution and institutionalization of privacy protections have followed a distinctly divergent trajectory. The processing and safeguarding of personal data have been elevated to the status of one of the fundamental pillars of the European legal order. In the European legal schema, privacy is accorded an equivalent degree of protection and reverence as other core human rights, underscoring its intrinsic significance in the contemporary regulatory landscape [9].

Notably, in its formative years, the European Community's legal framework—originally devised with the overarching objective of fostering economic integration—did not explicitly incorporate provisions addressing privacy within its foundational regulatory architecture. Issues pertaining to confidentiality and data protection were traditionally subsumed under the broader aegis of human rights jurisprudence and were consequently subject to a limited and often inadequate degree of legislative codification [10].

Over time, however, the Court of Justice of the European Union emerged as a pivotal actor in remedying this oversight. Through its jurisprudential contributions, the Court underscored and reinforced the primacy of fundamental human dignity and, in so doing, extended legal safeguards to privacy, thereby enshrining it as an indispensable component of individual rights within the European legal canon [11].

The formalization of a regulatory framework for the protection of personal data in Europe commenced with the promulgation of Directive 95/46/EC by the European Parliament and the Council—commonly referred to as the "Data Protection Directive" or the "Mother Directive" [12]. This legislative enactment represented a watershed moment in the annals of privacy law, providing a precise and meticulously articulated definition of personal data while establishing the foundational principles governing its lawful processing. The directive was conceived with a dual objective: first, to afford robust protections for the rights and liberties of natural persons in relation to data processing; and second, to facilitate the seamless and unimpeded flow of information across national and jurisdictional boundaries within an increasingly interconnected European economic space. It delineated key principles governing the legitimacy of data processing activities, striving to strike a judicious balance between the imperatives of individual rights protection and the operational exigencies of an evolving digital economy [13].

Despite successive amendments and legislative refinements introduced to keep pace with the relentless advancement of communication and information technologies, the European legal architecture governing privacy protections remained, in many respects, outdated and ill-suited to address the complexities of the digital age. This regulatory inertia persisted until the advent of the General Data Protection Regulation (GDPR), a landmark legal instrument that fundamentally redefined and modernized the landscape of data protection across the European Union, establishing a comprehensive and harmonized framework for the governance of personal data processing [14].

Against the backdrop of these historical and legal developments, the ensuing analysis will embark on an in-depth exploration of the intricate interplay between privacy and the emergent digital frontier represented by the metaverse. Furthermore, it will critically examine the mechanisms and regulatory frameworks through which data protection principles can be effectively upheld and operationalized within this innovative and immersive virtual environment, ensuring that fundamental privacy rights remain safeguarded in the face of rapidly evolving technological paradigms.

2 Safeguarding Privacy in the Metaverse: Approaches and Mechanisms

Within the boundless and intricately immersive digital expanse of the metaverse, the safeguarding of personal privacy assumes an indispensable and overarching significance. This paramount importance arises not only from the sheer scale and sophistication of digital interactions but also from the inherent necessity of data exchange—often frequent, persistent, and, in many instances, unavoidable—within this rapidly evolving virtual domain. The omnipresence of interconnected networks, coupled with the proliferation of intelligent systems and real-time digital engagement, renders privacy protection an imperative of utmost priority, requiring vigilant oversight and continuous adaptation to emerging risks and challenges [15].

Addressing the multifaceted obstacles associated with privacy protection in such an elaborately interconnected digital environment necessitates the implementation of a comprehensive, multi-tiered, and dynamically responsive approach. As technological innovation accelerates at an unprecedented pace, it becomes imperative to conceive, develop, and execute sophisticated regulatory frameworks and best practices. These measures must be meticulously designed to ensure the security, integrity, and confidentiality of user data while simultaneously demonstrating the flexibility required to counteract the increasingly intricate and adaptive nature of digital threats. Within this extensive and ever-expanding virtual ecosystem, an array of strategic methodologies has emerged, each tailored to confront the intricate challenges posed by the collection, processing, and dissemination of personal information [16].

From a technological standpoint, the deployment of state-of-the-art encryption protocols, alongside an arsenal of cutting-edge cybersecurity mechanisms, constitutes a fundamental bulwark against unauthorized access to sensitive personal data. These robust security architectures are meticulously engineered to align with the foundational principles of privacy-centric design, a paradigm frequently encapsulated under the concept of "privacy by design". This innovative approach emphasizes the seamless integration of privacy-enhancing features at the very core of digital infrastructure, embedding them intrinsically within the developmental framework of virtual environments rather than relegating them to peripheral considerations or post hoc modifications. The foremost objective of these proactive measures is to curtail the unnecessary collection, retention, and processing of sensitive information, thereby significantly mitigating the risk of data misuse, unauthorized exploitation, or inadvertent exposure. By prioritizing such preemptive safeguards, digital platforms can foster an environment that is inherently more secure, resilient, and conducive to preserving user trust and autonomy [17].

The architectural and developmental processes underlying the metaverse must be intrinsically infused with a rigorous commitment to privacy protection, ensuring that security measures are not mere ancillary add-ons but are instead enshrined as foundational tenets from the inception of digital platform design. This entails a meticulously structured and systematically executed approach, wherein privacy-compliant principles are woven into every stage of technological innovation. By embedding regulatory safeguards at the core of platform development, excessive data collection and indiscriminate processing can be preemptively curtailed, thereby reinforcing security and confidentiality as intrinsic, non-negotiable imperatives rather than mere afterthoughts or reactive measures implemented in response to regulatory scrutiny [18].

Several pivotal dimensions must be methodically considered in order to actualize this vision with efficacy and precision.

Foremost among these is the principle of data minimization, a cornerstone of contemporary privacy-preserving practices. This fundamental tenet advocates for the collection and retention of only the most essential and strictly necessary information required for the delivery of specific digital services. By rigorously adhering to this principle, the potential for unauthorized data access, security breaches, and systemic vulnerabilities is significantly diminished, thereby affording users a substantially heightened degree of digital safety and control. Furthermore, enabling individuals to navigate and engage within the metaverse without the obligation to disclose their real-world identities emerges as a critical design consideration. Providing users with the option to operate under pseudonymous or entirely virtual personas serves to fortify privacy protections, granting them the ability to participate securely and anonymously without compromising their overall digital experience or the functional efficacy of the virtual platforms they engage with [19].

Equally indispensable is the principle of user autonomy, which must be accorded a central and unwavering position in the design and implementation of privacy-enhancing mechanisms within the metaverse. This necessitates the development of user interfaces that are both intuitive and comprehensively transparent, ensuring that privacy settings and data management functionalities are readily accessible, easily comprehensible, and operable without undue complexity. Users must be unequivocally empowered with granular control over the dissemination and accessibility of their personal information, thereby allowing them to dictate the precise parameters governing data-sharing practices. The ability to seamlessly customize privacy preferences in alignment with individual comfort levels and contextual requirements is a fundamental prerequisite for fostering an environment of trust and digital self-determination [20].

Another indispensable pillar upon which a privacy-respecting metaverse must be constructed is the principle of transparency. The overarching framework of this virtual domain must be underpinned by an uncompromising commitment to openness and clarity, ensuring that users are provided with unambiguous, readily comprehensible insights into the mechanisms governing the processing, storage, and potential distribution of their personal data. This necessitates the establishment of rigorous informational policies that present data-handling procedures in an accessible and non-obfuscatory manner, thereby enabling users to make fully informed and conscientious decisions regarding the management of their digital footprint. The cultivation of user trust within the metaverse is inextricably linked to the accessibility, lucidity, and reliability of these privacy policies [21].

Moreover, the advancement of transparency mandates the formalization of explicit and legally binding contractual agreements with developers, service providers, and all entities operating within the metaverse. Such agreements must enforce stringent adherence to well-defined privacy standards, ensuring that all stakeholders are unequivocally committed to upholding the principles of data protection, confidentiality, and ethical responsibility. Developers and service providers must be held to rigorous accountability standards, with clearly delineated obligations designed to preclude any deviation from established data protection frameworks. The imposition of enforceable penalties for non-compliance, coupled with the establishment of regulatory oversight mechanisms, serves as an essential safeguard against potential privacy violations and ensures that all digital actors operate within the prescribed ethical and legal boundaries [22].

The overarching infrastructure of the metaverse must, therefore, be meticulously engineered to cultivate a digital culture that prioritizes, reveres, and diligently safeguards user privacy at every conceivable level. This necessitates a profound and unwavering cultural commitment, wherein every stakeholder—from platform architects and system developers to content creators and regulatory bodies—recognizes and internalizes the fundamental necessity of preserving personal data integrity.

Only through a holistic, unrelenting, and systemically embedded dedication to privacy principles can a truly secure, ethically responsible, and user-centric digital ecosystem be realized. Such an environment will not merely accommodate privacy considerations as secondary concerns but will instead enshrine them as non-negotiable imperatives, ensuring that the rights, freedoms, and digital self-sovereignty of individuals remain sacrosanct within this ever-expanding virtual domain. In doing so, privacy will cease to be an afterthought and will instead be elevated to its rightful status as an intrinsic and inviolable cornerstone of the metaverse's ethical and operational framework.

3 Artificial Intelligence and the Sanctity of Privacy

The ascendance of Artificial Intelligence (AI) marks the dawn of a new epoch in technological innovation, one that is reshaping industries, redefining societal norms, and posing significant challenges to the foundational principles of privacy and data protection [23]. As AI continues to permeate an ever-widening array of spheres—ranging from healthcare and education to finance, governance, and beyond—its transformative potential is inextricably linked to its unparalleled capacity to harness, analyze, and derive insights from vast and ever-growing repositories of data [24].

Yet, this unprecedented reliance on data raises profound questions about the delicate balance between technological advancement and the safeguarding of individual rights, particularly in relation to the rights to privacy and data protection [25].

At its core, AI thrives on the acquisition of extensive datasets, which are necessary to train sophisticated algorithms, refine predictive models, and generate outputs that are increasingly precise and reliable. This voracious appetite for data not only accelerates the erosion of traditional informational boundaries but also exacerbates the tension between technological innovation and the fundamental principle of data minimization [26]. While traditional data protection frameworks, such as the General Data Protection Regulation (GDPR) in Europe, advocate for the collection of data that is strictly necessary to fulfill the intended purpose, AI systems often require the aggregation of diverse and expansive datasets to achieve the level of accuracy and robustness that is characteristic of truly advanced systems [27].

This divergence between conventional data protection principles and the data demands of AI raises critical concerns about the proportionality of data collection and the legitimacy of processing activities, particularly when sensitive personal information is involved, such as health records or financial details. In such contexts, AI's expansive data requirements often challenge the very tenets that have long underpinned privacy laws and frameworks [28].

The issue of opacity further complicates the seamless integration of AI into systems that are governed by privacy norms. Many AI-driven technologies, particularly those that rely on complex machine learning algorithms, operate as "black boxes", wherein the inner workings of the algorithms remain inaccessible or incomprehensible to most stakeholders, including regulators, policymakers, and even the developers who deploy these systems [29]. This opacity creates a significant barrier to accountability, as it prevents individuals from understanding how their personal data is processed, whether ethical principles are consistently upheld throughout the AI system's operation, and whether the outcomes generated by these systems are free from bias or discrimination.

This lack of transparency is especially concerning in applications that have profound and far-reaching societal impacts, such as predictive policing, credit scoring, or medical diagnostics, where unexamined algorithmic decisions can perpetuate systemic inequities, entrench existing biases, or lead to significant harm [30].

In healthcare, biased algorithms could lead to unequal access to treatments or misdiagnosis, particularly for marginalized populations. The opacity of AI systems, therefore, is not just a technical concern but also an ethical one, with the potential to undermine public trust in both AI technologies and the institutions that adopt them [31].

Another critical risk exacerbated by AI is the potential for the re-identification of individuals within datasets that were presumed to be anonymized. Through the use of advanced pattern recognition and cross-referencing capabilities, AI systems can reconstruct detailed personal profiles from data that had previously been stripped of identifying information [32].

This phenomenon poses a grave threat to privacy protections traditionally afforded by anonymization techniques, rendering them inadequate in the face of increasingly sophisticated AI technologies. The ability of AI to "de-anonymize" data calls into question the efficacy of existing safeguards and raises urgent concerns about the integrity of anonymized datasets in protecting individuals' privacy [33].

Furthermore, the aggregation of data for the purpose of training AI systems often occurs without the explicit knowledge or consent of the individuals whose data is being utilized. This lack of informed consent not only strains the legitimacy of consent-based frameworks that lie at the heart of most data protection laws but also highlights a fundamental challenge in the regulatory oversight of AI technologies [34]. The use of personal data in AI training is frequently characterized by a lack of transparency about where the data originates from, how it is collected, and whether individuals have the opportunity to opt out of having their data used in this way [35].

Without proper consent mechanisms, AI systems risk perpetuating a cycle of data exploitation, where individuals are not given a say in how their personal information is used or monetized.

Beyond these technical and regulatory challenges, AI also introduces novel and often difficult ethical dilemmas. The creation and deployment of AI systems are frequently marked by asymmetries of power and knowledge, wherein the entities that design these technologies—be they corporations, research institutions, or governments—wield considerable influence over how data is collected, interpreted, and utilized [36]. This dynamic of unequal power distribution can lead to the commodification of personal information, reducing individuals to mere data points within larger economic and algorithmic ecosystems. In such environments, individuals may lose control over their own personal data, which becomes increasingly valuable in the marketplace, yet remains largely beyond their reach or control [37].

Moreover, AI's inherent propensity to perpetuate biases that are encoded in the training data poses significant risks to vulnerable and marginalized populations. These groups are often disproportionately impacted by automated decision-making, whether in the context of employment, loan applications, insurance premiums, or criminal justice [38].

The risk is that AI systems, which rely on historical data to inform their decisions, can inadvertently reinforce existing biases and inequalities, thereby perpetuating systemic discrimination. Addressing this issue requires a concerted effort to ensure that AI systems are developed in a way that actively mitigates bias and promotes fairness, equity, and inclusivity [39].

Tackling these multifaceted challenges requires a concerted and holistic approach that addresses both the technical and ethical dimensions of AI.

First and foremost, embedding privacy-enhancing technologies (PETs) into AI systems is essential for mitigating the risks to privacy while preserving the functionality and effectiveness of AI technologies. Techniques such as federated learning, differential privacy, and homomorphic encryption offer promising avenues for ensuring that sensitive data remains protected during the processing phase, thereby safeguarding individuals' privacy without stifling innovation [40].

Second, fostering algorithmic accountability through the development of robust auditing mechanisms and the promotion of explainable AI (XAI) methodologies is critical for enhancing transparency and building public trust in AI-driven processes. By ensuring that AI systems can be audited, understood, and explained, regulators and users alike can ensure that these systems operate ethically and justly [41].

Furthermore, the evolution of regulatory frameworks must keep pace with the rapid advancements in AI technology. Policymakers must grapple with the inherent complexities of regulating systems that defy traditional categories of oversight, ensuring that legal instruments remain flexible, adaptive, and enforceable [42].

Given the global nature of AI systems, harmonized international standards, informed by interdisciplinary expertise, are crucial for addressing cross-border data flows and the transnational challenges posed by AI technologies. This global approach will be necessary to ensure that AI systems do not exploit regulatory gaps across jurisdictions and that personal data is protected no matter where it is processed or stored [43].

Finally, it is essential to cultivate an ethical ethos within the AI community, ensuring that developers, organizations, and governments prioritize human rights, fairness, and accountability in the design, deployment, and regulation of AI technologies.

These principles must be integrated into every stage of AI system development, from initial design to post-deployment monitoring, to ensure that AI serves the broader societal good while respecting individual autonomy and privacy [44].

Only through such a comprehensive and proactive approach can we reconcile the tremendous promise of AI with the imperatives of privacy and data protection, ensuring that innovation flourishes without compromising the fundamental rights of individuals in an increasingly algorithmic world.

4 Strategies for Data Minimization: Navigating the Ethical Landscape

In the rapidly evolving and immersive domain of the metaverse, the principle of data minimization stands as a foundational pillar for the protection of user privacy. As we delve into this sophisticated digital realm, data minimization emerges as an indispensable strategy, one that revolves around the prudent and conscientious collection, processing, and retention of personal information. At its core, this approach dictates that only the most essential data—strictly those pieces of information necessary for the provision of the desired services—be handled, thereby eschewing the excessive accumulation of sensitive or superfluous data. Such a disciplined and methodical practice serves a dual purpose: it not only mitigates the risk of security breaches and unauthorized data access but also fosters an environment of trust, where users feel secure and confident in engaging with various metaverse platforms [45].

Central to the ethos of data minimization is the concept of selective and intentional data gathering. This concept underlines the significance of deliberate policies implemented by metaverse platforms to delineate the exact nature and extent of data collection. These platforms meticulously evaluate and determine which data points are indispensable for service delivery, purposefully avoiding the aggregation of extraneous or unnecessary information. This deliberate selectivity represents a thoughtful balancing act: it seeks to harmonize the demands of delivering highly engaging and interactive user experiences with the ethical imperative of preserving user privacy. The equilibrium between technological innovation and privacy protection is thus maintained through carefully crafted data governance practices that prioritize user autonomy and security [46].

Moreover, the principles governing data retention further fortify the architecture of data minimization. One of the key strategies involves the implementation of clear and finite data retention timelines. These pre-determined timeframes outline how long personal information will be retained before it is systematically and irrevocably erased from the system. By enforcing such structured data retention policies, platforms can significantly reduce the potential for data misuse, unauthorized access, or prolonged exposure to privacy risks. The automatic deletion of data, once it has outlived its utility, exemplifies a commitment to minimizing long-term vulnerabilities and ensures a more robust and resilient data security framework. The result is a minimized risk landscape that better protects user privacy in an environment that is inherently rich in data interactions [47].

In addition to these protective measures, advanced techniques like data masking and tokenization offer an additional layer of security in data management. These techniques are part of a broader arsenal of data obfuscation strategies designed to further safeguard personal information. Data masking involves the transformation of sensitive information into non-identifiable, surrogate data elements, effectively concealing the original details without hindering functional processes. Tokenization, on the other hand, replaces sensitive data with randomly generated tokens, which cannot be reversed without access to a secure, centralized token vault. Such methodologies are crucial in maintaining the operational efficacy of systems while simultaneously ensuring that real, sensitive data remains secure from exposure or exploitation. Thus, the employment of these sophisticated techniques illustrates an unwavering dedication to the principle of minimizing privacy risks while facilitating secure yet seamless interactions within the metaverse [48].

Despite the impressive arsenal of technological safeguards, the role of user agency and control remains an irreplaceable facet of the data minimization framework. In this context, empowering users to exercise greater control over their personal information is of paramount importance. It is not enough to implement robust data security measures; there must also be an emphasis on user empowerment through transparency and informed consent. Providing clear, accessible, and comprehensible information about data collection and usage practices is essential. Users must be equipped with intuitive tools and options that enable them to adjust their privacy settings and dictate the terms of their data sharing. By placing the reins of data control firmly in the hands of users, metaverse platforms can reinforce a sense of agency, autonomy, and trust. This approach ensures that individuals are actively engaged in the governance of their personal data, making deliberate and well-informed decisions about what they choose to share and under what circumstances [45].

However, it is crucial to acknowledge the inherent complexities that arise from the interplay between the ideals of privacy and transparency. The juxtaposition of these two values presents an ongoing challenge, especially within a societal construct that increasingly values openness and the free flow of information. In a "transparent" society, where information is readily accessible to a wide array of stakeholders, the sanctity of individual privacy can be severely compromised.

This tension gives rise to the observation that it is reasonable to associate the growing transparency of information with a corresponding decline in respect for the right to privacy. As transparency intensifies, the protections surrounding personal privacy often erode, necessitating a careful reconsideration of how these values can coexist within the digital and interconnected spaces of the metaverse [49].

From a practical standpoint, one viable method of addressing the privacy challenges exacerbated by the transparency of the metaverse involves the strategic use of advanced encryption technologies. End-to-end encryption, for instance, ensures that communication between parties remains confidential by employing a system of public and private cryptographic keys. Unlike simpler encryption models that use a single key for both encrypting and decrypting data, end-to-end encryption enhances security by ensuring that only the intended recipients can decrypt the transmitted data [50].

The underlying principle of this approach advocates for communication protocols to operate securely at the endpoints of a system, or as close to the controlled resource as possible, thereby minimizing exposure to potential interception or manipulation along the communication path [51].

Societal expectations regarding transparency have also evolved dramatically, especially in the realm of algorithmic accountability and governance. The demand for increased oversight and the democratization of automated decision-making systems has intensified. There is a growing emphasis on ensuring that these systems are not only transparent but also governed by principles of fairness, ethics, and legal compliance. Technological solutions, such as impact assessments for algorithms, regular audits, and certification protocols, are being called for to verify that automated decisions align with established standards of equity and justice. Embedding these accountability mechanisms into policy frameworks is seen as crucial for fostering a responsible and ethically aware digital ecosystem. Such measures become even more pertinent given the current developmental stage of the metaverse, which remains an emerging and evolving technological construct, yet to be fully realized and regulated [52].

Nevertheless, as we continue to navigate the complexities of the metaverse, one of the most formidable challenges remains the tension between the ideal of a truly open, global network and the assertion of state sovereignty over national territories and populations. The Internet—and by extension, the metaverse—illustrates this fundamental tension. On one hand, there is a compelling vision of a borderless, interconnected digital world; on the other, there is the legitimate desire of sovereign states to exercise jurisdiction over activities within their borders. Resolving this dichotomy requires more than mere goodwill or superficial measures. It necessitates a nuanced and collaborative approach to governance, acknowledging the legitimate interests of states while striving to preserve the openness and universality of the digital space [53].

From the perspective of state sovereignty, the metaverse can be analogized to a data transmission protocol—an intricate system for transferring digital packets across public networks. Legally, this protocol holds significance in the realms of intellectual property and regulatory frameworks. Yet, it remains an "entirely immaterial" construct, and thus, it cannot be subjected to claims of sovereignty in the same way a tangible object or physical territory might be. The analogy extends to viewing the metaverse as akin to a telecommunications standard or a measurement system like the metric decimal system, both of which transcend the bounds of territorial jurisdiction [54].

It is also important to recognize that, although the metaverse operates through intangible protocols such as the Internet, the physical infrastructure that underpins these digital networks is very much material. The servers, data centers, and transmission cables that facilitate these virtual interactions are, in fact, located on sovereign territories. The tangible nature of this infrastructure underscores the complex interplay between the virtual and the physical, where digital activities are inextricably linked to real-world jurisdictions [55].

Yet, we must also contend with the reality that contemporary governance has increasingly seen the rise of regimes and organizational structures that operate beyond the confines of traditional state systems. These new forms of governance may not be wholly detached from state authorities but often exist in a state of semi-independence or even outright opposition. These entities maintain connections to state systems to varying degrees, yet they also exhibit the capacity to act autonomously and, at times, in direct conflict with national regulations. It is a phenomenon that challenges conventional notions of state-centric governance and raises questions about the regulation and oversight of digital spaces like the metaverse [56].

However, to declare the existence of a singular, universal legal system governing these digital interactions would be a misrepresentation. The reality is far more complex. Various national and transnational legal frameworks continue to intersect and interact, often in intricate and unpredictable ways. These frameworks collaborate and sometimes clash in their efforts to regulate digital networks, demonstrating that governance in the metaverse is not a monolithic or universally codified system. Instead, it is a dynamic and multifaceted web of rules, reflecting the interplay between different jurisdictions and the collaborative efforts to regulate activities in the increasingly interconnected digital world [57].

5 Conclusive Remarks: Charting the Future of Privacy in the Digital Sphere

In light of the observations expounded upon in the preceding pages, it becomes undeniably clear that the protection of privacy is a cornerstone of the metaverse, a dimension both critical and foundational to its sustainable development. The analysis of personal data protection within this burgeoning virtual realm has brought to light an array of pressing concerns that demand immediate and decisive action. As the metaverse continues its rapid expansion, it introduces profound challenges to privacy, underscoring the urgent necessity for the conception and enforcement of a comprehensive and resilient regulatory framework.

Foremost among these considerations is the imperative for platforms to ensure transparency and foster user awareness. It is incumbent upon these virtual environments to articulate, with unequivocal clarity, their policies concerning the management, storage, and utilization of personal data. The security of such data must be treated as a matter of paramount importance, requiring the deployment of state-of-the-art technological measures. Advanced encryption methodologies, coupled with rigorous cybersecurity protocols, stand as essential bulwarks against the ever-present specter of data breaches and malicious exploitation. In this regard, an ongoing commitment to fortifying digital infrastructures is indispensable, as threats evolve in tandem with technological advancements. The implementation of adaptive security frameworks, capable of preempting and mitigating potential vulnerabilities, should be prioritized to ensure the continued integrity and resilience of data protection measures.

Equally vital to this endeavor is the collaboration among developers, regulators, and other key stakeholders in crafting and promulgating best practices for data protection. This collaborative approach must also extend to addressing the ethical dimensions that are intrinsically linked to the collection and processing of data within the metaverse. It is imperative to establish and enforce robust ethical guidelines, ensuring that the deployment of algorithms and automated decision-making systems adheres to principles of fairness, impartiality, and accountability. Such measures are crucial in averting discrimination, bias, and other forms of systemic abuse. Moreover, the development of ethical AI governance mechanisms should be accompanied by continuous auditing and monitoring processes to uphold standards of transparency and mitigate risks associated with algorithmic opacity.

Furthermore, an essential facet of privacy protection within the metaverse lies in the empowerment of users through enhanced digital literacy and control over their personal information. Providing individuals with intuitive and accessible tools to manage their data, coupled with clear, comprehensible consent mechanisms, is pivotal in fostering an environment of trust. The reinforcement of user agency should be a guiding principle in the design of virtual ecosystems, ensuring that privacy preferences are respected and upheld without undue complexity or ambiguity.

Looking toward the horizon, the challenges outlined herein demand nothing short of a concerted international effort. Governments, organizations, and individual users must come together in a spirit of shared responsibility to create a metaverse that is not only innovative but also secure, ethical, and respectful of individual privacy. Only through such a global coalition can the evolution of this immersive virtual reality unfold in a manner that safeguards the sanctity of personal data and upholds the fundamental rights of its users. A multilateral approach, encompassing international regulatory cooperation and the harmonization of data protection laws, is paramount to preventing jurisdictional fragmentation and ensuring a cohesive, standardized framework for privacy preservation.

Continuous research and the ceaseless pursuit of technological innovation will play an indispensable role in achieving this equilibrium. It is only by harmonizing the relentless march of technological progress with the inviolable principles of privacy protection that the metaverse can flourish as an inclusive and equitable space. In doing so, we ensure that this digital frontier evolves in a manner that enriches humanity, serving as a testament to our collective commitment to advancing technology while preserving the dignity and rights of every individual who ventures within its boundless realms. The metaverse, as a transformative paradigm, holds immense potential to redefine human interaction, commerce, and social engagement. However, its success hinges upon our ability to construct and maintain an infrastructure that upholds privacy, security, and ethical integrity. Thus, the path forward must be paved with diligence, foresight, and an unwavering dedication to ensuring that technological evolution remains inextricably linked to the fundamental values of human rights and digital sovereignty.

References

1. See on this point, Zambardino F (2022) La blockchain e la protezione dei dati personali: una tecnologia privacy compliant by design? Eur J Priv Law Technol 136. See also Borroni A, Zambardino F (2024) The data protection in the metaverse. Revista Româna de Drept al Afacerilor (Romanian Business Law Review), no 3, pp 17–32; Wolters Kluwer; Alpa G, Resta G (2019) Le persone e la famiglia. Le persone fisiche e i diritti della personalità. In Trattato di diritto civile, diretto da R. Sacco, pp 145–632, Utet Giuridica, Turin. In particular, a 'value-oriented' interpretation is favoured, which is considered the most appropriate line for dealing

with issues concerning the person and personality rights. In this scenario, the globalisation of markets and the evolution of technologies constitute, for the authors, complex challenges to the role of law

2. Rodotà S (1996) Tecnologia e diritti, p 19, Il Mulino, Bologna. Part of the doctrine affirms how, in the same way as intellectual property rights, the right to privacy is also closely linked to technology. The evolution of the concept of privacy is fruitfully read in terms of law and technology. Pascuzzi G, Izzo U, Macilotti M (eds) (2013) Comparative issues in the governance of research biobanks. Property, privacy, intellectual property, and the role of technology. Springer, Heidelberg-New York-Dordrecht-London. In argument, Pascuzzi G (2003) Il diritto dell'era digitale. Tecnologie informatiche e regole privatistiche, Il Mulino, Bologna
3. Warren S, Brandeis LD (1890) The right of privacy. Harvard L Rev 193 ff
4. For more on confidentiality, Rodotà S (2000) Riservatezza. Enciclopedia treccani. See also Pardolesi R (1974) Riservatezza: problemi e prospettive. In Spinelli M (ed) Responsabilità civile, vol II, p 316 ff, Cacucci, Bari
5. See in particular Griswold v. Connecticut, 381 U.S. 479 (1965) NAACP v. Alabama, 357 U.S. 449 (1958), Katz v. U.S. 347 (1967)
6. Miglietti L (2014) Profili storico-comparativi del diritto alla privacy. Diritti Comparati
7. Zambardino F (2022) La blockchain e la protezione dei dati personali: una tecnologia privacy compliant by design?. *op.cit.*, p 137. See also Borroni A, Zambardino F (2024) The data protection in the metaverse. *op. cit.*, pp 18–19
8. See amplius, Pagallo U (2008) La tutela della privacy negli Stati Uniti d'America e in Europa: modelli giuridici a confronto, Giuffrè, Milan
9. Miglietti L (2014) Profili storico-comparativi del diritto alla privacy. *op.cit.*, p 138. On the notion of data processing, see Lambo L (2003) La disciplina sul trattamento dei dati personali: profili esegetici e comparatistici delle definizioni. In Pardolesi R (ed) Diritto alla riservatezza e circolazione dei dati personali, p 75; Giuffrè M, Zambrano V (2019) Il Comitato europeo per la protezione dei dati, in I dati personali nel diritto europeo, pp 985–999, Giappichelli, Turin
10. For a careful reconstruction of the debate on the subject, which has animated doctrine since the early decades of the 20th century, see Niger S (2006) Le nuove dimensioni della privacy: dal diritto alla riservatezza alla protezione dei dati personali, Cedam, Padua, p 39 ff. In Italy, for example, close to the reform of the Civil Code, some jurists began to take an interest in the subject of 'confidentiality', framing it in the more general context of personality rights. From that time onwards, a wide-ranging doctrinal debate developed on the subject of the right to confidentiality, which involved illustrious jurists and which was caused, first and foremost, by the lack of an explicit and general rule that would provide a legal basis for the aforementioned right. On this subject, see ex multis Giampiccolo G (1963) La tutela giuridica della persona umana e il c.d. diritto alla riservatezza nel quadro dei diritti della personalità. Riv Dir Civ De Cupis A (1955) Teoria generale, diritto alla vita e all'integrità fisica, diritto sulle parti staccate dal corpo e sul cadavere, diritto alla libertà, diritto all'onore e alla riservatezza, Giuffrè, Milan. Carnelutti F (1955) Il diritto alla vita privata. Riv Trim Dir Pubb Ravà A (1938) Istituzioni di diritto privato, Cedam, Padua
11. Borroni A, Zambardino F (2024) The data protection in the metaverse. *op. cit.*, pp 19–20. For an in-depth analysis see Terrasi A (2017) Il rapporto tra diritto alla privacy e protezione dei dati personali tra Corte di giustizia dell'Unione europea e Corte europea dei diritti dell'uomo. In: Di Stefano M, La protezione dei dati personali ed informatici nell'era della sorveglianza globale. Editoriale Scientifica, pp 127–149; Naples. Caso R (2014) Misure tecnologiche di protezione: cinquanta (e più) sfumature di grigio della Corte di giustizia europea. Trento Law and Technology Research Group. Research Papers. It is interesting, in this regard, to observe the reconstruction made by Gambaro, who, with reference to the free circulation of personal data, implies the conception of the latter as legal goods subject to exchange. For an in-depth study of this thought, see Gambaro A (2012) I beni. In Cicu A, Messineo F (dir), Tratt. dir. civ. e comm., Giuffrè, Milan. In such a scenario, at the very moment when the concept of privacy became more prominent within institutions, considered as a right to be legally guaranteed, the term 'data protection' became widespread, and the original right to privacy was transformed into a

real specific control of data. On this subject, see Finocchiaro G (2012) Privacy e protezione dei dati personali. Disciplina e strumenti operativi, Zanichelli Editore, Bologna

12. Cardarelli F, Sica S, Zeno-Zencovich V (2004) Il codice dei dati personali. Temi e problemi, p 6, Giuffrè, Milan
13. Zambardino F (2022) La blockchain e la protezione dei dati personali: una tecnologia privacy compliant by design?. *op.cit.*, p 140
14. Cardarelli F, Sica S, Zeno-Zencovich V (2004) Il codice dei dati personali. Temi e problemi. *op.cit.*, p 9
15. Some authors, in this regard, refer to the emergence of a true virtual identity insofar as talking about personal identity, in today's context, also means coming to terms with a new dimension, namely that of information technology, in which personal identity is indispensable for the performance of a series of actions, mostly of a patrimonial nature—one thinks of the use of credit cards on the web, of the various commercial transactions, and even of social networks. On this point see, Rodotà S (2018) Quattro paradigmi per l'identità. In Vivere la democrazia, p 20 ff, Laterza, Bari; Alpa G (2017) L'identità digitale e la tutela della persona. Spunti di riflessione. Contr Impr, p 725 ff; Resta G (2007) Identità personale e identità digitale. Dir Infor
16. Cerrina Feroni G (2023) Il metaverso tra problemi epistemologici, etici e giuridici. Media Laws, n 1, p 27. The author states how in the metaverse there will be protection of personal data, but not privacy in the original sense of the term, as right to be let alone the problem that arises is no longer how to have access to the personal data of the inhabitants of this reality, but only how to manage their processing
17. Bholasing J (2022) How technological advances in the big data era make it impossible to define the 'personal' in GDPR's 'personal data.' Eur Data Prot L Rev 8(3):346–361
18. Bignami F, Resta G (2015) Transatlantic privacy regulation: conflict and cooperation. L. & contemporary problems, vol 78, n 4, pp 231–266. See also Caso R (2021) La società della mercificazione e della sorveglianza: dalla persona ai dati. Casi e problemi di diritto civile, Ledizioni, Milan
19. Borroni A, Zambardino F (2024) The data protection in the metaverse. *op. cit.*, pp 21–22. On the subject of of data minimisation and pseudonymisation, widely, Tosoni L (2022) Article 4(5). Pseudonymisation, in The EU General Data Protection Regulation: a Commentary. In Kuner C, Bygrave LA, Docksey C, Drechsler L (eds), pp 225–230, Oxford University Press, Oxford; Foglia C (2019) Il dilemma (ancora aperto) dell'anonimizzazione e il ruolo della pseudonimizzazione nel GDPR. In Panetta R (ed) Circolazione e protezione dei dati personali tra libertà e regole del mercato, pp 309–332, Giuffrè, Milan; Stalla-Bourdillon S, Knight A (2017) Anonymous data v. personal data–a false debate: an EU perspective on anonymization, pseudonymization and personal data. Wisconsin Int'l L. J., vol 34, pp 284–322
20. This aspect is also provided for in Article 2 of the GDPR ('Scope'). See Graziadei M (2021) Commento art. 2 GDPR. In D'orazio R, Finocchiaro G, Pollicino O, Resta G (eds.), Codice della privacy e data protection, Giuffrè, Milan, pp129–143
21. Noto La Diega G (2018) Against the dehumanization of decision-making—algorithmic decisions at the crossroads of intellectual property, data protection, and freedom of information. J Intellect Prop Infor Techn E-Commer L 9(3)
22. Palmieri A (2022) Diritto all'oblio, deindicizzazione e conclusioni non consequenziali alle premesse (nota a Cass., ord. 31 maggio 2021, n. 15160). Foro It., vol 147, fasc 1, pp 331–335; Palmieri A, Pardolesi R (2014) Diritto all'oblio: il futuro dietro le spalle (Nota a corte giust. 13 maggio 2014, causa C-131/12). Foro It, vol 139, fasc 6, pp 317–322; Palmieri A, Pardolesi R (2020) Polarità estreme: oblio e archivio digitali. Foro It., vol145, fasc 5, pp 1570–1576; Pardolesi R (2017) Oltre «Google Spain» e il diritto all'oblio Foro It, vol 142, n 4, pp 219–222
23. Falletta P, Marsano A (2024) Intelligenza artificiale e protezione dei dati personali: il rapporto tra Regolamento europeo sull'intelligenza artificiale e GDPR. Riv It Infor Dir, p 120
24. Frosini TE (2022) La privacy nell'era dell'intelligenza artificiale. DPCE Online, no 1, p 282. See also Alpa G (2021) L'intelligenza artificiale. Il contesto giuridico, p 71 ff, Mucchi Editore, Modena

25. Frosini TE (2020) Le sfide attuali del diritto ai dati personali. In: Faro S, Frosini TE, Peruginelli G (eds) Dati e algoritmi. Diritto e diritti nella società digitale, p 25 ff, Zanichelli, Bologna
26. On which see Peluso MG (2024) Intelligenza artificiale e tutela dei dati: prospettive critiche e possibili benefici per una governance efficace, p 93 ff, Giuffrè, Milan
27. Falletta P, Marsano A (2024) Intelligenza artificiale e protezione dei dati personali: il rapporto tra Regolamento europeo sull'intelligenza artificiale e GDPR. *op.cit*, pp 126–129
28. Casonato C, Fasan M, Penasa S (eds) (2022) Diritto e intelligenza artificiale. DPCE Online, n 1, p 178 ff
29. Rudin C, Radin J (2019) Why are we using black box models in AI when we don't need to? a lesson from an explainable AI competition. Harv Data Sci Rev 3 ff
30. In the case of predictive policing, for example, AI algorithms may unintentionally reinforce racial or socioeconomic biases, exacerbating existing injustices in the criminal justice system. On this point, see amplius Chiti E, Marchetti B (2020), Divergenti? Le strategie di Unione europea e Stati Uniti in materia di intelligenza artificiale. Riv. Regolazione Mercati, n 1; Cath C, Watcher S, Mittelstadt B, Taddeo M, Floridi L (2018) Artificial intelligence and the "Good Society": the US, EU and UK approach. Sci Eng Ethics, n 2
31. For an analysis of the healthcare Signorini E, Zambardino F (2024) La blockchain e la telemedicina: quali opzioni per la sicurezza e l'integrità dei dati sanitari. Federalismi.it, no 27, pp 205–231
32. Frosini TE (2022) La privacy nell'era dell'intelligenza artificiale. *op.cit*, p 283
33. For an in-depth analysis see Zambardino F (2024) The applications of the blockchain technology. A comparative law study, p 308, footnote 101, Edizioni Scientifiche Italiane, Naples. See also Resta G, Anonimato, responsabilità, identificazione: prospettive di diritto comparato. Dir infor, p 171
34. Finocchiaro G (2021) XVIII lezione: intelligenza artificiale, privacy e data protection. In: Ruffolo U (ed) XXVI Lezioni di diritto dell'intelligenza artificiale, p 338, Giappichelli, Turin
35. *Ibid*
36. On this point see amplius Zambardino F (2024) Artificial intelligence and law: hypothesis of civil liability. Opinio Juris in Comparatione, *passim*
37. Giraudo M, Fosch-Villaronga E, Malgieri G (2024) Competing legal futures—"commodification bets" all the way from personal data to AI. Germ L J 1–25
38. Scherer MU (2016) Regulating artificial intelligence systems: risks, challenges, competencies, and strategies. Harv J L & Tech 29(2):365 ff
39. Ciaralli CA (2023) Intelligenza artificiale, decisione politica e transizione ambientale: sfide e prospettive per il costituzionalismo. Federalismi.it, n 15, pp 41–42
40. For an in-depth analysis on the functioning of PETs for the artificial intelligence see D'Aliberti L, Gronberg E, Kovba J (2024) Privacy-enhancing technologies for artificial intelligence-enabled systems. IWSPA, Porto
41. Morandini S et al (2023) Assessing the transparency and explainability of AI algorithms in planning and scheduling tools: a review of the literature. Hum Interact Emerg Technol 111:609–619
42. Generally on regulation see Carantani I (2024) IA Act: l'Unione Europea approva la proposta di Regolamento sull'intelligenza artificiale. Ius in itinere; Finocchiaro G (2022) La regolazione dell'intelligenza artificiale. Riv Trim Dir Pubbl, n 4; Marchetti B, Parona L (2022) La regolazione dell'intelligenza artificiale: Stati Uniti e Unione europea alla ricerca di un possibile equilibrio. DPCE Online, n 1; Stradella E (2019) La regolazione della Robotica e dell'Intelligenza artificiale: il dibattito, le proposte, le prospettive. Alcuni spunti di riflessione. MediaLaws, n 1
43. Falletta P, Marsano A (2024) Intelligenza artificiale e protezione dei dati personali: il rapporto tra Regolamento europeo sull'intelligenza artificiale e GDPR. *op.cit*, p 126–129
44. Frosini TE (2022) La privacy nell'era dell'intelligenza artificiale. *op.cit.*, p 283 ff
45. Stalla-Bourdillon S, Knight A (2017) Anonymous data v. personal data–a false debate: an EU perspective on anonymization, pseudonymization and personal data. *op.cit.*, pp 284–322
46. On the subject see Burdon M (2020) Digital data collection and information privacy law. Cambridge University Press, Cambridge

47. Guarda P, Bincoletto G (2023) Diritto comparato della privacy e della protezione dei dati personali, p 66 ff, Ledizioni, Milan
48. Borroni A, Zambardino F (2024) The data protection in the metaverse. *op. cit.*, pp 24–25. For an analysis of the concept of data masking see Ahmed W (2024) Data masking & techniques. Int'l J Tech Res & Sci, p 13 ff
49. Zambardino F (2022) La blockchain nel mercato del lavoro italiano con particolare enfasi sulla questione privacy e il rapporto con il General Data Protection Regulation. TSL, pp 23–38. See also Mayes GR (2018) Privacy and Transparency, pp. 125–129, Sacramento State University
50. *Ibid*
51. Borroni A, Zambardino F (2024) The data protection in the metaverse. *op. cit.*, pp 24–26. With reference to risks see Gardella Tedeschi B, Thobani S (2020) Innovazione, diritto e tecnologia: temi per il presente e il futuro. Introduzione. Riv Dir Media. According to the authors, the risks arising from automatism, in particular, are exacerbated by the invasiveness of new technologies, which allow a high degree of intrusion into people's private lives
52. Zambardino F (2022) La blockchain e la protezione dei dati personali: una tecnologia privacy compliant by design?. *op.cit.*, p 145
53. Zeno-Zencovich V (2016) Intorno alla decisione nel caso Schrems: la sovranità digitale e il governo internazionale delle reti di telecomunicazione. In: Resta G, Zeno-Zencovich V (eds) La protezione transnazionale dei dati personali. Dai "safe harbour principles" al "privacy shield", Consumatori e Mercato, p 14, RomaTre-Press, Rome
54. *Ibid*. Again, the author, with specific reference to the Internet—but by analogy the same argument is also applicable to blockchain—states that there can be no more sovereignty over Internet protocol than there can be over the protocols used for Skype or WhatsApp services
55. *Ibid*. The author well exemplifies how the only case of non-material extra-territorial communication is that of a message coming from a satellite that can be received directly by the user (e.g. with a satellite mobile phone) without the need for a terrestrial infrastructure to distribute it
56. Poncibò C (2020) Il Diritto Comparato e la Blockchain. In Memorie del Dipartimento di Giurisprudenza dell'Università di Torino, p 226, Edizioni Scientifiche Italiane, Naples. Of the same author see also Poncibò C (2020) Blockchain and comparative law. In Cappiello B, Carullo G (eds) Blockchain, law and governance. Springer, Cham, pp 137–156
57. Zambardino F (2022) La blockchain e la protezione dei dati personali: una tecnologia privacy compliant by design?. *op.cit.*, p 146. In the same direction already Pascuzzi G (2020) Il diritto dell'era digitale, V ed., Il Mulino, Bologna; Pollicino O, Liguori L, Busia G (eds), Le nuove frontiere della privacy nelle tecnologie digitali, Aracne, Rome. On this point, some thought is given by Teubner G (2010) Ordinamenti frammentati e costituzioni sociali. In Rivista giuridica degli studenti dell'Università di Macerata, pp 45–57, who states that a semantic expansion of our concept of law is opportune, if not also desirable, so that it can also include norms that operate independently of the legal sources of state or international law

Quantum-Powered Metaverse Opportunities, Challenges, and Future Directions in Virtual Worlds

Mustafa El Bizri, Ali M. Haidar, Ahmad M. El-Hajj, and Layth Sliman

Abstract The metaverse is a virtual and physical world that integrates virtual and physical realities using the latest technologies such as artificial intelligence, blockchain, and extended reality. However, as metaverse environments grow and become more sophisticated, classical computing models are no longer able to keep up with the computational, security, and scalability demands of these virtual worlds. Quantum computing offers a breakthrough solution by using concepts such as superposition and entanglement to speed up data processing, security, and artificial intelligence-based automation. This review explores the integration of quantum computing with the metaverse, focusing on its envisioned applications, such as quantum-aided artificial intelligence, post-quantum cryptography, and ultra-high-speed, secure quantum networking. This study highlights key challenges such as quantum hardware constraints, compatibility with classical computing systems, and ethics of digital economies fueled by quantum. Through an analysis of recent developments, gaps in research, and future directions, this chapter offers a holistic guide to how quantum computing can be used to transform metaverse ecosystems. The insights presented aim to provide researchers, developers, and decision-makers on how to develop an optimized and secure quantum-enabled metaverse.

Keywords Quantum computing · Metaverse · Quantum cryptography · Artificial intelligence · Quantum networking · Virtual worlds

---

M. E. Bizri (✉) · A. M. Haidar
Electrical and Computer Engineering Department, Beirut Arab University, Beirut, Lebanon
e-mail: mostafa.bizri@bau.edu.lb

A. M. Haidar
e-mail: ari@bau.edu.lb

A. M. El-Hajj
Olayan School of Business, American University of Beirut, Beirut, Lebanon
e-mail: ae37@aub.edu.lb

L. Sliman
Efrei Paris, Panthéon Assas University, Paris, France
e-mail: layth.sliman@efrei.fr

A. Borroni et al. (eds.), *Metaverses: Reshaping Law, Economy, and Society*, PoliTO Springer Series, https://doi.org/10.1007/978-3-032-08664-8_11

1 Introduction

The metaverse is a fusion of digital and physical space, resulting in virtual ecosystems where users engage, socialize, and transact. This new digital frontier harnesses the power of advanced technologies, artificial intelligence (AI), blockchain, and extended reality (XR) to deliver experiences so immersive that they closely approximate or exceed in-person interactions [1]. However, with the metaverse evolving in complexity and size, it runs into issues of computational power, security, and scalability. Traditional computing infrastructures often cannot fulfil these necessities, which justifies the search for alternative solutions.

Quantum computing has the potential to efficiently solve such issues. Quantum computers utilize the fundamental concepts of quantum mechanics, viz., superposition and entanglement, to process larger sets of data with greater efficiency compared to conventional computers [2]. With the processing of data at unprecedented speeds, quantum computing is a game-changing development for the metaverse, supporting revolutionary use cases such as real-time data analysis, secure communication, and complex simulation scenarios.

This chapter focuses on investigating the role of quantum computers in the metaverse ecosystem. We will outline the basic principles of both fields, draw the limitations of existing technologies, and discuss where quantum destruction can overcome them. Moreover, the potential ethical, social and economic impacts of this innovation also provide an overall perspective for researchers and developers, political decision-makers seeking to construct the future of a virtual world.

1.1 Overview of the Metaverse

The term "metaverse" defines a common virtual space, created by the overlap of physically persistent virtual reality and virtually enhanced physical reality. The metaverse evolves through three phases, as depicted in Fig. 1: Digital Twins, where real-world entities are replicated virtually; Digital Natives, creating purely virtual environments; and Surreality, merging physical and digital realities for seamless immersion [3]. It comprises augmented reality (AR), virtual reality (VR), mixed reality (MR), and other immersive digital worlds [1]. AI is a key element of metaverse as it is used to improve the user experience through intelligent virtual agents and personalized content. Blockchain provides secure, decentralized real estate and transactions for transactions with digital assets and transparency genes and trust in virtual economy. XR, which includes AR, VR, and MR technologies, allows for immersive experiences for users by merging the digital and real worlds. All of these combination technologies allow virtual and dynamic world to reflect or increase the real world.

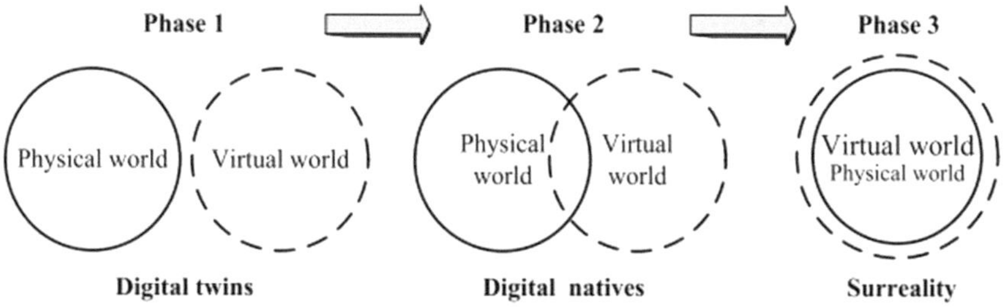

Fig. 1 Three phases of the development of the metaverse [3]

1.2 Limitations of Classical Computing in the Metaverse

As metaverse grows, many computational resources are required in order to maintain a smooth and responsive user interface. Classical computer systems encounter several limitations [4]. One significant constraint is arithmetic constraints because complex virtual worlds and real-time representation of simulations require processing power beyond what traditional CPU and GPU can handle. With the growing number of matched users and interactions, scalability becomes the classical architecture issues. Moreover, security vulnerabilities are also a major issue while protecting user information and digital properties from cyber-attacks because they are becoming increasingly vulnerable to highly advanced attacks [5]. All these factors emphasize the need for a more sophisticated computing paradigm to support the metaverse's evolution.

1.3 Quantum Computing as a Solution

Quantum computing provides a change in the model of our calculation capabilities that link the qubits that can exist simultaneously in multiple states, allowing the processing of complex calculations at unprecedented speed. This potential makes quantum itself a feasible solution for metaverse challenges [4]. Improved calculation power allows quantum algorithms to make the calculations more complex, facilitating the production and simulation in real time in Metaverse. Improved expansion ability allows quantum systems to manage and process larger datasets, by supporting the expansion of the virtual world. In addition, advanced security measures such as distribution of quantum locks provide strong encryption techniques to protect countries communicate against potential violations [6]. With the incorporation of quantum computing, the metaverse can expand beyond existing limits and become a safer, more advanced, and immersive role-playing environment.

1.4 Objectives and Structure

This chapter aims to provide a comprehensive view regarding how the metaverse can be changed by quantum computing. The aims are the changing of quantum and metaverse in building an early impression of two technologies. It shall acknowledge limitations of common computers in Metaverse, indicating restrictions that could be transcended through quantum computers. This chapter is going to enumerate quantum solutions by highlighting some of the applications and innovations that would improve the infrastructure of Metaverse. It would also assess the ethical, social and economic consequences of incorporating quantum technologies in Metaverse and would make recommendations regarding future development.

The chapter is structured as follows. Section 2 is a discussion of the fundamentals of quantum computing, looking at key concepts and the status of quantum technologies. Section 3 explores the metaverse's infrastructure and the current limits of classical computing in the metaverse. Section 4 sets out the role of quantum computing in the metaverse, with its potential applications and benefits. Section 5 conducts a technical exploration of embedding quantum computing in the metaverse, and pointing out significant challenges. Section 6 addresses the ethical, social, and economic implications of this integration. Section 7 identifies research gaps and probable future areas, and Sect. 8 concludes with an overview of key findings and consideration of the future landscape for a metaverse based on quantum.

Through this structured approach, the chapter aims at providing an encompassing understanding of the transformational potential at the confluence of quantum computing and the metaverse.

2 Fundamentals of Quantum Computing

Quantum computing represents a paradigm shift from classical computing, leveraging the principles of quantum mechanics to process information in fundamentally new ways. This section delves into the core concepts of quantum computing, contrasts it with classical computing, explores various quantum computing models, and discusses the current state of quantum hardware development.

2.1 Basic Principles of Quantum Mechanics

At the heart of quantum computing are several basic principles that emerged from quantum mechanics. One of them is superposition, whereby qubits may exist in many states simultaneously rather than a single binary state of 0 or 1, as in traditional computing. The unique characteristic allows quantum computers to perform operations in parallel, significantly enhancing processing power and efficiency [7].

The other fundamental principle is entanglement, in which two or more qubits become entangled such that the state of one qubit immediately influences the state of another regardless of the distance between them [8]. This property allows the establishment of effective quantum networks and safe media channels, which are the basis of quantum cryptography and quantum teleportation.

Quantum interference is also a basic phenomenon of quantum mechanics that contribute to quantum computation [9]. Probability margin of states in quantum system can be added together with creation and annihilation and to create interference patterns. Scientists can increase precise outputs with less error and hence making quantum more efficient on some problems for problem-solving through the construction of interference-based quantum algorithms [10].

2.2 *Quantum Versus Classical Computing*

The inherent differences between quantum computing and classical computing lie in their information representation, processing capability, and vulnerability to errors [11]. In classical computing, information is encoded as bits that can exist in one of two states, 0 or 1 as shown in Fig. 2a. In contrast, as illustrated in Fig. 2b, Quantum computing employs qubits that leverage superposition, i.e., they can exist in multiple states at the same time. This feature allows quantum computers to perform multiple calculations at once, greatly improving computational efficiency.

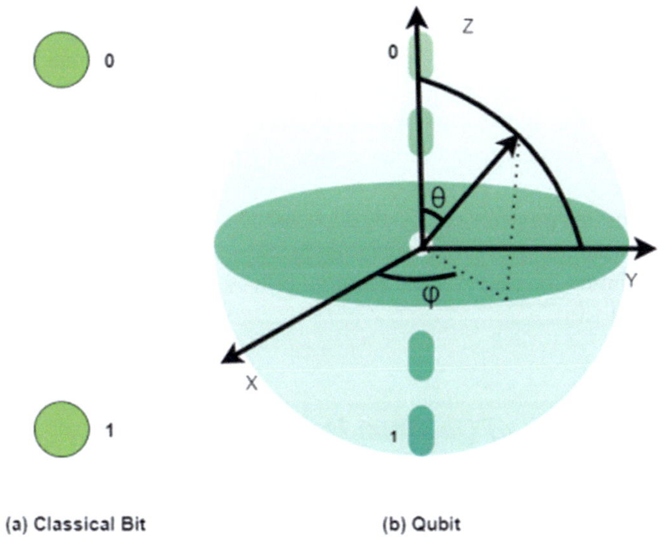

Fig. 2 Classical bits "0 or 1"—qubits, represented by the Bloch sphere, exist in superposition between $|0\rangle$ and $|1\rangle$ [12]

Processing power is the second significant differentiating factor. Classical computers execute problems in sequential or parallel steps, while quantum computers can tackle several possibilities all at once using quantum superposition and entanglement. This aspect provides exponential speedups for complex problems, such as factorizing large numbers for cryptography and optimizing large-scale simulations [13].

Despite its advantages, quantum computing is riddled with significant challenges, particularly susceptibility to errors. Quantum systems are very sensitive to external disturbances and therefore prone to decoherence, where the fragile quantum state turns into a classical state [14]. Quantum computing error correction methods are still in their early stages, requiring significant breakthroughs before quantum devices can reliably outperform classical supercomputers in actual applications.

2.3 Types of Quantum Computing Models

Several models of quantum computing have been proposed, each with unique approaches to performing quantum operations. The Quantum Gate Model is the most widely researched and is conceptually similar to classical logic gates. In this model, qubits are manipulated through unitary quantum gates to form quantum circuits that execute specific computations [15, 16]. This approach is used in quantum processors developed by IBM and Google [17].

Quantum Annealing is another approach that focuses on solving optimization problems. Rather than performing general-purpose computations, quantum annealers exploit quantum tunneling and superposition to find the minimum energy configuration of a system, making them particularly useful for tasks such as logistics optimization and machine learning applications [17].

A more advanced and less developed model is Topological Quantum Computing, which relies on exotic quasiparticles called anyons [18]. These particles can be braided in specific ways to perform computations. The major advantage of this approach is its inherent resistance to decoherence, making it a promising candidate for fault-tolerant quantum computing. Microsoft has been actively researching this model, aiming to create robust quantum processors with fewer error correction challenges [19, 20].

2.4 Current State of Quantum Hardware

Recent advancements in quantum hardware indicate significant progress toward scalable and practical quantum computing. Microsoft recently unveiled the Majorana 1 Chip, which employs a novel material called a topoconductor to generate reliable qubits based on Majorana particles [21]. This chip has the potential to scale up to

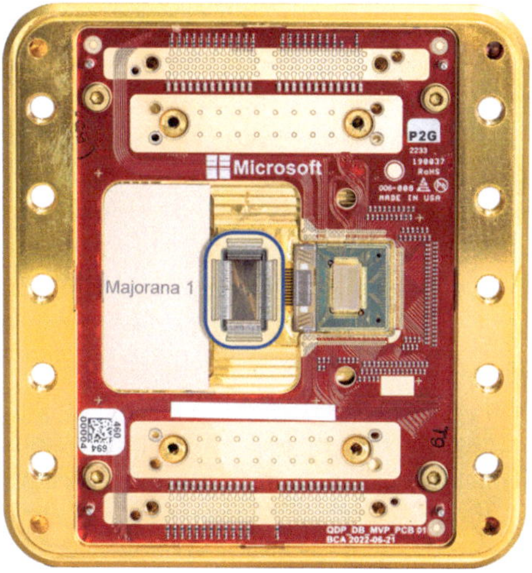

Fig. 3 Majorana 1—a new quantum chip [21]

one million qubits, marking a revolutionary step in quantum computing hardware development (Fig. 3).

Google has introduced the Willow Quantum Processor, which is capable of performing complex quantum calculations in just minutes—tasks that would take classical supercomputers an impractical amount of time to complete [21]. This processor demonstrates Google's ongoing efforts to achieve quantum supremacy and make quantum computing a commercially viable technology [22].

China has also made significant strides in quantum computing with the launch of the Wukong Quantum Machine, a quantum computer featuring 72 working qubits that is accessible via cloud computing [23]. Developed by Origin Quantum Computing Technology, this machine underscores the global competition in quantum research and its increasing accessibility to researchers worldwide.

These advancements reflect the rapid evolution of quantum computing, with major technology companies and research institutions making significant progress toward practical and scalable quantum processors. While quantum computing is still in its early stages, the continuous development of hardware innovations suggests that the field is steadily moving toward widespread adoption and real-world applications.

3 The Metaverse: Infrastructure and Challenges

The metaverse is built on a fusion of technologies that create a persistent, shared virtual world. Core components include immersive interfaces (virtual reality and augmented reality) and decentralized platforms (e.g. blockchain) that support 3D avatars, digital assets, and virtual economies [24]. AI engines drive content generation and personalized experiences, while wireless networks and cloud/edge computing provide the massive data processing and storage needed to render rich, interactive environments in real time. Together, these elements form the infrastructure that blurs physical and digital boundaries, enabling users to work, play, and socialize in a parallel digital universe [25].

As metaverse worlds grow more complex and populated, they face significant scalability challenges. Large-scale simulations and live interactions demand *enormous computational power and low latency* to run smoothly. Currently, the limits of classical computing present a bottleneck—conventional servers and GPUs struggle to handle the scale and complexity of fully immersive 3D environments with millions of concurrent users [26]. This leads to lag, bandwidth strain, and an inability to simulate realistic physics or AI-driven events at planetary scale. Ensuring *high reliability and performance* (ultra-low latency, high bandwidth) in these vast virtual spaces is an ongoing challenge [27]. In short, today's computing models often cannot meet the metaverse's demands for real-time responsiveness and massive parallel processing, highlighting the need for more powerful computational paradigms.

In addition to technical performance issues, *security and privacy* emerge as critical concerns in the metaverse. Immersive platforms collect a vast amount of personal data on users—far beyond what traditional web services gather [28]. Biometric identifiers like face movements, gaze, voice, and heart rate can be tracked continuously in virtual experiences [29]. If not properly protected, this sensitive data could be misused for surveillance, identity theft, or intrusive targeted advertising. Likewise, the metaverse could become a breeding ground for cybercrime and fraud if robust safeguards aren't in place [30]. Users need assurance that their digital identities and assets (e.g. avatar appearances, virtual property) are shielded from hacks and unauthorized access. Privacy regulations and advanced encryption will be essential to maintain trust as people engage in these all-encompassing digital worlds.

Another major challenge is interoperability across the multitude of emerging metaverse platforms. Currently, many virtual worlds are siloed ecosystems run by different providers, making it hard for users to move assets or identities between them. Without common standards, we risk a patchwork of incompatible or isolated metaverses, which would slow down innovation and limit the seamless "universe" experience [31]. Open interoperability—allowing avatars, items, and data to traverse different virtual spaces—is technically complex but crucial for a thriving metaverse. Achieving it requires consensus on data formats, communication protocols, and governance models among companies that are often competitors [32]. In summary, the metaverse's promise depends on overcoming these challenges in scalability, security/privacy, and interoperability. Addressing the limitations of classical computing and establishing robust standards will be key to unlocking a truly immersive, interconnected virtual world at scale.

4 The Role of Quantum Computing in the Metaverse

Quantum computing has the potential to *dramatically enhance metaverse performance* and overcome the bottlenecks of classical systems. Unlike traditional bits, quantum bits (qubits) leverage superposition and entanglement to process vast amounts of information in parallel. This means a quantum computer can perform complex calculations at unprecedented speed, which could be a game-changer for metaverse applications that require real-time computation [17]. For example, rendering a highly detailed virtual city or running a physics simulation for thousands of interacting objects could be accelerated by quantum algorithms that evaluate many possibilities simultaneously. Improved processing power translates to more responsive virtual environments, enabling richer graphics, more realistic physics, and larger numbers of simultaneous users without slowing down the experience [12]. In essence, quantum computing's massive parallelism can boost the scalability and realism of the metaverse beyond the limits of classical CPUs and GPUs [33, 34].

Another critical contribution of quantum technology is in security through quantum cryptography. Today's metaverse platforms rely on cryptographic protocols (like RSA and elliptic curves) to secure transactions, identities, and assets. However, a sufficiently powerful quantum computer could crack much of this classical encryption—for instance, breaking a 2048-bit RSA key, which would take a classical computer billions of years, might be done in mere seconds by a quantum computer [9, 35]. This looming threat has driven interest in both post-quantum cryptography and quantum-based encryption methods. Quantum computing offers tools like *quantum key distribution (QKD)*, which uses quantum states of particles to exchange encryption keys with theoretically unbreakable security shown in Fig. 4 [36].

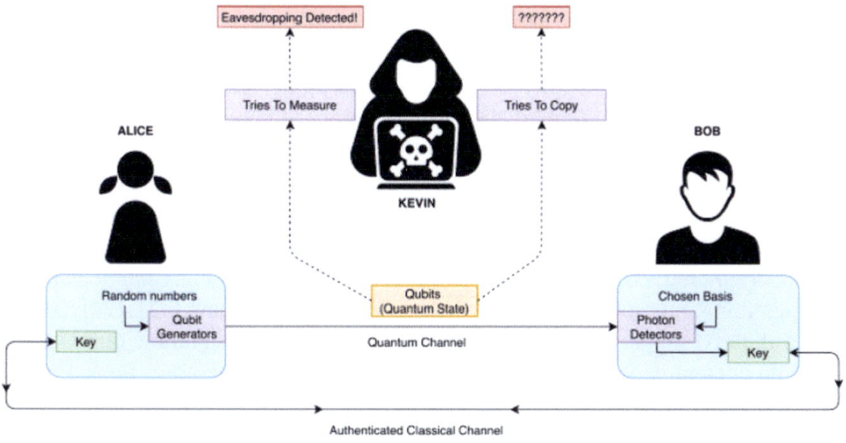

Fig. 4 Basic workflow of quantum key distribution

Incorporating QKD or other quantum cryptographic schemes into metaverse communications can ensure that sensitive data (financial transactions, personal info, etc.) remains secure even in the era of quantum attacks [37, 38]. In a quantum-powered metaverse, users could interact and trade with confidence that their communications are protected by physics itself, not just computational hardness. Indeed, experts anticipate that quantum-resistant security will be required for every virtual trade and asset transaction, and that *quantum-resistant blockchains* will become the foundation of metaverse economies [4]. By deploying quantum cryptography alongside classical measures, virtual worlds can stay one step ahead of cyber-adversaries and safeguard the burgeoning digital economies. A key enabler of such physics-based security is quantum entanglement, which further enhances trust, authentication, and privacy in metaverse interactions. *Quantum entanglement* is a phenomenon in which two or more qubits become so deeply connected that the state of one instantly affects the state of the other, even across large distances. This unique property enables secure information exchange that is fundamentally protected by the laws of physics.

Recent contributions have proposed quantum signature protocols that use multi-qubit entangled states to bind users to their avatars and enable secure interactions between virtual identities. As displayed in Fig. 5, these schemes ensure critical security properties such as authentication, unforgeability, and traceability within the metaverse [39]. Additionally, entanglement-enhanced frameworks for multiparty authentication have been developed, integrating quantum secret sharing and decentralized blockchain systems to enable user verification and access control, even in environments with untrusted authorities. Such entanglement-driven mechanisms offer robust protection against both classical and quantum threats, establishing a foundation for trusted identity management and communication integrity in quantum-powered virtual worlds [40].

Beyond security and raw speed, quantum computing can elevate AI and data processing in the metaverse. Machine learning drives many aspects of the metaverse—from intelligent NPC behavior to personalized content curation—but training and running advanced AI models is computationally intensive. Quantum machine learning (QML) algorithms offer a way to accelerate these processes by leveraging qubits to explore complex data patterns more efficiently than classical algorithms. For instance, a quantum computer might rapidly optimize a recommendation system for virtual events, or improve real-time speech translation between users of different languages in a virtual conference [12]. As metaverse data complexity grows, such capabilities become vital. Researchers predict that as the *"metaverse's complexity increases, machine learning will play an increasingly important role"* and quantum computing will be key to scaling those AI efforts [41]. Early studies in Quantum AI suggest potential improvements in pattern recognition, optimization, and even content generation—all relevant to richer metaverse interactions. Furthermore, quantum computers could analyze the vast streams of data generated within the metaverse (user movements, transactions, sensor inputs) far faster, enabling on-the-fly analytics and adaptive environments. This synergy is expected to stimulate innovation in graphics, AI, and data analysis for virtual worlds [42]. In summary, quantum computing stands to bolster the metaverse on multiple fronts: boosting

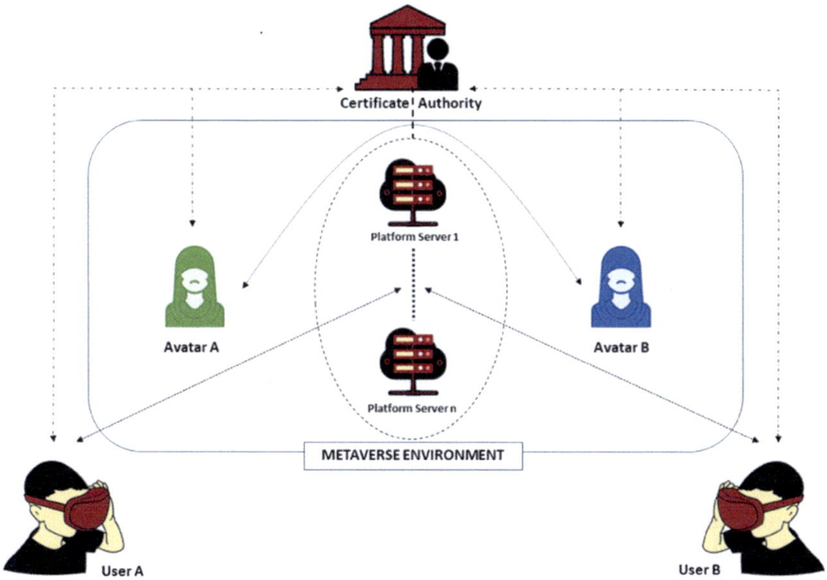

Fig. 5 System model proposed in metaverse environment

computational performance, providing next-generation security, and enhancing the intelligence and interactivity of virtual environments. These enhancements collectively push the metaverse closer to its full vision—a seamlessly immersive, secure, and intelligent digital universe.

5 Technical Implementation of Quantum Computing in the Metaverse

Integrating quantum computing into the metaverse's infrastructure will require hybrid models that combine quantum and classical resources. In practice, near-term quantum computers will function as specialized co-processors accessed via the cloud, working alongside traditional servers to handle tasks that benefit from quantum speed-ups. For example, a metaverse platform might offload an optimization-heavy computation (like global pathfinding for thousands of avatars or cryptographic key generation) to a quantum cloud service, then integrate the result back into the virtual world. Given the limited qubits and early-stage nature of current quantum hardware, a fully quantum-powered metaverse is not yet feasible—instead, hybrid quantum–classical computing is seen as the key approach [43]. In such a model, classical systems manage most routine rendering and game logic, while quantum processors are invoked for specific hard problems (optimization, large-scale simulation, cryptography) where they have an edge. This division of labor maximizes the strengths of each paradigm. Early research on next-generation networks and the Internet of Everything indicates that hybrid quantum–classical algorithms can significantly improve service management, even with today's quantum devices. Still, making this integration seamless will be challenging. Metaverse developers must learn to interface with quantum APIs and languages, meaning there is a knowledge gap to bridge between quantum computing experts and 3D application engineers [12]. Developing middleware and standards for quantum-enhanced cloud services (accessible through familiar game engines and platforms) is an active area of research. Overall, the integration model likely to emerge is one of cloud-based quantum acceleration, where metaverse applications tap into remote quantum hardware on-demand, analogous to how AI cloud services are used today.

A variety of *quantum algorithms* are being explored for metaverse scenarios, each targeting different application needs. One class of algorithms focuses on optimization problems, which abound in virtual worlds—from optimizing network resource allocation to orchestrating thousands of interactive objects efficiently. Quantum algorithms such as QAOA (Quantum Approximate Optimization Algorithm) or quantum annealing techniques can tackle NP-hard optimization tasks faster or more effectively than heuristics on classical computers, improving things like server load balancing and content streaming in the metaverse [44]. Another active area is quantum machine learning for the metaverse: algorithms like quantum neural networks or quantum support vector machines could be used to train AI agents or personalize experiences more quickly by exploiting quantum parallelism. For instance, a quantum-enhanced recommender system might analyze user behavior across a vast virtual world to suggest relevant activities in real time. Quantum cryptography protocols (e.g. QKD, discussed earlier) are also part of the algorithm toolkit, ensuring secure communications within and between virtual environments. Even quantum simulation algorithms have a role—metaverse developers could use quantum computers to simulate complex phenomena (like realistic fluid dynamics or material properties) that would be infeasible to compute in real time on classical hardware [45]. Each of these algorithms must be tailored to work within a hybrid system and often requires novel formulations of the problem. Researchers note that implementing quantum solutions in the metaverse demands expertise in both domains and careful problem encoding for quantum hardware. Progress is being made: for example, experimental work has applied quantum optimization to metaverse resource management, showing improvements in service scalability. As quantum processors grow, we expect a growing library of quantum algorithms purpose-built for enhancing virtual world operations. The challenge is to integrate them in a way that is transparent to metaverse users—the experience should simply become faster, smarter, and more secure, without users needing to know a quantum computer solved a part of it in the background.

The integration of quantum computing into the metaverse involves multiple technical aspects, each with its unique challenges and opportunities. Table 1 summarizes these key aspects, their applications, and the technical hurdles to be addressed.

As shown in Table 1, while quantum computing holds immense promise for optimizing metaverse operations, its practical implementation is currently constrained by hardware limitations, error rates, and scalability issues.

Table 1 Summary of key technical aspects in quantum computing for the metaverse

Aspect	Applications in the metaverse	Challenges
Hybrid Integration Model [12]	Quantum cloud computing for handling complex tasks alongside classical systems	Interoperability issues and the need for middleware
Optimization Algorithms [41]	Solving NP-hard problems like resource allocation and server load balancing	Quantum algorithm adaptation and scalability limitations
Quantum Machine Learning [42]	AI-driven personalization and faster training for metaverse AI agents	Limited hardware capabilities and data privacy concerns
Quantum Cryptography [43]	Secure metaverse communications using Quantum Key Distribution (QKD)	High deployment costs and integration with classical systems
Quantum Simulations [43]	Enhancing real-time rendering and realistic physics modeling	Computational demands and error correction challenges
Hardware Limitations [44]	Improving qubit stability, coherence time, and operational scalability	Extreme conditions (near absolute zero), decoherence, and limited qubit count
Scalability Issues [44]	Expanding quantum computing capabilities for large virtual worlds	Need for thousands/millions of stable qubits and reliable error correction techniques
Quantum Cloud Services [44]	On-demand quantum computing to enhance metaverse applications, similar to AI cloud services	Bandwidth limitations, latency concerns, and security risks
Quantum Networks [45]	Enabling quantum-based data exchange and distributed quantum processing	Developing stable quantum internet and ensuring compatibility with classical networks
Middleware Development [12]	Creating APIs and development frameworks for quantum-metaverse integration	Bridging the knowledge gap between quantum experts and 3D application developers

The hardware requirements and limitations of quantum computing are a central factor in its metaverse deployment. Current quantum computers (as of mid-2020s) are mostly experimental devices with tens or hundreds of qubits that must operate in specialized conditions (such as near absolute zero temperatures for superconducting qubits). They are extremely sensitive to external noise and disturbances; even slight vibration or heat can cause qubits to lose their quantum state in a phenomenon known as decoherence. This fragility means quantum computations are error-prone, and error correction protocols are still in early stages of development. For metaverse applications, which might require long and complex computations, these errors pose a serious challenge—a computation must remain stable long enough to be useful. Scalability of hardware is another issue: to handle the full scope of a massive virtual world, we may eventually need quantum computers with thousands or millions of reliable qubits, far beyond what exists today. Engineering such systems will take significant breakthroughs in quantum error correction, qubit design, and manufacturing. In the near term, the limited availability of quantum hardware means only certain tasks can be offloaded, and there may be queue times or throughput constraints when many requests hit a quantum cloud simultaneously. Metaverse architects must design around these limits, perhaps by scheduling quantum jobs during moments where a slight delay is acceptable, or using approximate/hybrid algorithms that mitigate quantum hardware's constraints [46]. There is also the consideration of hardware interoperability—how to connect quantum processors with classical data centers and user devices. This might involve high-speed classical links to quantum cloud data centers, and eventually quantum networks that can transmit qubit states between nodes. Early proposals for a "quantum internet" suggest it could be used to link quantum computers over distance for tasks like QKD or distributed quantum processing, which could directly benefit a geographically distributed metaverse. However, such networks are in infancy [47]. In summary, while quantum hardware is rapidly advancing, its current limitations mean that a quantum-powered metaverse will start modestly, focusing on specific high-value uses within a hybrid framework. Overcoming qubit stability and scaling challenges remains an open technical frontier, but steady progress is being made to turn quantum computing into a reliable component of future metaverse infrastructure.

6 Ethical, Social, and Economic Considerations

The convergence of quantum computing and the metaverse brings not only technical advances but also a host of ethical questions and societal implications. One major ethical concern is the potential for abuse or misuse of quantum's immense power when applied to virtual worlds. Just as social media raised issues of data misuse and manipulation, a quantum-enhanced metaverse could amplify those risks. Quantum computing could be used to break encryption or analyze personal data at unprecedented speeds—if wielded by malicious actors or oppressive regimes, it might enable unprecedented surveillance or cyber attacks in virtual spaces. These "known unknowns" of quantum technology include serious ethical considerations stemming from abuse, misuse, or unintended consequences. For example, quantum-boosted AI could generate hyper-realistic deepfakes and bots in the metaverse that are indistinguishable from real users, potentially deceiving people or spreading disinformation. The metaverse already blurs reality; with quantum-enhanced AI, the line between genuine and fake could become nearly impossible to discern, raising concerns about trust and consent in virtual interactions. Ensuring responsible use of quantum capabilities will be critical—stakeholders may need to establish ethical guardrails and oversight so that these tools augment user experiences without infringing on rights or wellbeing. Transparency in how quantum algorithms influence the metaverse (for instance, in content moderation or behavioral nudging) will also become important to avoid hidden biases or manipulations. In short, as we embrace quantum innovations, we must proactively address their ethical landscape, developing policies to prevent misuse and to promote fairness and user autonomy in the quantum-powered metaverse [48].

On the social front, one key issue is how quantum technology might impact digital equality and inclusion. Advanced quantum infrastructure could become concentrated in the hands of a few large tech companies or governments, potentially creating a "quantum digital divide." If only wealthy organizations can afford or access quantum computers, they could gain disproportionate control over metaverse platforms and data, exacerbating power imbalances between corporations and individuals. In such a scenario, smaller developers and under-resourced communities might be locked out of fully participating in or benefiting from the quantum-enhanced metaverse. Ensuring broad, democratic access to quantum tech is thus a social imperative; some experts argue that citizens should have access to quantum computing so it does not become an exclusive tool of the powerful. Additionally, if the metaverse becomes increasingly lifelike and essential for work, education, and social life (a trend that quantum could accelerate by making it more immersive), there are concerns about its impact on human behavior and society. Hyper-real virtual worlds could affect mental health or social skills, for instance, if individuals prefer virtual interactions over real ones or if the virtual experiences become too absorbing. Communities might need to grapple with questions of identity (when quantum-AI avatars can mimic anyone), digital addiction, and the cultural changes that come with a deeper integration of virtual existence. On the positive side, a quantum-powered metaverse might enable new forms of social connection across distances and more powerful tools for education and collaboration, which could benefit society if deployed equitably. The challenge is to maximize these benefits while mitigating risks like exclusion, loss of privacy, or erosion of trust. Policymakers and ethicists are starting to explore frameworks to guide this development, emphasizing fairness, accessibility, and user rights in parallel with the technical innovation [49, 50].

There are also significant *economic considerations* surrounding a quantum-enhanced metaverse. Virtual worlds already host booming digital economies (with virtual goods, services, and currencies), and quantum computing could both boost and disrupt these economies. On one hand, quantum computing might unlock new economic value by enabling more complex virtual market simulations, faster financial transactions, or robust smart contract platforms. It's projected that the metaverse could contribute trillions of USD to the global economy by 2040, and quantum tech could amplify this growth by increasing efficiency and trust in virtual commerce. For instance, quantum optimization could improve logistics in virtual trade or enable more sophisticated economic models in simulations [51]. On the other hand, quantum computing poses a threat to the security of digital assets that underpin metaverse economies. As discussed, quantum attacks could break the cryptographic safeguards of cryptocurrencies and other blockchain-based assets used in virtual worlds. Without proactive measures, this could erode confidence in virtual property ownership—a cornerstone of the metaverse economy. To protect digital asset value, the community will need to transition to post-quantum or quantum-safe cryptography well before large-scale quantum computers come online. In fact, planners are already calling for quantum-resistant security for "any trade and transaction" in the metaverse to future-proof virtual asset ownership. Another economic factor is the cost associated with quantum infrastructure: quantum computers and networks are expensive to develop and maintain [52]. This could influence the business models of metaverse providers (e.g. offering premium quantum-powered features at higher prices, or consortiums of companies sharing quantum resources). It also raises questions about monetization and ROI–companies will invest in quantum tech for the metaverse only if it yields clear economic benefits or competitive advantages. Lastly, the introduction of quantum capabilities might spur new digital business models and jobs. We may see the rise of "quantum content creators" or analysts who specialize in leveraging quantum tools to design better virtual services. Entire new sectors could emerge at the intersection of quantum and virtual reality (for example, quantum-driven virtual healthcare or finance). Economies will need to adapt to these innovations, updating regulations (taxation of virtual quantum services, intellectual property rights for quantum-generated content, etc.) accordingly. Overall, the quantum-powered metaverse promises economic growth and transformation, but careful consideration is needed to ensure stability, security, and equitable opportunity in these emerging digital markets.

7 Future Directions and Research Gaps

As the idea of a quantum-powered metaverse is still in its infancy, there are numerous open research problems and uncertainties that future work must address.

7.1 Growth Trends in Quantum Computing and the Metaverse

The rapid market expansion shown in the Chart 1 emphasizes growing industry and research interest in quantum computing and the metaverse. Quantum computing is expected to grow exponentially, driven by technological advancements, increased investment, and broadening applications. Similarly, the metaverse market is experiencing robust growth due to increasing adoption across various sectors such as gaming, education, healthcare, and commerce. Understanding these trends highlights the importance of addressing existing research gaps and preparing for emerging opportunities [53, 54].

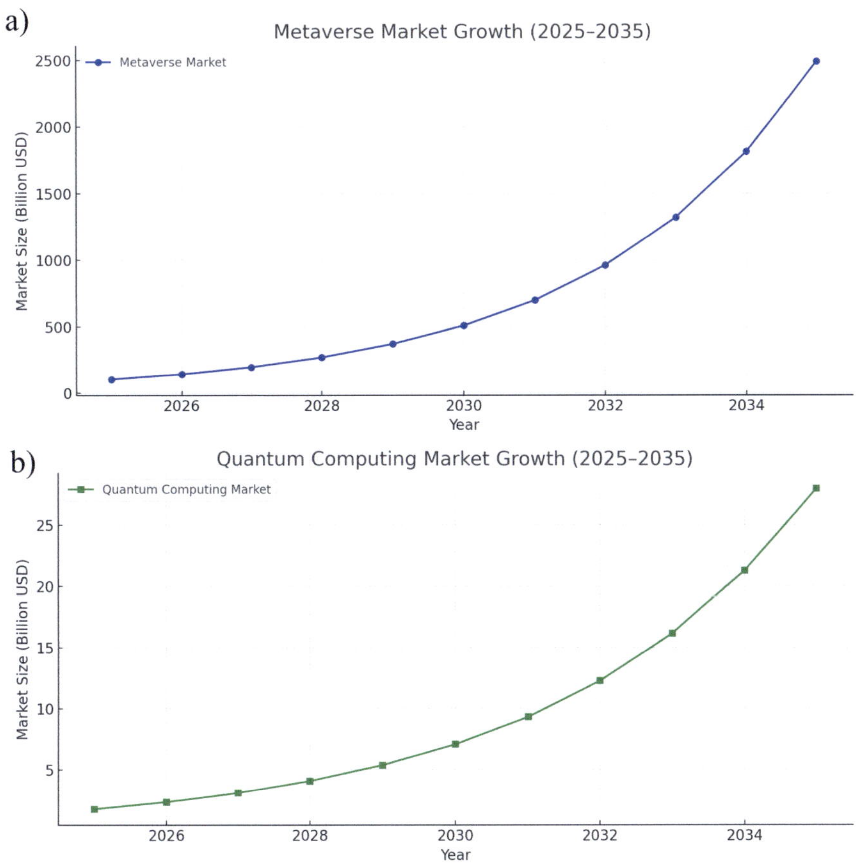

Chart 1. Market growth of metaverse (**a**) and Quantum computing (**b**) (2025–2035)

7.2 *Scalable Quantum Hardware for Virtual Worlds*

Developing quantum computers with far more qubits and longer coherence times is essential for handling the massive computational tasks of the metaverse. Research in error-corrected, fault-tolerant quantum computing aims to push beyond the current noisy intermediate-scale quantum (NISQ) devices. The goal is to eventually have stable, large-scale quantum processors that could run complex metaverse simulations continuously without errors or crashes [55].

7.3 Hybrid Architectures and Middleware

Because the metaverse will likely use quantum and classical computing in tandem, new architectures are needed to seamlessly integrate the two. This involves creating scheduling algorithms to decide which tasks go to quantum versus classical resources, and developing middleware that allows game engines or VR platforms to invoke quantum computations easily. Ensuring low latency in this handoff is a challenge—research is ongoing into minimizing the delay between a metaverse application requesting a quantum computation and receiving the result (possibly by predictive pre-fetching or parallelism).

7.4 Quantum Networking and Communication

Quantum Networking and Communication: To fully leverage quantum tech, the metaverse may eventually incorporate quantum communication networks. These could enable secure transmission of quantum keys (for encryption) or even entangled states between servers, reducing certain latency or security issues. Research in quantum networking (the "quantum Internet") and its interface with classical internet infrastructure is a crucial gap. For example, how might a VR headset someday use a quantum communication channel for ultra-secure connectivity? Recent advances, like experimental quantum communication techniques to protect Web 3.0 transactions, hint at the possibilities, but practical metaverse applications are still long-term [55].

7.5 Domain-Specific Quantum Algorithms

There is a need for quantum algorithms tailored to metaverse use cases. While general algorithms (optimization, machine learning, etc.) exist, researchers are exploring how to adapt them to problems like real-time physics simulation in games, dynamic resource scaling for virtual events, or even content generation. One open question: what new quantum algorithms might be invented specifically for virtual world management? This includes looking at algorithms for things like collision detection, social network analysis within the metaverse, or quantum-based search engines for virtual content.

7.6 Post-quantum Cryptography and Identity

While quantum cryptography holds promise, it's not yet widely adopted across the metaverse. As a result, post-quantum cryptography (PQC)—classical algorithms resistant to quantum attacks—remains essential. Key research gaps include integrating PQC into blockchain systems for asset ownership and designing secure, quantum-resistant identity frameworks. Another challenge is planning a smooth transition from today's encryption to quantum-safe methods before attackers exploit vulnerabilities.

Future breakthroughs could accelerate the fusion of quantum computing and the metaverse. Achieving fault-tolerant quantum computing would allow stable, large-scale quantum processing, enabling quantum-powered data centers. Quantum AI algorithms could revolutionize lifelike virtual agents and simulations, while quantum networking (e.g., long-distance entanglement) could connect dispersed VR environments with secure, low-latency links. Advances in hardware—like photonic or room-temperature quantum systems—might lead to compact, affordable chips for edge devices, boosting local metaverse computations [55].

These developments, though uncertain in timeline, could bring transformative industry applications. In healthcare, doctors might use quantum simulations in VR for treatment planning. Engineering and manufacturing could use quantum-enhanced digital twins for faster prototyping and optimization. Finance might adopt quantum algorithms for secure, real-time trading in virtual environments. In education, quantum simulations could power immersive science lessons. Gaming and social media could use quantum AI for dynamic, personalized content. Smart cities, defense, and disaster response simulations might also benefit from this powerful fusion [56].

While many ideas remain speculative, research is steadily progressing. By addressing current gaps, we move closer to a secure, intelligent, and quantum-powered metaverse.

8 Conclusion

The exploration of quantum computing's role in the metaverse reveals significant promise alongside notable challenges. This chapter highlighted how quantum computing could address the metaverse's infrastructure limitations, enhancing computational scalability, security, and interoperability. Quantum algorithms promise improved real-time simulations and personalized experiences, significantly advancing immersion within virtual environments.

Despite this potential, practical hurdles remain. Current quantum hardware is limited and error-prone, requiring hybrid quantum–classical solutions and phased implementations. Ethical guidelines, regulatory frameworks, and equitable access are also essential considerations for a safe and inclusive quantum-powered metaverse.

While writing this book, recent developments emphasize quantum computing's rapid progress, notably Microsoft's announcement of a new "topological qubit," representing a novel state of matter. Such breakthroughs highlight quantum computing as an evolving and transformative research area.

Looking forward, quantum technology could profoundly impact the next-generation internet, enabling virtual worlds of extraordinary realism and scale. Achieving this vision will demand continued innovation, interdisciplinary collaboration, and careful attention to ethical and policy considerations.

In conclusion, although a fully quantum-enabled metaverse remains an ambitious goal, current trends and breakthroughs indicate this vision is increasingly achievable. By addressing quantum computing challenges proactively and responsibly, developers and stakeholders can ensure a secure, equitable, and beneficial virtual future.

References

1. Huynh-The T, Pham QV, Pham XQ, Nguyen TT, Han Z, Kim DS (2023) Artificial intelligence for the metaverse: a survey. Eng Appl Artif Intell 117:105581
2. Lu Y, Sigov A, Ratkin, L., Ivanov, L. A., & Zuo, M. (2023). Quantum computing and industrial information integration: a review. J Ind Inf Integr 100511
3. Bojic L (2022) Metaverse through the prism of power and addiction: what will happen when the virtual world becomes more attractive than reality? Eur J Futures Res 10:22. https://doi.org/10.1186/s40309-022-00208-4
4. Gurung D, Pokhrel SR, Li G (2023) Decentralized quantum federated learning for metaverse: Analysis, design and implementation. arXiv preprint arXiv:2306.11297
5. Chehimi M, Hashash O, Saad W (2023) The roadmap to a quantum-enabled wireless metaverse: beyond the classical limits. In: 2023 fifth international conference on advances in computational tools for engineering applications (ACTEA), Zouk Mosbeh, Lebanon, pp 7–12. https://doi.org/10.1109/ACTEA58025.2023.10194017.
6. Using quantum technologies to make communications secure. https://www.nato.int/cps/en/natohq/news_207634.htm. Last accessed 19 February 2025
7. Quantum computing fundamentals. https://learning.quantum.ibm.com/course/quantum-business-foundations/quantum-computing-fundamentals. Last accessed 20 February 2025
8. Kumar V, Gupta S, Gill FS (2025) An overview on quantum computing for technology. Challenges Inf Commun Comput Technol 696–701
9. Rayhan A (2024) The quantum computing revolution: challenges and opportunities. https://doi.org/10.13140/RG.2.2.12467.85283
10. Lu Y, Yang J (2024) Quantum financing system: a survey on quantum algorithms, potential scenarios and open research issues. J Ind Inf Integr 100663
11. Valavanidis A (2024) Quantum Computing. A revolutionary computing capability to sift through huge numbers of possibilities and extract potential solutions to complex problems 1:1–25
12. Tuli EA, Lee JM, Kim DS (2024) Integration of quantum technologies into metaverse: applications, potentials, and challenges. IEEE Access 12:29995–30019
13. Sonko S, Ibekwe KI, Ilojianya VI, Etukudoh EA, Fabuyide A (2024) Quantum cryptography and US digital security: a comprehensive review: investigating the potential of quantum technologies in creating unbreakable encryption and their future in national security. Comput Sci IT Res J 5(2):390–414

14. Shimizu A, Miyadera T (2002) Stability of quantum states of finite macroscopic systems against classical noises, perturbations from environments, and local measurements. Phys Rev Lett 89(27):270403
15. Volpe D, Orlandi G, Turvani G (2025) Improving the solving of optimization problems: a comprehensive review of quantum approaches. Quant Rep 7(1):3
16. Ravindran D, Revathi S, Sowndharya V, Farzhana I, Sathya V, Girija P, Subramanian S (2025) Unraveling the quantum computing frontier: advancements, challenges, and future prospects. Integr AI, Quantum Comput Semiconduct Technol 139–158
17. Tripathi R, Tomar S, Kumar S (2025) A comprehensive survey on quantum annealing: applications, challenges, and future research directions. Authorea Preprints
18. Long J, Zhong J, Meng L (2025) Topological quantum compilation of metaplectic anyons. arXiv e-prints, arXiv-2501
19. Microsoft Azure Quantum, Aghaee M, Alcaraz Ramirez A et al (2025) Interferometric single-shot parity measurement in In As–Al hybrid devices. Nature 638:651–655. https://doi.org/10.1038/s41586-024-08445-2
20. Youvan DC (2025) Microsoft's Majorana 1: a paradigm shift toward scalable and fault-tolerant quantum computing
21. Chhimwal A, Singhal P, Singh M (2025) Majorana-1 and neuromorphic quantum systems: pioneering healthcare transformation. Available at SSRN 5222138
22. Besta M, Gerstenberger R, Iff P, Sonawane P, Luna JG, Kanakagiri R, Mahony AO (2024) Hardware acceleration for knowledge graph processing: challenges & recent developments. arXiv preprint arXiv:2408.12173
23. He J, Kan Y, Xue C (2024) Training quantum self-attention model in near-term quantum computer. In: 2024 16th international conference on wireless communications and signal processing (WCSP). IEEE, pp 139–144
24. Dwivedi YK, Hughes L, Baabdullah AM, Ribeiro-Navarrete S, Giannakis M, Al-Debei MM, Wamba SF (2022) Metaverse beyond the hype: Multidisciplinary perspectives on emerging challenges, opportunities, and agenda for research, practice and policy. Int J Inf Manage 66:102542
25. Chehimi M, Hashash O, Saad W (2023) The roadmap to a quantum-enabled wireless metaverse: beyond the classical limits. In: Proceedings of the fifth international conference on advances in computational tools for engineering applications (ACTEA), Zouk Mosbeh, Lebanon, pp 7–12. https://doi.org/10.1109/ACTEA58025.2023.10194017
26. Lee LH, Braud T, Zhou PY, Wang L, Xu D, Lin Z, Hui P (2024) All one needs to know about metaverse: a complete survey on technological singularity, virtual ecosystem, and research agenda. Found Trends® Hum–Comput Interact 18(2–3):100–337
27. Genç S (2023) The impact of quantum technology on the metaverse: future possibilities and challenges. J Quant Technol Inf Res 1(1):17–27. https://doi.org/10.70447/ktve.2222
28. Zhao R, Zhang Y, Zhu Y, Lan R, Hua Z (2023) Metaverse: security and privacy concerns. J Metaverse 3:93–99. https://doi.org/10.57019/jmv.1286526
29. Agarwal A, Ramachandra R, Venkatesh S, Prasanna SM (2024) Biometrics in extended reality: a review. Discov Artif Intell 4(1):81
30. Wu J, Lin K, Lin D, Zheng Z, Huang H, Zheng Z (2023) Financial crimes in web3-empowered metaverse: taxonomy, countermeasures, and opportunities. IEEE Open J Comput Soc 4:37–49
31. Karuppasamy K, Puram V, Johnson S, Thomas JP (2025) A comprehensive review of quantum circuit optimization: current trends and future directions. Quant Rep 7(1):2–18. https://doi.org/10.3390/quantum7010002
32. Li T, Yang C, Yang Q, Lan S, Zhou S, Luo X, Zheng Z (2023) Metaopera: a cross-metaverse interoperability protocol. IEEE Wirel Commun 30(5):136–143
33. Emu M, Choudhury S, Salomaa K (2023) Quantum computing empowered metaverse: an approach for resource optimization. In: ICC 2023—IEEE international conference on communications, Rome, Italy, pp 2412–2418. https://doi.org/10.1109/ICC45041.2023.10278858

34. Willsch D, Willsch M, Jin F, Michielsen K, De Raedt H (2022) GPU-accelerated simulations of quantum annealing and the quantum approximate optimization algorithm. Comput Phys Commun 278:108411
35. The Future of Cryptography and the Rise of Quantum Computing. https://www.genre.com/us/knowledge/publications/2023/september/the-future-of-cryptography-and-quantum-computing-en. Last accessed 15 March 2025
36. Aji A, Jain K, Krishnan P (2021) A Survey of Quantum Key Distribution (QKD) network simulation platforms. In: 2021 2nd global conference for advancement in technology (GCAT), Bangalore, India (2021), pp 1–8. https://doi.org/10.1109/GCAT52182.2021.9587708
37. Xu M, Niyato D, Xiong Z, Kang J, Cao X, Shen XS, Miao C (2022) Quantum-secured space-air-ground integrated networks: concept, framework, and case study. IEEE Wirel Commun 30(6):136–143
38. Lu Y, Yang J (2024) Quantum financing system: a survey on quantum algorithms, potential scenarios, and open research issues. J Ind Inf Integr 37:100663. https://doi.org/10.1016/j.jii.2024.100663
39. Kumar P, Bharmaik V, Prajapat S, Thakur G, Das AK, Shetty S, Rodrigues JJ (2024) A secure and privacy-preserving signature protocol using quantum teleportation in metaverse environment. IEEE Access
40. Tuli EA, Lee JM, Kim DS (2024) Leveraging quantum blockchain for secure multiparty space sharing and authentication on specialized metaverse platform. Sci Rep 14(1):25776
41. The Role of Artificial Intelligence (AI) in the Metaverse. https://metaversereality.ieee.org/publications/articles/the-role-of-artificial-intelligence-in-the-Metaverse. Last accessed 21 May 2025
42. Ramezani SB, Amirlatifi A (2024) Bridging the worlds of quantum computing and machine learning. Collections 57(03)
43. Geda MW, Tang YM (2025) Adaptive hybrid quantum-classical computing framework for deep space exploration mission applications. J Ind Inf Integr 100803
44. Chen KC, Burt F, Yu S, Liu CY, Hsieh MH, Leung KK (2025) Resource-efficient compilation of distributed quantum circuits for solving large-scale wireless communication network problems. arXiv preprint arXiv:2501.10242
45. Pirandola S, Andersen UL, Banchi L, Berta M, Bunandar D, Colbeck R, Wallden P (2020) Advances in quantum cryptography. Adv Opt Photon 12(4):1012–1236
46. Golec M, Hatay ES, Golec M, Uyar M, Golec M, Gill SS (2024) Quantum cloud computing: trends and challenges. J Econ Technol
47. Pfaendler SML, Konson K, Greinert F (2024) Advancements in quantum computing—viewpoint: building adoption and competency in industry. Datenbank Spektrum 24:5–20. https://doi.org/10.1007/s13222-024-00467-4
48. Possati LM (2023) Ethics of quantum computing: an outline. Philos Technol 36:48. https://doi.org/10.1007/s13347-023-00651-6
49. Gohar S, Choudhary R (2024) Ethical and social implications of the metaverse: an overview of emerging challenges. J Virtual Worlds Res 17(1):1–16. https://www.itexchangeweb.com/blog/ethical-and-social-implications-of-metaverse
50. World Economic Forum (2023) Bridging the quantum divide: why it matters. World Economic Forum Report. https://www.weforum.org/stories/2023/01/the-world-quantum-divide-why-it-matters-davos2023
51. Boston Consulting Group (BCG) (2024) Quantum computing market forecast: economic impacts through 2040. BCG Insights Report. https://www.bcg.com/press/18july2024-quantum-computing-create-up-to-850-billion-of-economic-value-2040
52. Fernandez-Carames TM, Fraga-Lamas P (2020) Towards post-quantum blockchain: a review on blockchain cryptography resistant to quantum computing attacks. IEEE Access 8:21091–21116
53. The Rise of Quantum Computing. https://www.mckinsey.com/featured-insights/the-rise-of-quantum-computing. Last Accessed 21 March 2025
54. Metaverse—Worldwide. https://www.statista.com/outlook/amo/metaverse/worldwide. Last Accessed 21 March 2025

55. Ren X, Xu M, Niyato D, Kang J, Xiong Z, Qiu C, Wang X (2025) Building resilient web 3.0 infrastructure with quantum information technologies and blockchain: an ambilateral view. Proceedings of the IEEE
56. Tang A (2025) Safeguarding the future: security and privacy by design for AI, metaverse, blockchain, and beyond. CRC Press

Evaluating the Potential of the Metaverse: The Chinese Perspective

Elena Martina Paone

Abstract This paper explores the evolution of the Metaverse in the People's Republic of China, analysing its development through the lens of national policy and regulation and international strategy. More specifically, it will be outlined how, through the implementation of the Cybersecurity Law, the Data Security Law, and the Personal Information Protection Law, this country is facing issues related to digital sovereignty, data control, and cybersecurity. Emphasizing the distinctively state-centric nature of China's approach, the study also outlines how the government has introduced the e-CNY (or digital yuan) and what role blockchain and infrastructure technologies have in constructing a domestically governed digital ecosystem. Special attention is given to the Digital Silk Road in the context of the Belt and Road Initiative, as an extension of China's ambition to foster interconnectivity in the Eurasian region. Finally, the paper discusses the challenges facing China's so-called 'Chinaverse', including issues of technological dependence, platform interoperability, and international perception.

Keywords Metaverse · Chinaverse · Cybersecurity · Digital Silk Road · e-CNY · New infrastructures

1 Introduction

The metaverse has been envisioned as a seamlessly interconnected network of 3D virtual worlds, rendered and brought to life in real time: a dynamic and persistent ecosystem, designed to simultaneously host a (potentially) unlimited number of users, offering them a persistent and synchronous experience. From identity and personal history to rights, communications, payments, and digital assets, the metaverse ensures that the continuity of every participant's essential data remains intact [1]. Thus, this system holds the potential to further revolutionize the already vast technological landscape, a transformative power that countless leading companies—from

E. M. Paone (✉)
University of Campania, "Luigi Vanvitelli", Caserta, Italy
e-mail: elenamartina.paone@unicampania.it

A. Borroni et al. (eds.), *Metaverses: Reshaping Law, Economy, and Society*, PoliTO Springer Series, https://doi.org/10.1007/978-3-032-08664-8_12

social networks to online platform providers—have swiftly taken notice of. Through significant investments, they are vying to secure a substantial share of this rapidly growing market, whose estimated potential value in China alone has reached 52 trillion yuan (approximately 8 trillion USD) [2].

It is therefore unsurprising that China is strategically targeting key markets for technologies that are closely linked to the metaverse (5G, VR and AR) [3]; more so, driven by—at least in part—new demands arising from the global health crisis, domestic investments and funding in the virtual reality industry alone have been reported to reach 20.709 billion yuan [4]. Furthermore, data indicate that in 2021, the Chinese AI market exceeded $23 billion, an estimated that is projected to triple its value by 2025 [5]. In addition to this, a stated objective of the Chinese government in its Next Generation Artificial Intelligence Development Plan is to ensure that the AI industry generates an annual revenue of 1 trillion RMB (approximately 1.546 trillion USD) per year [6].

It is not a coincidence, therefore, that 2021 has been defined for China as the "year one of the metaverse with extensive investment by local government, tech companies, major conferences, and studies" [7] all related to this topic. Among the major Chinese economic operators and tech giants that, also thanks to the support of the government's policies, are able to make their country shine as one of the driving forces for the development of the future Metaverse, Baidu is the one that has taken the first steps in this new reality: it was precisely in 2021 that this society has launched a VR platform known as XiRang ("Land of Hope") where it is possible to create an avatar to interact with other participants and carry out different activities [8].

Following that, ByteDance—the company behind the development of the TikTok video-sharing platform—has acquired Pico, the startup that has become famous for producing the multifunctional VR headset carrying the same name and developing two metaverse apps: Party Island, tailored for the Chinese market, and Pixsoul, developed for the Southeast Asian market [9]. Among the major players in the Metaverse sector, Tencent [10] stands out prominently. The company, which created WeChat [11], has incorporated a feature called Super QQ Show into its QQ messaging app, an interactive 3D space where users can socialize and customize their avatars [12].

However, the emergence of the metaverse represents not only the dawn of a new social model based on digital interconnectedness but, more significantly, a new economic paradigm in which technologies such as cryptocurrencies and blockchain underpin the vast majority of transactions taking place within it. Accordingly, as will be discussed throughout this work, China is complementing its investments in the sector with an increasingly proactive legislative agenda, aimed at regulating and exerting control over a domain that, while rapidly approaching mainstream integration, remains in a state of ongoing evolution.

2 The Chinese National Strategy on the Metaverse

There's no denying that interest in the Metaverse is gaining significant traction in China. This is evidenced by the inclusion of Metaverse-related initiatives in the five-year development plans of cities such as Shanghai, which are actively promoting its integration across a wide array of sectors, including public services and commercial venues, as well as manufacturing, security, gaming, and beyond [13]. Numerous conferences on the topic have been held across the country, and concrete promotional initiative have been launched—among those, the establishment of the Metaverse Innovation Alliance in Wuxi, of a special committee for the Metaverse in Hangzhou, of an innovation laboratory in Shenzhen, and of the Internet Education Innovation Center in Zhongguancun, Beijing's tech hub, aimed at exploring possible synergies between the Metaverse and the education sector [7].

This enthusiasm appears to be bolstered by the rhetoric of President Xi Jinping himself [14] and was already detectable as early as 2016 in the context of the drafting of the National Cyberspace Security Strategy by the Cyberspace Administration of China (CAC). This document emphasized the importance of sovereignty and legislative action in relation to the digital economy and cyberspace. The latter, in particular, is defined as "a new area for important human activity of equal importance to land, sea, air and space, national sovereignty has extended and stretched into cyberspace, sovereignty in cyberspace has become an important component part of national sovereignty" [15].

Given that the Metaverse represents a new frontier of cyberspace, it logically follows that China aspires to become a global leader in this technology—also as a means of more effectively safeguarding its sphere of sovereignty. In this regard, in October 2021, the China Institute of Contemporary International Relations (CICIR), a bureau under the Ministry of State Security, published a report—in which the expression 'Chinaverse initiative' appeared for the first time [16]—addressing the national security issues posed by the metaverse [17].

More specifically, as emerges from this document, China identifies a number of specific threats to national security that can be traced back to three kinds of vulnerabilities: (1) in terms of technological hegemony; (2) in terms of technological security, and (3) in terms of a technological paradigm shift [18]. This means that, in the first place, an overly rapid development of the metaverse risks exacerbating the so-called digital divide between advanced and developing systems, which may face difficulties in accessing such technologies. Second, the metaverse is expected to give rise to new and previously unknown types of cyberattacks—economically damaging and potentially targeting not only users' sensitive data but also digital service providers themselves [19]. Lastly, additional concerns relate to social impacts, such as the risk of alienation and dependency on excessively immersive experiences [20].

Being aware of these vulnerabilities, the Chinese government has adopted a dual approach to the metaverse: one of great interest, yet also of caution. This attitude has been well summarized by Zhang Ying, deputy director of the Shanghai's Economic and Information Technology Commission and member of the Chinese People's Political Consultative Conference, who, referring to the risks of speculative investments fuelled by excessive enthusiasm toward the entertainment industry, stated: "use the virtual to enhance the real, use the virtual to strengthen the real" [21].

A first national policy document specifically focused on the development of metaverse-related technologies was issued in 2022 by China's Ministry of Industry and Information Technology (MIIT). Titled 'Virtual Reality and Industry Application Integration Development Action Plan (2022–2026)', it sets out five key tasks to be pursued over the plan's duration: (i) promoting innovation in core technologies; (ii) strengthening the supply capacity of the industrial chain; (iii) accelerating the implementation of these technologies across different sectors (manufacturing, tourism, education, sports and healthcare, entertainment, disability assistance, etc.); (iv) establishing an "industrial public service platform"; and (v) creating a system of application standards to ensure the safety and reliability of such technologies [22].

To reach these objectives, the Action Plan proposes several measures that the government has to adopt, including the promotion of the development of small and medium-sized enterprises, the encouragement of R&D investments, the support of higher learning institutions, and the expansion of international exchange and cooperation channels [23]. However, an explicit reference to the Metaverse appeared only the following year, with the release of the 'Three-Year Action Plan for the Industrial Innovation and Development of the Metaverse (2023–2025) [24]. In a similar manner, this document is structured around five pillars: (i) the development of advanced technologies and the creation of metaverse-related industrial systems; (ii) the design of a three-dimensional, interactive industrial metaverse; (iii) the implementation of "immersive interactive digital life application"; (iv) the deployment of global-level support infrastructure (such as optical networks and 5G); and (v) the establishment of a secure and reliable "industrial governance system" [25].

The emphasis on the need to support industries and to foster the development and dissemination of foundational technologies for the Metaverse stems from the government's intent to achieve technological self-sufficiency and reduce its dependence on foreign technologies [26]. Strengthening domestic industries could, in turn, facilitate the work of legislators and regulatory authorities. However, this does not necessarily imply an intention to restrict foreign access to the Chinaverse. On the contrary, international technological cooperation and foreign investment in the related sectors are strongly encouraged by the Action Plan itself, which is intended as an initial guiding document to be followed by more targeted and precise policy instruments in the future [27].

3 Regulatory Approaches to Digital Innovation—The Same Old Story

As previously stated, cyberspace surveillance represents one of China's national priorities. In this regard, three legislative pillars form the regulatory framework governing cybersecurity and privacy issues [28]: the Cybersecurity Law (CSL), the Data Security Law (DSL), and the Personal Information Protection Law (PIPL) [29].

As stated by Article 1, the purpose of the CSL, which came into effect on June 1, 2017, is to "ensure cybersecurity; safeguard cyberspace sovereignty and national security, and social and public interests; protect the lawful rights and interests of citizens, legal persons, and other organizations; and promote the healthy development of the informatization of the economy and society" [30]. Furthermore, as a law of general scope and relatively vague language, its provisions necessarily require integration through additional legislative instruments [31].

Complementing the CSL, both the DSL and the PIPL entered into force in 2021, on September 1 and November 1, respectively. The DSL focuses on the storage and classification of data collected within China and its implications for national security. In particular, with reference to legal liability, it provides for financial penalties in cases where "data processing outside the territory of People's Republic of China harms the national security, public interests, or the lawful rights and interests of individuals or organizations of the People's Republic of China" [32].

The PIPL, on the other hand, is aimed at the protection of individual rights and interests related to personal data, the regulation of personal data processing, and the promotion of their reasonable use (as stated in Article 1) [33]. Regarded as China's equivalent of the European GDPR [34], this law "increases compliance costs, restricts what data can be stored, where data can be stored, and how data is authorized to leave the PRC's borders. Enforced by the newly created Cyberspace Administration of China (CAC), the PIPL imposes harsh financial penalties of up to fifty million CNY (approximately $7.8 million USD) or five percent of the offending company's revenue from the previous year" [35].

Moreover, consistent with the CSL, the PIPL established specific rules for the cross-border data transfer. Article 3 stipulates that this instrument "shall apply to the processing of personal information of natural persons within the territory of the People's Republic of China" but also "to the activities carried out outside the territory of the People's Republic of China to process the personal information of natural persons within the territory of the People's Republic of China under any of the following circumstances: (I) where the purpose is to provide products or services to domestic natural persons; (II) where the purpose is to analyze and evaluate the activities of domestic natural persons; and (III) other circumstances provided by laws and administrative regulations" [36].

Finally, with specific reference to technologies related to algorithms and their use within the metaverse, it is worth briefly mentioning a measure introduced by the CAC and titled 'Administrative Provisions on Deep Synthesis in Internet-Based Information Services'—a draft regulation released in 2022 (and entered into effect on January 10, 2023) with the aim of curbing the spreading of illegal content created through deep synthesis technologies, which rely on deep learning and AI and are capable of generating "text, images, audio, videos, or virtual scenens" [37].

3.1 The Raise of a "Block-China" Model

As already noted, the use of distributed ledger technologies and cryptocurrencies in China is progressively increasing under the vigilant oversight of the government—an attention level that has translated into rigorous and at times restrictive monitoring measures.

With specific regard to blockchain technology, for instance, the 14th Five-Year Plan of 2021 highlights the need for a "major correction" to its implementation. While this technology is acknowledged as one of the cornerstones of innovation, its nature is essentially decentralized; despite this, Chinese regulations aim at ensuring state control over blockchain development, application, and governance [38].

Among the best-known features of blockchain are its capacity to guarantee anonymity and the immutability of information, as each transaction is identified, validated, stored, and shared [39]. However, these characteristics stand in stark contrast with the Chinese government's tendency to censor sensitive content and with its previously discussed assertions regarding the protection of sovereignty in the cybersphere [40].

China's regulatory approach to blockchain thus rests on a fundamental paradox: on one hand, "the decentralised nature of the technology" and, on the other hand, "the highly centralised nature of the Chinese political system" [41]. A first step toward a regulatory framework aligned with this approach was taken in 2019, when the Cyberspace Administration of China (CAC) issued a set of provisions—effective from February of the same year—intended to regulate blockchain-based information services [42]. Among the most significant measures is Article 8, which, in accordance with the Cybersecurity Law, requires blockchain information service providers to verify the identity of their users. Those who fail to complete such identification will be denied access to these services [43].

From the perspective of monitoring activity, the government's Blockchain-based Service Network (BSN) project is noteworthy. It aims at enabling business, government entities, financial institution, and other actors to develop decentralized applications (dApps) on blockchain, thereby ensuring better interoperability between various blockchains and protocols, while also providing the necessary infrastructure to support widespread adoption of this technology [44].

Although blockchain is primarily associated with the use of cryptocurrencies, the Chinese government appears to pursue a much broader agenda since its promotion of this technology spans numerous arenas, "ranging from energy conservation to urban management and law enforcement" [45]. With specific reference to the latter, for instance, in the case *Hangzhou Huatai Yimei Cultural Media Co., Ltd. c. Shenzhen Daotong Technology Development Co., Ltd.* (2018), the Hangzhou court ruled that electronic evidence stored on a third-party blockchain-based platform is to be deemed legally valid and reliable [46]. The value of blockchain in the judicial system has been further recognized by the Supreme People's Court, which, in a May 2022 document titled 'Opinions of the Supreme People's Court on Strengthening Blockchain Application in the Judicial Field' outlines a series of standards for the development of blockchain-based platforms aimed at improving court operations. The Opinions also specify the objective of blockchain use in the judiciary: namely, increasing the credibility of the institution—also thanks to the immutability of data records stored on distributed ledgers—and enhancing the efficiency, coordination, and interoperability of the judicial system as a whole [47].

In conclusion, the distinctive features of China's one-party system are reflected in the creation of a more centralized and controlled alternative type of blockchain [48], and, through the systematic testing of its application, this country is beginning to gain a lead in the global race toward digital innovation.

3.2 The Ban on Cryptocurrency Usage and the Emergence of the Digital Yuan

The use of cryptocurrencies [49] in economic transactions is strictly related to the issue of blockchain. Like many other governments [50], the Chinese one considers cryptocurrencies a highly volatile form of investment, raising concerns related to anti-money laundering regulations and the support of illicit activities, as well as posing significant environmental challenges due to the high energy costs of mining operations [51].

The adoption of increasingly stringent measures in this sector began in 2021 and culminated in the issuance, by the People's Bank of China (PBOC) of the 'Notice on further preventing and disposing of the risk of speculation in virtual currency transaction' [52] which declares that "virtual currencies do not have the same legal status as legal tender […] and should not and cannot be circulated as currency in the market" and that, consequently, "[v]irtual currency-related activities are illegal financial activities. […] Such activities are unlawful and strictly prohibited by law and shall be forcibly banned in accordance with the law. Any person that has committed such illegal financial activities that constitute a crime will be prosecuted for criminal liabilities" [53]. The online sale of cryptocurrencies from abroad to residents within China is also deemed illegal [54].

At the same time, however, the Chinese government has developed its own digital currency: the digital yuan (also referred to as e-CNY), issued by the PBOC [55] and thereby granted legal tender status within China, where it is used for payments as an official digital version of the Renminbi (RMB) [56]. This currency operates under a centralized management model: the right to issue e-CNY belongs to the State; the PBOC distributes the digital yuan exclusively to authorized intermediaries (i.e., commercial banks), which in turn circulate it to the public. The digital yuan is therefore positioned as a mere alternative to physical cash, with which it will continue to exist "as long as there is demand for the physical RMB" [57]. Moreover, compared to other cryptocurrencies, its main advantage lies in the enhanced security of transactions, ensured by technologies such as "digital certificate system, digital signature, and encrypted storage to make double-spending, illegal duplication and counterfeit, transaction falsification, and repudiation unfeasible" [58].

Currently, the circulation of e-CNY is experiencing considerable momentum [59]: as recently stated by the former PBOC governor Yi Gang, transactions conducted using the digital yuan had reached 1.8 trillion (approximately USD 249.33 billion) by the end of June 2023—a massive increase compared to the 100-billion-yuan figure recorded in August of the previous year [60]. Nonetheless, while the Chinese experiment may serve as a useful model for regulating transaction in the metaverse in a manner that ensures a certain degree of State monitoring and traceability, it also raises the question of how far it can go in fulfilling the need for interoperability between Chinese and foreign platforms [61], also in light of China's long-standing strategic objective to internationalize its own currency [62].

4 New Connectivity Horizons: The Digital Silk Road and New Infrastructure (NI)

China's growing presence and relevance on the global stage also stem from the development of the well-known Belt and Road Initiative (BRI), an infrastructure development strategy aimed at enhancing connectivity, announced in 2013 by President Xi Jinping. It is based on intergovernmental cooperation and, in terms of geographical scope, is structured around two main pillars: the Silk Road Economic Belt and the 21st-Century Maritime Silk Road [63]. The former primarily encompasses a land-based route starting from Xi'an in central China, crossing Asia and extending to Europe and North Africa; the latter, instead, consists of a maritime network of "navigation routes, ports, and other coastal infrastructures, connecting South and Southeast Asia, East Africa, and the northern Mediterranean Sea" [64].

According to the 2015 document 'Vision and Actions on Jointly Building Silk Road Economic Belt and 21st-Century Maritime Silk Road' [65], the BRI must be strengthened across five key areas: (i) policy coordination; (ii) facilities connectivity; (iii) unimpeded trade; (iv) financial integration; (v) people-to-people bond [66].

In the context of constructing a system entirely based on the concept of connectivity, the digital dimension cannot be overlooked. In fact, the most recent extension of the BRI, announced in 2015 during the China-EU Digital Cooperation Roundtable in Belgium, is the Digital Silk Road (DSR), which has been described as the combination of the BRI and digital economy [67]: "the Digital Silk Road is where digital technology supports the Belt and Road Initiative. Relying on internet technology and cross-border e-commerce platforms to promote digital infrastructure, smart payment and logistics systems, the Digital Silk Road aims to become a new engine of the

Belt and Road international cooperation, boosting the establishment of cooperation mechanisms" [68].

The DSR is therefore a project aimed at improving connectivity in order to facilitate the "exchange of data, information, technology, and knowledge", while also promoting cross-border e-commerce and cultural exchange [69]. More specifically, the challenges posed by the recent developments in e-commerce relate to increasing demands in the logistics sector, and in particular to the development of the so-called New Infrastructure (NI)—a concept that was first introduced during the 2018 Central Economic Work Conference of the Central Committee of the Chinese Communist Party and the State Council [70]. More precisely, this term refers to infrastructures that diverge from those traditionally encompassed by BRI projects (so-called traditional physical infrastructure or TPI) [71], as they are based on technological innovation and digital networks [72].

Initially, NI encompassed infrastructures linked to four specific sectors: 5G, Artificial Intelligence (AI), the Industrial Internet, and the Internet of Things (IoT). Later, with the continued progress of technological development, a press conference by the National Development and Reform Commission (NDRC) in April 2020 replaced this sector-based classification with a more general reference to three broad categories: (i) information infrastructures [73], referring to services representative of the new generation of digital technologies (5G, AI, IoT, blockchain, etc.) [74]; (ii) integrated infrastructures, i.e. traditional physical infrastructures that have been upgraded through the employment of new technologies (such as ultra-high-voltage (UVH) or intercity high-speed railways); and (iii) innovative infrastructures, referring to those that are meant to support "scientific research, technology development, and product development, which includes the shared work-space, incubator, high tech laboratory, industrial technology innovation platform, etc." [75].

With particular regard to the NI relevant to digital finance and trade, in 2021, China established its first sectoral working group on the Metaverse—the 'China Mobile Communications Association Metaverse Committee' (CMCAMC). In partnership with the Asia Digital Bank (ADB), this body promotes international cooperation in this field with other BRI countries, aiming at integrating "both domestic and international digital and financial resources to offer diversified, customized, international and digital services, and promote metaverse partnerships among China, ASEAN members and the rest of the world" [76]. In addition, these institutions aim to connect China's economy with the global digital economy through the implementation of blockchain technology, which—together with the progressive rise of the digital yuan—may prove instrumental in fostering a form of cooperation within the BRI that can unfold without any physical barriers.

5 Diverging Metaverse Regulations and Applications: A Comparison with South Korea's Approach

The governance of the Metaverse is shaped and reshaped by national political, economic, and cultural tenets. As previously discussed, the Chinese government is designing a legal framework that's characterized by strict norms regarding data usage and transparency standards, and by heavy restrictions on cryptocurrencies and NFTs. Thus, while the Chinaverse appears strongly regulated, state-driven, and primarily rooted in industrial and regulatory objectives, other Asian economies have pursued different strategies.

Notably, amongst the so-called Asian Tigers, South Korea has manifested its intent to develop a new dedicated regulation for the emerging reality, rather than applying and adapting existing laws (for instance, gaming laws, internet driving principles, or virtual *lex electronica*) that could soon be outdated [77]. By pursuing a 'softer' digital transformation strategy (the so-called Digital New Deal 2.0 initiative) [78], this country has decided to mainly structure its Metaverse around ethical principles [79] aimed at regulating users' conduct and has established the Metaverse alliance in 2021, involving and putting together more than 200 companies, with the goal to accelerate its development [80]. Most importantly, though, South Korea was positioning itself at the forefront of the industry from what concerns accessibility: Seoul was (one of) the world's first city to see the Metaverse as a virtual communication channel between the local government and its citizens. With an investment of 2.8 billion USD, the South Korean capital had set out to develop 'Metaverse Seoul' [81]—a project (originally scheduled for completion by 2026) that envisioned a fully digital urban environment where residents, through virtual reality headsets, would be able to interact with municipal officials represented by avatars, but also file complaints, participate in public events, visit iconic cultural landmarks and engage in civic consultations [82]. Nevertheless, even if it initially appeared to have gathered large media attention, the project was officially shut down on October 16, 2024 [83]. The decision came in the wake of gradually declining user engagement, persistently high maintenance costs, and growing uncertainty among investors regarding the long-term sustainability of the platform at this stage of technological development.

Still, South Korea strongly encourages and continues to explore metaverse applications in sectors like education, tourism, and art while also planning to regulate cross-border virtual asset transactions starting in the second half of 2025 [84]. The upcoming MVEX 2025 that will be held in Seoul from 18 to 20th June 2025 [85] is a concrete manifestation of the Country's commitment, even if with a more prominent preference for industrial applications and private sector collaborations [86].

Although 'Metaverse' is often used as an umbrella term encompassing different experiments of virtual environments, it is clear that no one single model dominates the digital landscape. Ranging from immersive social platforms related to gaming to government-led digital infrastructure projects, the most successful ones still appear to be those rooted in industrial applications and gaming ecosystems, especially when they allow the proliferation of user-generated content that can be monetized, granting a somewhat stable and continuous level of user participation.

On the other hand, administrative or civic-oriented Metaverse initiatives—such as Seoul's virtual city services—often struggle to gain traction, despite their potential to bring socially beneficial results. This may depend not only on the absence of entertainment or economic value but also, somewhat paradoxically, on the overly "cartoonish" visual language of these platforms, which many users are unable to see as a reliable and professional *longa manus* of public institutions.

Regardless, it is important to emphasize how these "administrative metaverses" could enhance social inclusivity. At least in theory, virtual public spaces could enable or facilitate access to services for the elderly and persons with disabilities, who are often marginalized in physical real-world settings. But even though they could be able to mitigate barriers related to mobility and geographic isolation, a major obstacle still remains: that of a persistent digital divide, which particularly affects older adults and economically disadvantaged populations.

Without targeted digital inclusion policies and strategies, those lacking internet access, digital skills, or adequate devices are effectively excluded, leaving the leap into the Metaverse an unrealized prospect. With appropriate inclusion measures, however, such projects may begin to mitigate—rather than reproduce existing disparities and evolve into tools that support greater equity.

6 Concluding Remarks: Key Obstacles to the Full Development of a Metaverse with Chinese Characteristics

The development of the Metaverse in the Chinese context is a phenomenon that is difficult to interpret, as it is shaped by the specific socio-political framework in which it is taking place. Moreover, it represents a process that still requires time to reach its full expression. As stated by Du Zhengping, head of the already mentioned China Mobile Communications Association Metaverse Committee, "[t]raditional Chinese internet businesses developed first and were then regulated. Industries like the metaverse will be regulated as they are built" [87]. However, as emerges from the analysis conducted thus far, China appears determined to develop a Metaverse whose defining features can be summarized as "clean, censored, compliant and crypto-less" [88], and which is therefore strictly monitored and grounded in a robust regulatory framework [89].

The rise of the so-called Chinaverse in undoubtedly facilitated by several factors: thanks to the Country's vast demographic base, the extensive reach of its 5G infrastructure [90], and a government-led industry promotion approach, China clearly possesses the potential to achieve its declared goal of becoming a global leader in the tech sector—and particularly in AI, without which the very conception of the Metaverse would be impossible—by the year 2030 [91].

Nevertheless, this ambition is hindered by some systemic limitations. First, China remains dependent on foreign markets for certain technological components—a dependency that may prove difficult to overcome, especially in light of U.S. restrictions that have been in place since 2022 and ban the export of advanced microchips to the PRC [92]. A second issue is the "limited appeal of the made-in-China metaverse" [93], which is primarily tied to competition with far more popular American social media alternatives [94], to the low degree of interoperability with platforms developed in other countries, and to the tights governmental controls imposed on products, transactions, and, more generally, on the content produced within this ecosystem [95].

These factors may hinder not only the tangible development of the Chinese metaverse but also its international reception, particularly given the current limitations in cross-platform compatibility with foreign virtual ecosystems. Ultimately, beyond the domestic economic constraints it faces, the PRC must navigate the complex challenge of reconciling its intrinsic need for extensive state oversight with openness as an essential and imperative condition for fostering a digital ecosystem that meets the expectations of an increasingly interconnected society. Otherwise, the Chinaverse risks becoming a paradoxically isolated system, an insular construct mirroring the broader divide that already separates China from the global internet and social media landscape.

References

1. Ball M (2022) The metaverse: and how it will revolutionize everything. Liveright Pub Corp, New York, p 29
2. Chittum M (2022) Morgan Stanley Sees $8 Trillion Metaverse Market—in China Alone, https://blockworks.co/news/morgan-stanley-sees-8-trillion-metaverse-market-eventually, last accessed 2025/04/18; Kharpal A (2022) China's tech giants push toward an $8 trillion metaverse opportunity—one that will be highly regulated, https://www.cnbc.com/2022/02/14/china-metaverse-tech-giants-latest-moves-regulatory-action.html, last accessed 2025/04/18. Over 10,000 metaverse-related trademarks were registered in China in 2021 alone—an exponential increase compared to the fewer than 1,000 trademarks registered in the years 2019 and 2020. Baughman J (2022) Enter the Battleverse: China's Metaverse War. Military Cyber Affairs 5(1)
3. This technological push is paralleled by developments in the legal domain, where Chinese institutions are beginning to adapt dispute resolution mechanisms to this new reality. In fact, in July 2022, the Guangzhou Arbitration Commission of China (GZAC) announced the creation of the Meta City Arbitration Court (Yuanbang), In November of the same year, this court resolved its first case, which involved both the real world and the metaverse: the dispute concerned the application of a digital image, received from a NFT development company, to a clothing line intended for offline sale. The case was concluded with the signing of a settlement agreement between the parties "[b]ut questions arise about what would happen if an arbitration award was made? Would the winning party be able to enforce the arbitration award and how?". Ermakova EP (2023) Blockchain, metaverses and NFT in civil procedure and arbitration in Russia, China and USA. RUDN J Law 27(1):162–163
4. As highlighted in a 2021 white paper on the development of the virtual reality industry, jointly published by the China Institute of Electronics and Information Technology and the Virtual Reality Industry Alliance. Cheng S (2022) Metaverse: concept, content and context. Springer,

Cham, pp 196–199. This sector has experienced significant growth in recent years: notably, while approximately 1,9000 VR-related companies were registered in 2016, this figure had enormously increased by 2021, culminating in a total of 3,022 registrations of companies within the same domain. Ivi, p 190.

5. Casarotto F (2022) La Cina punta sul metaverso per battere gli Stati Uniti. Domino 9:23. See also China Briefing (2022) L'intelligenza artificiale in Cina: normativa, opportunità, e sfide per gli investitori, https://www.china-briefing.com/news/intelligenza-artificiale-in-cina-opportunita-sfide/. Last accessed 16 April 2025.
6. China Briefing, L'intelligenza artificiale in Cina… cit
7. Baughman J, Enter the battleverse: China's metaverse War… cit
8. Casarotto F, La Cina punta sul metaverso per battere gli Stati Uniti… cit., p 22
9. China Briefing (2022) China's debut in the metaverse: trends to watch, https://www.china-briefing.com/news/metaverse-in-china-trends/. Last accessed 16 April 2025
10. Ma Huateng, the CEO of Tencent, in publicly addressing investors' fears regarding the metaverse, has stated: "[t]he metaverse is an exciting topic, and I believe Tencent has a lot of technology and capability to explore and develop the metaverse, for example, in areas related to games, social media, and artificial intelligence". Cheng S, Metaverse: Concept, Content and Context… cit., p 197
11. WeChat, that has approximately 1.2 billion active users, is an instant messaging platform often compared to WhatsApp but, more than its western counterpart, it has been integrated with features traditionally associated with social media. Piccinali M (2023) Interoperatibilità e geopolitica: le scelte cinesi. In: Piccinali M, Puccio A, Vasta S (eds) Il metaverso. Modelli giuridici e operativi. Giuffrè, Milan, p 37
12. China Briefing, China's Debut in the Metaverse: Trends to Watch… cit
13. China is also advancing the concept of the so-called "virtual factory" aimed at eliminating the reliance on physical prototypes during the pre-production phase, an approach intended to significantly decrease "the consumption of energy, water, and raw materials", thereby aligning industrial processes with broader sustainability objectives. Kshetri N (2023) National metaverse strategies. Computer 56(2):138
14. In the words of Xi Jinping, "[i]f China is to grow strong, prosperous, and rejuvenated; we most certainly need to develop science and technology in a big way. […] We are closer than any time in history to the objective of the great rejuvenation of the Chinese nation, and we need more than any time in history to build China into a world science and technology superpower"; to that end "the strategy of strengthening the country through the internet is an indispensable key step". *Ibidem*
15. An English translation of the national cyberspace security strategy is available at https://chinacopyrightandmedia.wordpress.com/2016/12/27/national-cyberspace-security-strategy/. Last accessed 16 April 2025
16. Kshetri N, National metaverse strategies… cit., pp 137–138
17. It also contains the first official statements of the Chinese Communist Party (CCP) on the subject: "[i]t [the metaverse] will have a major social, political and economic impact on countries […] and on political and cultural security […] it could bring about major changes in the social structure". *Ibidem*
18. Vanorio F (2021) Metaverso e sicurezza nazionale. Internet 3.0 e nuovo ordine mondiale digitale, pp 24–25. https://www.strategicstudies.it/wp-content/uploads/2021/12/Edizioni-Machiavelli-Metaverso-e-Sicurezza-Nazionale.pdf. Last accessed 16 April 2025
19. In this regard, given the enduring nature of data retention in virtual environments, the overall cost of the damage sustained would exceed that of the current digital ecosystem Furthermore, the unauthorized appropriation of 'avatar' assets and user information within synthetic worlds would entirely nullify the value of the user base. Damage to information storage systems would also substantially diminish the aggregate value of the virtual environment, resulting in severe economic losses. *Ibidem*
20. *Ibidem*

21. This has been reiterated multiple times by various academics, such as during the 2022 world metaverse conference held in Beijing or in occasion of the creation of the metaverse and virtual-reality interchange and interconnection Research Institute of Shanghai. Liu Z (2022) China says no to speculative all-virtual metaverse, Jing Daily, https://jingdaily.com/posts/china-metaverse-real-economy-industrial-power. Last accessed 16 April 2025
22. Huld A (2022) Building China's virtual world—the new action plan for metaverse technology¸ China briefing. https://www.china-briefing.com/news/virtual-reality-in-china-new-action-plan-for-developing-industry/. Last accessed 16 April 2025
23. *Ibidem*
24. In Chinese language https://www.gov.cn/zhengce/zhengceku/202309/content_6903023.htm. Last accessed 16 January 2025
25. Interesse G (2023) China releases three-year action plan for metaverse industry development, China briefing. https://www.china-briefing.com/news/china-releases-three-year-action-plan-for-metaverse-industry-development. Last accessed 16 April 2025
26. As it will be discussed later, Beijing is not self-sufficient in the production of semiconductors (essential components for any project involving artificial intelligence); therefore, the dependence from foreign countries is a factor that weighs heavily in the long term, representing an unsustainable vulnerability for the People's Republic. Casarotto F, La Cina punta sul metaverso per battere gli Stati Uniti… cit., p 25
27. Interesse G China releases three-year action plan for metaverse industry development… cit
28. For a general overview of privacy-related issues in the context of the metaverse, see Tortorano P (2025) From virtual realms to tangible reality: privacy at the nexus of artificial intelligence and the metaverse, in this volume
29. Chaskes W (2022) The three laws: The Chinese communist party throws down the data regulation gauntlet. Wash Lee Law Rev 79(39):1169 ff
30. More specifically, the CSL applies to both the so-called Networks Operators and the operators of Critical Information Infrastructure (CII): pursuant to the combined provisions of Articles 40 and 41, both are required to "maintain the confidentiality of user information they collect, and establish and complete user information protection systems" and to collect and use personal information in compliance with the principles of "legality, propriety, and necessity; they shall publish rules for collection and use, explicitly stating the purposes, means, and scope for collecting or using information, and obtain the consent of the persons whose data is gathered". Operators must not collect data unrelated to the services they provide. In the event of the violation of legal provisions, administrative regulations, or agreements between the parties on the collection and use of personal data, the affected individuals have the right to "demand the network operators delete their personal information; where discovering that personal information gathered or stored by network operators has errors, they have the right to demand the network operators make corrections" (Article 43). An English translation of the CSL is available at https://digichina.stanford.edu/work/translation-cybersecurity-law-of-the-peoples-republic-of-china-effective-june-1-2017/. Last accessed on 16 January 2025
31. Piccinali M Interoperatibilità e geopolitica: le scelte cinesi… cit, p 26
32. As established by Article 2 of the DSL. For further details on the sanctioning system, please refer to Chapter VI—legal liability. An English translation of the Data Security Law is available at http://www.npc.gov.cn/englishnpc/c2759/c23934/202112/t20211209_385109.html. Last accessed on 16 January 2025
33. The English translation of the Chinese Personal Information Protection Law is available at https://personalinformationprotectionlaw.com/. Last accessed on 16 January 2025
34. Chaskes W The three laws: the Chinese communist party throws down the data regulation gauntlet… cit., p 1175
35. *Ibidem*
36. The PIPL also includes provisions regarding the use of algorithms and other automated systems that have been observed to result in the discrimination of certain consumers. In such cases, pursuant to Article 24 of the PIPL, personal data controllers who use this information to

carry out automated decision-making processes must ensure that transparency of the decision-making process itself, as well as the fairness and impartiality of the outcomes, and may not apply unreasonable differential treatment to individuals in terms of pricing or other contractual conditions. On the subjects of algorithms, see also THE Provisions on the management of algorithmic recommendations in internet information services of sarch, 2022. Piccinali M, Interoperatibilità e geopolitica: le scelte cinesi… cit., pp 30–31
37. Pashentsev E (2023) The Malicious Use of Deepfakes against psychological security and political stability. In: Pashentsev E (ed) The Palgrave handbook of malicious use of AI and psychological security. Palgrave Macmillan, Cham, p 55. In particular, pursuant to Article 13 of the Provisions, providers of deep synthesis services bear the responsibility of employing adequate measures to address the violation of relevant state provisions. Similarly to what the cybersecurity law provides for blockchain service providers, it also states that deep synthesis service providers must verify the identity of their users (Article 9 of the Provisions)
38. Due to this peculiar approach, Piccinali M, Interoperatibilità e geopolitica: le scelte cinesi… cit., pp 19–24 pointedly refers to "block-chain vs. block-China"
39. In relation to the characteristics, potential uses and regulation of the blockchain, see Borroni A (2019) Blockchain: uses and potential value. In: Borroni A (ed) Legal perspective on blockchain. Theory, outcomes, and outlooks. Edizioni Scientifiche Italiane, Napoli; Zambardino F (2019), Blockchain technology and regulatory issues. In: Borroni A (ed) Legal perspective on blockchain. Theory, outcomes, and outlooks. Edizioni Scientifiche Italiane, Naples. On the topic of privacy, see also Finocchiaro G (2012) Privacy e protezione dei dati personali. Disciplina e strumenti operativi. Zanichelli Editore, Bologna; Pascuzzi G, Izzo U, Macilotti M (eds) (2013) comparative issues in the governance of research biobanks. Property, privacy, intellectual property, and the role of technology. Springer, Heidelberg-New York-Dordrecht-Londra; Resta G, Zeno-Zencovich V (eds) (2016) La protezione transnazionale dei dati personali. Dai "safe harbour principles" al "privacy shield". RomaTre-Press, Rome; Seghesio M (2019) Blockchain and privacy. In: Borroni A (2019) Legal perspective on blockchain. theory, outcomes, and outlooks. Edizioni Scientifiche Italiane, Naples; Zambardino F (2022), La blockchain e la protezione dei dati personali: una tecnologia privacy compliant by design? Eur J Priv Law Technol 2; Zambardino F (2021) La blockhain nel mercato del lavoro italiano con particolare enfasi sulla questione privacy e il rapporto con il General Data Protection Regulation. Tutela e sicurezza del lavoro 3 and Zambardino F (2024) The applications of the blockchain technology. Comp Law Study. ESI, Naples
40. Ekman A (2021) China's blockchain and cryptocurrency ambitions: the first-mover advantage, BRIEF/15 of the European Union Institute for Security Studies, p 1. https://www.iss.europa.eu/content/chinas-blockchain-and-cryptocurrency-ambitions. Last accessed 16 April 2025
41. *Ivi*, p 2
42. The Administrative Provisions on Blockchain Information Services are based on the cybersecurity Law of 2017, on the "Measures for the Administration of Internet Information Services" of 2000, and on the "Notice of the State Council on Authorizing the State Internet Information Office to be responsible for the management of Internet information content" of 2004 (see also Article 1 of the Administrative Provisions). Piccinali M, Interoperatibilità e geopolitica: le scelte cinesi… op. cit., p 20. The text of the Administrative Provisions is available at http://www.cac.gov.cn/2019-01/10/c_1123971164.htm. Last accessed 16 April 2025
43. Moreover, pursuant to the combined provisions of Articles 16 and 17 service providers are required to censor posts and to retain user data, intervening by imposing restrictions on user account functionalities or even proceeding with account deletion in cases of unlawful conduct by the users themselves. In order to prevent the dissemination of illegal information content, providers must record users' posts and logs for a period of no less than six months and make such records available upon request by law enforcement authorities. Piccinali M, Interoperatibilità e geopolitica: le scelte cinesi… op. cit., p 21
44. *Ibidem*
45. Ekman A, China's blockchain and cryptocurrency ambitions: the first-mover advantage… cit., pp 2–3

46. Based on the inherent features of blockchain technology, the Chinese Supreme Court published the 2021 Rules of Online Litigation of the People's Court, which state that electronic evidence stored on a blockchain is presumed to be authentic. The introduction of this new evidentiary presumption into Chinese procedural law could serve as a model for the evolution of court practices worldwide. Ermakova EP, Blockchain, metaverses and NFT in civil procedure and arbitration in Russia, China and USA… cit., pp 161–162
47. Piccinali M, Interoperatibilità e geopolitica: le scelte cinesi… cit., p 22. See also Todaro D (2020) Efficienza e controllo: l'uso della tecnologia nei tribunali cinesi. Mondo Cinese XLVII 167(1)
48. According to some scholars, the implementation of a blockchain system designed in this manner has the potential to consolidate the political apparatus of the State, particularly in domains such as law enforcement and the control of dissent. In a sense, for less democratic regimes, "China could emerge as a model […] in developing blockchain with surveillance and censorship capabilities". Ekman A, China's blockchain and cryptocurrency ambitions: the first-mover advantage… op. cit., p 6
49. See, in general, Borroni A, Seghesio M (2019) Bitcoin and Blockchain: Un'analisi comparatistica dalla nascita alla potenziale regolamentazione. Ianus 19 and Borroni A (2016) Bitcoins: regulatory patterns. Bank Finance Law Rev 32(1)
50. In addition to China—whose stance will be examined in greater detail later in this section – countries such as Algeria, Bangladesh, Egypt, Nepal, and Tunisia have also imposed bans on cryptocurrencies; meanwhile, strict regulations on crypto mining are currently in force in Iran. For further insights on the matter, see Shapiro J (2022) Il futuro delle criptovalute è una scommessa. Domino 9:44
51. Piccinali M Interoperatibilità e geopolitica: le scelte cinesi… op. cit., p 22
52. The text of the Notice on further preventing and disposing of the risk of speculation in virtual currency transactions is available at http://www.pbc.gov.cn/en/3688253/3689012/4353814/index.html. Last accessed 16 January 2025. A few days prior to the issuance of the Notice by the People's Bank of China, the Chinese National Development and Reform Commission had already released a Notice on rectifying virtual currency "mining" activities, officially prohibiting any further investment in prospective cryptocurrency mining projects. Piccinali M, Interoperatibilità e geopolitica: le scelte cinesi… cit., p 22
53. The Notice on further preventing and disposing of the risk of speculation in virtual currency transaction enumerates the commercial activities related to cryptocurrencies that are deemed unlawful: "[e]xchanging legal tender for virtual currencies or vice versa, exchanging one virtual currency for another, buying and selling virtual currencies as a central counterparty, providing information and pricing services for the trading of virtual currencies, issuing tokens to raise funds, trading virtual currency derivatives, and engaging in other virtual currency-related activities are potentially instances of illegal issuance of coupons as substitutes for the legal tender, unauthorized public offering of securities, illegal conduct of futures business, or illegal fundraising"
54. This framework has significant implications for key activities associated with the metaverse: the purchase of NFTs—denominated in RMB or yuan—is restricted to the primary market and must be carried out using traditional payment methods, since they can be almost assimilated to cryptocurrencies. Piccinali M, Interoperatibilità e geopolitica: le scelte cinesi… op. cit., p 23
55. On Central Bank Digital Currencies see Borroni A (2025) Banks in the metaverse: the legal framework between evolution and revolution, in this volume
56. As stated by PBOC, "E-CNY is the digital version of fiat currency issued by the PBOC and operated by authorized operators. […] e-CNY is the fiat currency issued by the central bank [and] has all the basic functions of money […]". Working Group on E-CNY Research and Development of the People's Bank of China, Progress of Research & Development of E-CNY in China, 2021. http://www.pbc.gov.cn/en/3688110/3688172/4157443/4293696/2021071614584691871.pdf?utm_source=aposto. Last accessed 16 April 2025
57. *Ibidem*
58. *Ibidem*

59. In promoting the use of the e -CNY, PRC has begun to pay the salaries of public sector employees using the digital currency. Demertzis M, Martins C (2023) The value added of central bank digital currencies: a view from the euro area. Bruegel Policy Brief 13:9
60. Wee R (2023), China's digital yuan transactions seeing strong momentum, says cbank gov Yi, Reuters. https://www.reuters.com/markets/asia/chinas-digital-yuan-transactions-seeing-strong-momentum-says-cbank-gov-yi-2023-07-19/. Last accessed 16 April 2025
61. Piccinali M, Interoperatibilità e geopolitica: le scelte cinesi… cit., p 24
62. An important step in this process happened in October 2015, when the International Monetary Fund (IMF) included the RMB in the basket of currencies that comprise the Special Drawing Rights (SDRs), alongside the U.S. dollar, the euro, the British pound and the Japanese yen
63. The Silk Road Economic Belt and the 21st-century Maritime Silk Road represent a direct evolution of China's transformation that began with its rapid rise in the 1980s, during which the country directed its productive capacity toward building an "ever-expanding network of industrial hubs and ports" along its coastline—thus establishing what has been defined as "its geo-economic interface with the international market". Holslag J (2010) China's Road to Influence. Asian Survey 50(4), p 641. Indeed, with the sixth Five-Year Plan (1981–1985), China introduced the concept of three economic belts, corresponding to the eastern, central, and western regions of the country. These were formally institutionalized in the seventh Five-Year Plan, which came into effect in 1986. The reform strategy outlined the following: in the eastern region (the so-called "first belt"), the construction of export-oriented industries; in the central region ("second belt"), the development of agriculture and energy production activities; and in the western areas ("third belt"), livestock farming and the establishment of extractive industries. This new urban and territorial planning framework identified the coastal zone as a national priority. Basso M (2015) L'Agenda Urbana Nazionale della Cina: una Prima Esplorazione. Working papers (Urban@it) 1, p 6. While China was transforming its coastal areas into "Asia's new center of commerce and manufacturing", the government also started to develop and valorise the hinterland. In this sense, the ninth Five-Year Plan (1996–2001) marked a pivotal moment in China's policy trajectory, as it was the first comprehensive national strategy to articulate a vision for expansive infrastructure development extending beyond the traditionally prioritized coastal provinces. It set forth the objective of establishing an integrated network of roads, canals, and railways to enhance connectivity with neighbouring Asian countries. Holslag J China's road to influence… cit., pp 654–655
64. Yu Y, Chang YC (2018) The 'One Belt One Road' initiative and its impact on shipping law in China. Mar Policy 87:292. For further insights into the economic corridors and the blue economic passages that unfold along these routes, reference should be made to Rimmer PJ (2020) China's global vision and actions: reactions to belt, road and beyond. Edward Elgar Publishing, Cheltenham and Rimmer PJ (2020) China's Belt and Road Initiative: underlying economic and international relations dimensions. Asian-Pac Econ Lit 32(2)
65. National Development and Reform Commission, Ministry of Foreign Affairs, and Ministry of Commerce of the People's Republic of China, with State Council authorization (2015) Vision and actions on jointly building silk road economic belt and 21st-century maritime silk road. http://2017.beltandroadforum.org/english/n100/2017/0410/c22-45.html. Last accessed 16 April 2025
66. Participating BRI countries are expected to establish ties through the creation of mechanisms aimed at promoting cultural and academic exchanges, international tourism, and cooperation with medias. In this context, it is worth noting that UNESCO, with the support of Azerbaijan, China, Germany, Kazakhstan, and Oman, has launched the Silk Road Online Platform with the objective of fostering people-to-people connectivity and safeguarding the cultural heritage of the historical Silk Road. The initiative promotes research through the creation of archives and the organization of cultural activities such as photography contests (see https://en.unesco.org/silkroad/. Last accessed 16 April 2025). This type of virtual and 3D projects is particularly relevant to cultural heritage preservation: while the mere act of digitalization already serves to broaden access to archives and museums, the emergence of the metaverse could further

enhance this potential by enabling the creation of digital equivalents of events, museums, or art exhibitions

67. The concept of digital economy refers to "new economy that had been developing under influence of digital technologies including, among other examples, broadband Internet as well as the Internet of Things (IoT), advanced robotics, big data and advanced analytics, artificial intelligence or machine learning". Majcherczyk M, Shuqiang B (2019) Digital silk road—the role of cross-border e-commerce in facilitating trade. J WTO China 9(2):113
68. Haiyan W (2020) China and Central Asian countries jointly building the digital silk road: foundations, challenges and paths. China Int Stud 82:142
69. Majcherczyk M, Shuqiang B, Digital silk road—the role of cross-border e-commerce in facilitating trade… cit., p 115
70. On this occasion, the term was initially proposed as a concept of national economic development and later reframed as "a new focus on economic interactions with overseas countries under the BRI framework". Zhang C, Zhang M, Xiao C (2022) From traditional infrastructure to new infrastructure: a new focus on China's belt and road initiative diplomacy? Eurasian Geogr Econ 63(3):424–425
71. The term 'infrastructure' has an interdisciplinary nature and generally refers to "matter that enables the movement of other matter". Larkin B (2013) The politics and poetics of infrastructure. Annu Rev Anthropol 42: 329. For the Chinese system, the main differences between Traditional Physical Infrastructure (TPI) and New Infrastructure (NI) can be summarized as follows: regarding the participating entities, a TPI is primarily controlled by governments and SOEs, while a NI may involve the additional collaboration of private enterprises. A TPI requires more significant and necessarily long-term investments and is realized in traditional physical spaces, whereas a NI integrates both physical and digital spaces. Finally, TPI has a well-defined function and produce effects limited to the specific geographical area, while NI can impact various aspects of the economy and social life, generating indirect benefits as well. Zhang C, Zhang M, Xiao C From traditional infrastructure to new infrastructure: a new focus on China's belt and road initiative diplomacy?... cit., pp 428–429
72. See Zhang C, Zhang M. Xiao C, From Traditional Infrastructure to New Infrastructure: a new focus on China's belt and road initiative diplomacy?... cit., p 425 and Huang Y, Huppenbauer N, Mayer M (2022) Infrastructuring cyberspace. Exploring China's imaginary and practices of selective connectivity. Int Quart Asian Stud 53:422
73. The advantage of information infrastructures, which are the primary foundation for the metaverse, lies in their nature as virtual assets. Unlike physical ones such as bridges, railways, etc., these infrastructures are less susceptible to "political turbulence in recipient countries". Zhang C, Zhang M, Xiao C, From Traditional Infrastructure to New Infrastructure: a New Focus on China's Belt and Road Initiative Diplomacy?... cit., p 432
74. Several CSL articles are targeted at operators of Critical Information Infrastructures—i.e., those infrastructures leveraging on informatic tools that "if destroyed, suffering a loss of function, or experiencing leakage of data" might impact in a negative way "national security national welfare, the people's livelihood, or the public interest" (Article 31). These operators are subject to specific obligations, including those set forth in Article 34: "(1) Set up specialized security management bodies and persons responsible for security management, and conduct security background checks on those responsible persons and personnel in critical positions; (2) Periodically conduct cybersecurity education, technical training, and skills evaluations for employees; (3) Conduct disaster recovery backups of important systems and databases; (4) Formulate emergency response plans for cybersecurity incidents, and periodically organize drills; (5) Other duties provided by law or administrative regulations". Moreover, they are subject to a national security review conducted by the State cybersecurity and informatization departments, as well as by the relevant departments of the State Council (Article 35)
75. Zhang C, Zhang M, Xiao C, From Traditional Infrastructure to New Infrastructure: a New Focus on China's Belt and Road Initiative Diplomacy?... cit., p 428. See also National Development and Reform Commission. See (in Chinese language) https://m.mofcom.gov.cn/article/i/jyjl/e/202004/20200402957398.shtml. Last accessed 16 April 2025

76. As stated by Yu Jianing, executive director of the CMCAMC. Global Times (2022) China's metaverse industry group signs strategic partnership with Asia Digital Bank. https://www.globaltimes.cn/page/202202/1252898.shtml. Last accessed 16 April 2025
77. As reported, in 2022 the Ministry of Science and ICT (MSIT) declared: "[w]e will not make the mistake of regulating a new service with existing law". See https://www.forbesindia.com/article/cryptocurrency/south-koreas-msit-reveals-plans-to-enact-special-metaverse-laws/79929/1. Last accessed 16 April 2025
78. See https://www.msit.go.kr/eng/bbs/view.do?sCode=eng&mId=4&mPid=2&pageIndex=&bbsSeqNo=42&nttSeqNo=621&searchOpt=ALL&searchTxt=. Last accessed 16 April 2025
79. Above all, those of "equity, inclusion, and diversity", along with the "establishment of self-regulatory standards". The discussion has also focused on the proposal to create a "pan-governmental consultative body based on self-regulation, minimum regulation, and pre-emptive regulation of users' virtual property". Nonetheless, regulatory efforts to date have remained largely circumscribed to issues such as privacy. Karimov E (2024) Meta-morphosis of copyright and user-generated content: can East asia's emerging policies navigate through the metaverse? Asian J Law Soc 11(4):483
80. Kshetri N, National Metaverse Strategies… cit., p 139
81. Giovanardi E (2022) I diversi modi di utilizzo dei token non fungibili, p 34. In: Deotto D, De Simone D, Giovanardi E, Rampone F, NFT e Metaverso. Cosa sono, come funzionano, caratteristiche e aspetti tecnici. Guida operativa alle regole giuridiche e fiscali, Il Sole 24 Ore Professional
82. Giovanardi E (2022) Introduzione alle piattaforme del Metaverso, diversi modi di utilizzo dei token non fungibili, p 48. In: Deotto D, De Simone D, Giovanardi E, Rampone F, NFT e Metaverso. Cosa sono, come funzionano, caratteristiche e aspetti tecnici. Guida operativa alle regole giuridiche e fiscali, Il Sole 24 Ore Professional
83. See the official Notice of Termination available at https://news.seoul.go.kr/gov/archives/560968. Last accessed 16 April 2025. See also https://www.fnnews.com/news/202408130507185408?ref=blog.copilot.kr. Last accessed 16 April 2025
84. See https://www.reuters.com/markets/asia/south-korea-regulate-cross-border-trade-virtual-assets-2024-10-25/. Last accessed 16 April 2025
85. See https://metavexpo.com/eng/. Last accessed 16 April 2025
86. Among South Korea's metaverse key players there are major tech conglomerates such as Samsung, LG Electronics, and Naver (developer of Zepeto, the popular 'social universe' inhabited by users' avatars). Some of them will be present at the MVEX 2025, which is aimed at exploring 'the latest trends in interactive technology'—from extended, virtual and augmented reality technologies to Artificial Intelligence and Web3. *Ibidem*
87. Baptista E (2022) A metaverse with Chinese characteristics is a clean and compliant metaverse, Reuters. https://www.reuters.com/markets/funds/metaverse-with-chinese-characteristics-is-clean-compliant-metaverse-2022-01-25/. Last accessed 16 April 2025
88. *Ibidem*
89. China's experiments in regulating metaverse-related products may contribute to the development of shared practices, rules, and standards, thereby shaping the very concept of the metaverse itself. For example, following the launch of the aforementioned digital yuan, the European Union is also exploring the possibility of introducing a digital euro. For further details, see https://www.ecb.europa.eu/paym/digital_euro/html/index.it.html. Last accessed 16 April 2025
90. It has been observed that 60% of the world's 5G infrastructure is located in China. Piccinali M, Interoperatibilità e geopolitica: le scelte cinesi… op. cit., p 38
91. See New Generation Artificial Intelligence Development Plan (2017), available in English at https://digichina.stanford.edu/work/full-translation-chinas-new-generation-artificial-intelligence-development-plan-2017/. Last accessed 16 April 2025

92. Restrictions that, through the so-called foreign direct product rule, extend event to microchip companies that do not fall under U.S. jurisdiction, yet whose products still contain U.S. technology. See Casanova GA (2023) Microchip: sfida cinese senza licenza, ISPI, https://www.ispionline.it/it/pubblicazione/microchip-sfida-cinese-senza-licenza-154243. Last accessed 16 April 2025. See also Morelli E (2023) La guerra dei semiconduttori. Domino 9 and Herczegh V, Friedman G (2023) Xi non può rianimare l'economia cinese. Domino 3
93. Casarotto F, La Cina punta sul metaverso per battere gli Stati Uniti… cit., p 25
94. In this regard, it has been noticed an actual geopolitical dimension of the metaverse, that is characterized by the opposition between the Chinese model and the Western model, primarily identified with the U.S. model, especially in sectors such as Artificial Intelligence. A competition that "is likely to be as much about differences of approach as technological dominance (already access to the Chinese-owned Tik Tok social network service varies depending on whether the consumer is Chinese or not) and could eventually lead to the emergence of distinct "Regional metaverses". See Council of the European Union General Secretariat (2022), metaverse-Virtual World, Real Challenges, p 9, https://www.consilium.europa.eu/media/54987/metaverse-paper-9-march-2022.pdf. Last accessed 16 April 2025
95. It should be noted that China's entertainment industry is already subjected to extensive monitoring, regulation, and censorship by the government

Navigating the Law of the Sea and Ports—the Impact of Technologies in the Maritime Sector and Navigation Law

Federica De Robertis

Abstract The maritime sector has historically been at the forefront of technological advancements, continuously adapting to innovations that enhance efficiency, safety, and sustainability. With the emergence of the Fourth Industrial Revolution, digital technologies such as artificial intelligence (AI), blockchain, smart contracts, and the metaverse are reshaping maritime operations. This article explores the legal challenges and adaptations required to accommodate these transformations, focusing on the autonomy of maritime law and its intersection with emerging technologies. A key focus is the rise of Maritime Autonomous Surface Ships (MASS), which present new regulatory and liability issues, particularly in the absence of clear international legal frameworks. Additionally, the increasing adoption of the metaverse in training and operational simulations highlights its potential to revolutionize maritime workforce development and risk management. The study also addresses the development of smart ports, leveraging AI and digital tools to optimize logistics, environmental impact, and overall efficiency. While technological advancements promise significant benefits, they also challenge existing legal principles, requiring a balance between innovation and regulatory oversight. The new scope of the maritime operators is to achieve a harmonized legal approach to ensure that the integration of intelligent navigation and digital tools aligns with international maritime conventions while maintaining security, efficiency, and environmental sustainability.

Keyword New technologies · Maritime sector · Sustainability

1 Introduction

Since ancient times, the transport sector has been one of the most adaptable industries when it comes to integrating innovative solutions and technological advancements to improve efficiency, combing it with the solutions required by the domain of the law.

F. De Robertis (✉)
Università Telematica "Leonardo Da Vinci", Torrevecchia Teatina, Italy
e-mail: f.derobertis@drblex.it

A. Borroni et al. (eds.), *Metaverses: Reshaping Law, Economy, and Society*, PoliTO Springer Series, https://doi.org/10.1007/978-3-032-08664-8_13

Early literature identified transportation as a key indicator of civilization's progress, linking human development to the evolution of communication and transportation methods [1].

In this perspective, the transfer of people and/or goods from one place to another, and respect for safety and security meant that shipping—more than other sectors—became the testing ground for technological innovations of the time [2], due to its inherent need for safety and reliability. With the digital transformation sweeping across various industries, the legal framework has had to evolve in response to these changes, adapting to the challenges of what is now considered the Fourth Industrial Revolution [3]. This modern phase of industrialization builds upon previous revolutions—mechanization in the 1780s, mass production in the 1870s, and computerization in the 1970s—by integrating ICT (Information and Communication Technologies), blockchain, smart contracts, artificial intelligence, and other digital tools into industry practices [4].

The maritime sector has long been at the forefront of engineering advancements. Historically, innovations such as the transition from rowers to sails, the adoption of mechanical propulsion, and the shift from coal to liquid fuels significantly transformed ship operations. Presently, the industry faces two primary challenges: the ecological transition toward more sustainable energy sources and the integration of intelligent navigation technologies [5].

In fact, if with the first industrial revolution in 1780, there was the beginning of a process of mechanization of production, aimed at industrializing the entire manufacturing sector, and with the second in 1870, there was a reorganization of work aimed at massifying production thanks to the use of electricity, with the third in 1970, there was the start of computerization, aimed at further automating production processes.

In this respect, Industry 4.0—seen as the fourth industrial revolution—can be said to be characterized by the use of ICT (Information and Communication Technologies) technologies in various areas of production through the concept of interconnection and dematerialization of the production chain, through multiple digital tools and the latest blockchain, smart contracts, metaverse, cryptocurrencies, and artificial intelligence [6].

Generally speaking, the world of navigation was the first to experience the impact of engineering innovations. More specifically, with reference to navigation by water, the introduction of the sail rather than rowing by rowers was already a change that influenced the type of traffic and cargoes; and, again, the introduction of mechanical propulsion allowed ships to be freed from the uncertainty of winds and weather conditions to the advantage of greater certainty regarding the duration and success of the voyage itself [7].

Similarly, the introduction of liquid fuels in place of coal, with the consequent reorganisation of the ship's technical spaces and the reduction of the rooms used to manage the propulsion systems, together with the greater energy efficiency and the relative smaller volume occupied by fuels and the significant reduction in the vessel's tonnage, also led to the transformation of crew photography (one thinks, immediately, of the disappearance of the need to embark stokers) [8].

Today, navigation is facing new challenges: on the one hand, the ecological transition, which makes it necessary to find traction solutions with the lowest possible environmental impact, and on the other hand, the implementation of new technologies, which will mark a point of no return from so-called intelligent navigation.

This study examines the legal adaptations required to accommodate these transformative changes in the maritime industry.

2 The Specialty of the Maritime Sector

Maritime law has always operated with a degree of autonomy, following its own specialized legal framework rather than general laws [9]. This principle is enshrined in legal systems such as the 1942 Italian Code of Navigation, which establishes maritime law as a *lex specialis* that prevails over general civil law when applicable [10]. This autonomy is justified by the unique risks and technical factors associated with navigation, necessitating distinct legal principles tailored specifically for the maritime domain.

Maritime law—and by logical connection also maritimc law—has always been considered as such: an area of the legal system, characterised by the presence of special institutes in comparison to their common law counterparts. The reason why the law of navigation was initially considered a special law lies in the fact that the ship, once it had set sail from the port, set out on its voyage and for a long time was never heard from again, and on this ship lived a travelling community that, living for a long time detached from the mainland, needed special rules. Suffice it to mention, by way of example, that the ship's captain is allowed to perform marriages in extremis. This faculty is not granted to the driver of land vehicles because the duration of maritime transport tends to be much longer than that of other modes of transport.

In fact, the Italian institutes of maritime and aviation law have been brought together in the text of the 1942 Code of Navigation which, in its opening, outlines the hierarchy of sources as follows: '[i]n matters of navigation, maritime, inland and air, this Code, the laws, regulations, corporate rules and customs relating thereto shall apply. Where there are no provisions of navigation law and there are none applicable by analogy, civil law shall apply' [11].

It still represents today, a legislative monument of enormous importance for its originality and completeness, an emblem of the unity of the subject. Over the years, the Navigation Code has been supplemented—and even surpassed—by a wide range of other sources of navigation law. As a result, a true "poly-system" has emerged, meaning that today, the Code can only be considered a reference point and a starting framework for understanding the legal discipline currently in force within the Italian legal system. Particularly clear and insightful in this regard are the observations of important doctrine, who states: "Within the realm of economic law, the process of legal creation occurs at such an intense pace that traditional sources, almost eroded by reality, reveal their insufficiency, while new forms of regulation emerge or intertwine with the legal framework of economic activities" [12]. The economic factor plays a significant role in the formation of law. Other factors, such as ethical and sociological considerations, also contribute, but with less intensity. On closer inspection, contractual regulation and uniform rules come into force when there is no legislative text regulating a particular phenomenon or when existing rules are subject to derogation. For instance, legal relationships concerning chartering and transport derive their regulatory framework from specific standard contracts. In no other field is the gap between codified law and living law as pronounced as in this one. It often happens that the law guiding us cannot be entirely traced back to the cold and generic provisions written in codes but is instead implemented according to the ever-changing reality of facts [13].

This approach is linked to the so-called theory of the autonomy of maritime law: according to which, in the light of the specificities and risks to persons and property that characterise the technical fact of navigation, institutes of their own, detached from the common order, are applicable [14].

It should also be emphasised that military navigation does not fall within the scope of (Italian) navigation law. The navigation code emphasises its non-reference to military navigation, sometimes expressly confirming it (art. 748), sometimes establishing the cases in which its rules apply exceptionally to warships (art. 589) or to military aircraft (art. 965) [15].

Indeed, merchant shipping has totally different purposes from those pursued by military shipping, which are characterised by functions and interests that transcend private ones. However, a legal relationship falls within the scope of navigation law when it relates for one of its subjects to merchant shipping and for the other to military shipping.

3 The Role of the New Technology in the Maritime Sphere from (AI) Artificial Intelligence and Metaverse

In Shipping continues to dominate global trade, handling over 90% of the $15 trillion spent annually on transportation. With over 800 active ports worldwide, China leads as the busiest hub, managing 1.43 billion tons of cargo per year. Given its economic significance, the maritime sector has pioneered technological innovations, prompting legal frameworks to evolve alongside these advancements [16].

Among the innovations available to maritime operators stands out artificial intelligence, which in the words of the European Commission in its Communication to the European Parliament, the Council, the European Economic and Social Committee and the Committee of the Regions of 25 April 2018: 'is not science fiction: it is already part of our lives' [17].

The use of unmanned maritime instruments to support naval operation is a phenomenon not new at the US navy. In fact, during the second world war unmanned surface vessels were used during strategical operations for testing the effects of nuclear weapons [18].

Furthermore, this innovative technology has grown exponentially in the 21st century, both in terms of the number and complexity of the missions assigned to them. Between 2000 and 2014, the US Department of Defence increased the number of unmanned aerial systems from 50 to over 10,000 [19]. This trend extends globally, with over 90 nations and non-state groups, including ISIS, using such technologies for surveillance or combat operations.

The adoption of these systems is due to their ability to improve situational awareness, reduce human workload and increase mission effectiveness, while reducing risk to civilian and military personnel. The US Department of Defence predicts that by 2035, more than 70 per cent of air assets will be unmanned systems, while the US Navy will integrate more autonomous surface ships and submarines to support the manned fleet [20].

These developments will radically alter the way nations conduct military operations, enabling commanders to meet new challenges on the battlefield.

The capabilities of unmanned systems will continue to evolve as technological advances improve their autonomy, stealth, mobility, versatility and survivability. Such technologies prove particularly useful for operations considered too tedious, dangerous or risky for human personnel, such as surveillance, reconnaissance and explosive ordnance clearance.

Unmanned systems are set to become a key component of armed forces, contributing to greater operational efficiency and reduced risks for personnel. However, their integration raises legal and regulatory questions, especially regarding their classification as ships or aircraft, their compatibility with international conventions and the management of liability in the event of accidents or attacks [21].

The US Navy is working to expand the use of unmanned surface vehicles, which can be used for anti-submarine warfare, maritime surveillance and mine's countermeasures. At the same time, autonomous submarines capable of operating in hostile environments, collecting hydrographic and oceanographic data and supporting intelligence missions are being developed [22].

From a legal point of view, the classification of these vehicles remains complex [23]. The UNCLOS Convention does not provide a clear definition of unmanned ships, but some conventions, such as SOLAS and MARPOL, recognise autonomous vehicles as part of the ship category [24].

It should also be noted that artificial intelligence (AI) is emerging as a transformative tool in the maritime industry, offering significant enhancements in hydrography and navigational safety. AI's potential is vast and still largely unexplored, much like the ocean itself. The scientific community has been investigating AI for decades, driven by its ability to improve hydrographic mapping and ensure safer navigation [25].

One critical aspect of maritime operations is Maritime Domain Awareness (MDA), which involves understanding all elements associated with the maritime domain that could influence security, safety, economic activities, and environmental conditions [26]. Geospatial intelligence (GEOINT) plays a vital role in MDA by analyzing and utilizing spatial data to assess physical characteristics and track maritime activities [27]. Both governmental and private entities leverage GEOINT through remote sensing and direct observations to enhance their comprehension of oceanic processes and potential risks.

AI-driven maritime systems can be classified into three categories based on the degree of human oversight: (i) Human-in-the-Loop (HITL)**:** AI assists human decision-making, but final actions require human intervention. This model is commonly used in ship navigation where AI provides recommendations, but the captain makes the ultimate decision; (ii) Human-on-the-Loop (HOTL): AI operates autonomously, with human oversight available to intervene when necessary. This system is beneficial in highly automated ship operations where AI controls routine tasks; (iii) Human-out-of-the-Loop (HOOTL)**:** AI functions independently without real-time human involvement. This model is still in its early stages for maritime navigation but is being explored for fully autonomous vessels [28].

Basically, Unmanned Surface Vessels (USVs) or Maritime Autonomous Surface Ships (MASS) refer to all those ships that present one of the following four degrees of autonomy: (1). 'Ship with automated processes and decision support', (2). 'Remotely controlled ship with seafarers on board', (3). 'Remotely controlled ship without seafarers on board', (4). 'Fully autonomous ship'.

Autonomous navigation technology has the potential to revolutionize shipping by reducing human errors, enhancing safety, minimizing pollution, and lowering operational costs.

A few significant examples may make the state of the art in this sector clear: (i) the YARA Birkeland, which for the time being is only allowed to sail within twelve nautical miles from the coast—already ready to set sail from the port of Heroya—constitutes the first fully electric and autonomous container ship in the world; (ii) the first test site for autonomous vehicles in Europe has been positioned in the Norwegian fjord of Trondheim; (iii) since 2015, the European Commission has launched a research project on unmanned ships, called Maritime Unmanned Navigation through Intelligence in Networks (MUNIN), to investigate the possibility of short sea shipping with autonomous vehicles in order to achieve electrification and fully autonomous driving [29].

The European Union has invested over €150 billion in AI research, emphasizing autonomous navigation as a priority [30]. Unmanned Surface Vessels (USVs) or Maritime Autonomous Surface Ships (MASS) exemplify this technological shift.

Potentially, Unmanned Surface Vessels (USVs) or Maritime Autonomous Surface Ships (MASS) constitute a major achievement in terms of technological innovation, cost reduction—to the extent that the human component is eliminated or greatly reduced—as well as increased efficiency and safety.

What motivates maritime operators to pursue this ship revolution is the expectation that the use of autonomous ships will significantly reduce the number of accidents at sea caused by human error, reduce pollution by virtue of the reduced possibility of accidents involving the spillage of toxic liquids into marine waters and the reduction in the number of voyages, due to improved logistics resulting from the intervention of artificial intelligence applied to the stowage and loading of goods in ships [31].

The lack of crew on board will help to compress costs and specific risks, such as piracy at sea deprived of the possibility of hostage-taking. In addition, unmanned ships are more manageable for dealing with all those situations identified in the relevant literature as dangerous and complex, from scientific experimentation to rescue at sea [32].

For the time being, design studies are aimed exclusively at the construction of autonomous cargo ships, although the step towards passenger transport is a short one: in that case, the presence of seafarers would initially be limited to monitoring the proper functioning of automatic control systems and other on-board equipment.

Although the aim of autonomous navigation is to eliminate harmful consequences due to human error with all the associated legal issues, the use of autonomous driving will also entail new challenges and scenarios for the law, starting with the possible harmonization of existing and forthcoming legislation, as well as the reconsideration of certain cardinal concepts such as the cardinal concept of the ship.

According to the strict wording of the supranational rule, the notion of 'ship' does not require the presence of a crew on board as one of its constituent elements, and Article 1(d) of the Hague-Visby Rules defines ship as any vessel used for the carriage of goods by sea [33].

Moreover, even the Italian Navigation Code, in Article 136, disregards the presence of a crew on board when identifying the concept of a ship [34].

With foresight, by omitting any mention of the command of the vessel, the legislator allows the rule to be adapted to the technical evolution of the vessel and in particular to the hypothesis of conduction entrusted to artificial intelligence.

Although the statutory definition of a ship lends itself to being adapted to the new concept of autonomous driving, new interpretative scenarios open up on the applicable legal framework, especially with regard to liability in the event of collisions between ships [35].

There is a part of English doctrine that is oriented towards considering ships per se, as autonomous centre with legal personality [36]. This follows the line of interpretation of certain case law which, in settling the fate of a maritime accident, established that the vessel enjoyed its own legal personality, different from that of its legitimate owners and therefore could well be held liable for damages caused to third parties.

This personification of the naval unit would help to consider it as a centre of imputation of interests capable of generating rights and obligations even in the management entrusted to artificial intelligence.

However, there is other English doctrine that, in a majority form, has already shown the first perplexities, considering that the parties contractually exposed, and therefore the owner of the vessel, are not exempt from the burden of compensation for damage caused to third parties [37].

On this point, we refer to the granite principles expressed in the famous case of Donoghues v Stevenson on the subject of negligence and the legal duty on the part of the manufacturer to avoid damage to the health of the end user. Even with regard to self-driven vessels, in the case of maritime accidents, the success of any action for damages resulting from a collision at sea depends on the plaintiff's discharge of the burden of proof at trial that the damage suffered is the consequence of the owner's breach of that legal duty to safeguard the third party.

The significant theme is to establish what the true cause of the damaging episode is and whether it can be attributed to the owner of the vessel, i.e. whether or not it embodies the breach of a legal duty attributable to him.

This becomes difficult when the chain of relations between human beings is interrupted when dealing with a synthetic being who performs the most sophisticated manoeuvres by the means of a virtual calculation of algorithms, resulting in the lack of direct power of control on the part of the float owner.

In the absence of specific legislation in this respect and hoping that the issue of liability in the event of maritime accidents will not be left solely to the workings of case law, three potential scenarios can be identified as of now: (i) strict liability of the shipowner, accompanied by the obligation of adequate risk insurance and direct action by the injured party; (ii) strict liability of autonomous ships; (iii) fault-based liability of software and hardware developers, system engineers and autonomous ship builders.

The first hypothesis is one that is already familiar to the maritime environment, since under uniform international rules, there are Conventions that impose an objective type of liability on the shipowner with the obligation of insurance cover, e.g. the International Convention on Civil Liability for Oil Pollution Damage concluded in Brussels on 29 November 1969 and amended by the Protocol signed in London on 27 November 1992 (CLC) [38].

The second hypothesis could be considered as being studied by the European Commission as a result of the European Parliament Resolution of 16 February 2017 containing recommendations to the Commission concerning civil law rules on robotics. Indeed, the recitals of the aforementioned resolution literally state: 'the more autonomous robots are, the less they can be considered as mere tools in the hands of other actors (such as the manufacturer, the operator, the owner, the user etc.)' [39].

Under the current legal framework, robots (including autonomous vehicles and artificial intelligence) cannot be held liable in their own right for acts or omissions that cause damage to third parties [40]. This is because the existing legislation must necessarily assign liability to a human agent, be it the manufacturer or owner or user [41].

These objective regulatory shortcomings can only be remedied by specific legislation at transnational level, and the first questions that the European Commission, at the instigation of the European Parliament, is called upon to resolve are precisely the mechanisms for the objective imputation of liability and whether these can be extended to artificial intelligence [42].

The third scenario represents the current legal solution whereby the manufacturer/developer of the autonomous vehicle can be held liable for damages caused to third parties on the basis of product liability principles [43].

One should, therefore, partly draw on the rules of liability in general (civil code), the regulations on compulsory insurance for vehicles and the principles of the consumer code for defective products.

Another technology used in the maritime sector that deserves further investigation is the use of the metaverse [44].

The metaverse aims to reproduce the performance of any daily activity in a virtual dimension, through a digital alter ego or avatar, which, however, acts and interfaces with other users by means of the personal data referable to the real person.

In concrete terms, in the metaverse, the virtual twin recreates its own experience through an endlessly interconnected parallel community and can work, socialize, shop, and play there using special technologies. Some of these technologies include virtual reality visors, augmented reality glasses, and special applications for smartphones and/or other devices that allow for an even more real-world-like experience [45].

The latest technological advances tend towards a gradual disintegration of the unity of space and time, and this is having inevitable and severe repercussions on the world of work as it is traditionally understood [46].

Among the aspects that distinguish maritime employment from the common employment relationship are the pre-eminent public interest in the safety of navigation and the peculiarity of the bodies in charge of the seafarers' employment function [47].

While in common labour relations the legislator's aim is to protect the worker and more rarely to safeguard company assets and productive activity, in the nautical sector the objective of the safety of navigation and the marine environment enjoys the same relevance as the protection of the worker himself [48].

Undoubtedly, then, the use of the metaverse would also represent a method to curb the phenomenon of pollution, thus reducing sea trips for training purposes, an aspect that could be partially replaced by realistic replicas in the seas of a virtual reality.

Indeed, one of the main areas of the navy is simulation and training. The use of avatars instead of a real crew, i.e. the creation of a parallel reality where crews can actively participate in exercises and simulations, face combat scenarios, and foil acts of piracy, would make it possible to raise the level of safety at sea, reduce costs and avoid risks to people's safety, without sacrificing the training effect of the exercise. Then, by means of faithful virtual reproductions of maritime infrastructure, shipping lanes and navigation conditions, captains and maritime personnel could test and develop different tactical and strategic options quickly and effectively, with the possibility of repeating the exercise ad infinitum.

The imaginative universe created by the metaverse becomes a valuable decision-making, experiential and training platform for creating defensive and preventive strategies for acts of piracy and warfare. The immateriality of the meta-verse allows the military to collaborate and communicate with different units and levels of command, participating in meetings, planning sessions useful for planning movements—think of the possibility of navigating in different and adverse weather conditions, dealing with emergency situations and improving one's skills in coordinating the actions of a ship within a fleet or countering acts of piracy [49].

The integration of new technologies into maritime navigation presents a revolutionary shift in how ships operate, improving safety, efficiency, and compliance with international legal frameworks. However, as AI technologies or metaverse evolve, regulatory bodies must address the ethical and legal challenges associated with autonomous decision-making at sea. Ensuring that AI aligns with existing maritime conventions while fostering innovation will be essential for the sustainable development of AI-driven navigation systems.

4 Conclusions

The technologies going around the maritime sector combine different epiphenomena ranging from virtual and augmented reality (side to the metaverse) to the IA, IOT, and the Blockchain net. This leads potentially to an interactive, immersive experience and the generation of an increasingly realistic and all-encompassing three-dimensional virtual universe [50].

The innovations, on the basis of these premises, actively contribute to the creation of structured training paths, capable of bringing out the criticalities and potentialities not only of the new modes of employment, but also of the more traditional work expressions.

One of the prerequisites for being able to carry out navigational activities is safety, which is expressed both in terms of the protection of the seafarer and the protection of the marine space.

The use of the news technologies could favour a process of gradual reduction of human error in the case of navigation, which, while not integrating the perfect calculability of the unexpected and the fortuitous event, would represent an investment in the direction of a better quality of service rendered, through the combination of technological innovation, transparency and training and in addition, has the concrete potential to bring about a useful improvement in terms of the governance of vessel traffic, work sequences and the management of unforeseen emergencies.

Furthermore, the development of smart ports in terms of improved performance, greater efficiency in the storage and stowage of goods is already part of our present.

In this regard, the metaverse service offers key features across three perspectives: (i) Predicts estimated arrival times, monitors port operations, and optimizes workflow efficiency; (ii) Tracks and forecasts ship fuel consumption while monitoring emissions from port equipment; (iii) Assesses risks along ship routes, detects and monitors hazardous navigation paths, and oversees safety incidents involving workers and equipment.

Port operations optimization encompasses the enhancement of the entire port process, starting from a vessel's arrival, through berthing procedures, yard management, and the final transfer of goods beyond the port gates. This optimization serves multiple objectives, including efficient equipment allocation, scheduling improvements, and minimizing rehandling. It can focus on refining individual tasks or streamlining the entire workflow. The metaverse leverages the sensor data and Terminal Operating System (TOS) information to track various operations, assess congestion and delays, and recommend optimization strategies to mitigate inefficiencies [51].

The fusion of emerging technologies with maritime operations is poised to redefine industry standards, requiring corresponding adaptations in maritime law.

AI, autonomous ships, and virtual training simulations are reshaping the sector, enhancing safety and operational efficiency. However, legal frameworks must evolve to address liability concerns, labor market transformations, and the broader implications of these advancements.

A harmonized approach—integrating technology with specialized maritime legal principles—will be crucial for ensuring a smooth transition into this new era of intelligent navigation. The future of maritime law will depend on its ability to adapt to rapid technological progress while maintaining core legal principles to safeguard industry stakeholders.

References

1. Tajani F (1943), I trasporti sotto l'aspetto economico, ed. III, Università Ambrosiana, 6, where it is stated that it was not «[...] esagerato affermare che il progresso umano s'identifica nel progresso dei mezzi di comunicazione». Cfr. anche Abello L, (1927), Trattato della locazione, Torino, p. 3; Scialoja A, (1933), Sistema del diritto della navigazione, I, ed. III, Roma; Irti N, (2007) Tecno-diritto, in Il diritto nell'età della tecnica, Naples, pp. 11 ss.
2. Pardo Zaragoza C (2016), 'Análisis de la evolución jurídica del derecho aeronáutico desde 1911 a 1955 a través de las organizaciones aéreas internacionales' in 'Revista europea de derecho de la navegación marítima y aeronáutica', n. 33, pp. 31 ss.
3. See also, in historical perspective: Bastier J, (1985) Le temps des machines volantes. Une histoire du droit aérien français des origines à 1914, in 'Ann. dr. mar. aér'., p. 267.
4. Mantoux P (1928), The industrial Revolution in the eighteenth Century, (Jonatan Cape Thirty Bedford Square London, pp. 92 ss; Gray H L (1924) The Production and Esportation of English Woollens in Fourteenth Century, Historical Review, XXXIX, p. 32; Deande P (1957) The Output of the British Wollen Industry in the Eighteenth century in J. Econ. Hist. XVII, p. 220.
5. Tonizzi ME (2014) Lavoro e lavoratori del mare nell'età della globalizzazione, in Riv. di Storia dell'800 e del 900, 4, p. 691.
6. On the terminological roots of the term Industry 4.0, it should be noted that this term was first used in 2011 at the Hannover Messe in Germany, by intellectuals: Henning Kagermann, Wolfgang Wahlster and Wolf - Dieter Lukas, three representatives from economics, politics and science, wanting to refer to the ZuKunftsProjekt Industrie 4.0. In this regard, read the following consultation: https://wayback.archive-it.org/12090/20170401061823/; https://ec.europa.eu/digital-single-market/en/blog/implementation-industry-40-strategy-german-plattform-industrie-40; https://merics.org/en/report/made-china-2025;
7. In doctrine, see also: Wingen M – Scuttrumpf H, (2022) Water sector 4.0' in Frenz W, Handbook Industry 4.0 – Law, Technology, Society, (Springer), 681 ss,; F Munari, (2019) Blockchain and smart contracts in shipping and transport. A legal revolution is about to arrive?, in Soyer B, Tettenborn A (edited by.), New Technologies, Artificial Intelligence and Shipping Law in the 21st Century, London, p. 3.
8. Tonizzi ME (2014), Lavoro e lavoratori del mare nell'età della globalizzazione, in Riv. di Storia dell'800 e del 900, n. 4, p. 691; Flinn MV (1975) Economic and Social History of Britain Since 1700, ed. II, (Basingstoke), pp. 110 ss.; M Fusaro (2008) Reti commerciali e traffici globali in età moderna, in Quaderni Storici, Riv. Quad., n. 2, pp. 11 ss.
9. Scialoja A (1933) Sistema del diritto della navigazione, Rome, for the author, the object of the law of navigation is to be sought in the totality of social relations, of whatever kind, that is attributable to navigation in itself. In this sense, the subject matter is not to be limited strictu sensu to the phenomenon of navigation but extends to the environment, to the persons who take part in it, and to the things that take part in it. This is an autonomous vision of navigation

law. See also Carbone SM (1975) La c.d. autonomia del diritto della navigazione: risultati e prospettive, in Dir. mar., pp. 17 ss.

10. The navigation code, approved by Royal Decree of 30 March 1942, was drafted on the basis of the studies of Antonio Scialoja. In relation to the scientific value of the latter's work see, generally: Carbone SM (1983) Il Diritto Marittimo in Italia al Momento della sua Codificazione ed agli Inizi degli Anni '80, in Dir. Mar., fasc. 1–4 e da Pescatore G (1985) Antonio Scialoja: il Sistema e la Codificazione del Diritto della Navigazione, in Foro it, vol. 108, n. 12.
11. By Royal Decree No. 327 of 30 March 1942, the definitive text of the navigation code18 was published of navigation18 that finally came into force on the following 21 April, the day of Christmas in Rome, for a precise legislative political choice suggested by the propaganda purposes of the fascist regime. It represented, and still represents today, a legislative monument of enormous importance for its originality and completeness, an emblem of the unity of the subject. As stated above, the navigation code regulates all relations pertaining to shipping, maritime, inland and air navigation in an organic and complete manner. Moreover, it brings together in a unified system both private and public institutions. On this point read: Giannini T.C., Il nuovo codice marittimo nell'opinione delle Corti e delle Università italiane, in Dir. mar. 1937–1938, in commenting on the draft recognises it to be 'an excellent basis for review and discussion for a future project'. For a more detailed reconstruction of doctrinal positions on the 1931 Maritime Code Project see: Giannini A.,(1931) Il sistema del progetto del codice marittimo, in Dir. mar., p. 452.
12. Dominedò FM (1950), Le fonti del diritto della navigazione, in Riv. dir. nav., parte I, p. 273; Gaeta D (1965), Le fonti del diritto della navigazione, Milan.
13. According to the thinking of Scialoja A, this expression is identified with so-called autarkic transport. In particular, the author states: «[i]l rischio della navigazione che accomuna in una formidabile solidarietà di interessi le persone che alla nave hanno affidato i loro beni; la lontananza della nave necessariamente affidata, così per le cose private come per le pubbliche, all'autorità ed alla perizia di uno solo, di fronte ad ogni evento; questi elementi di fatto, così forti e diversi, è inevitabile che diano alle loro norme regolatrici, di creazione originale o di adattamento, una comune precisa caratteristica impronta». Scialoja A (1933) Sistema del Diritto della Navigazione, Rome, p. 12.
14. Cfr. Scialoja A, (1929), Sistema del diritto della navigazione, vol. I, ed. 2, Rome; Scialoja A, (1943) Corso di diritto della navigazione, vol. I, Rome.
15. See. Lefebre D'Ovidio A, Pescatore G, Tullio L(2022) Manuale di diritto della navigazione, Milan, p. 4.
16. Kraska J, Pedrozo R, (2022) Unmanned Maritime Systems', Disruptive Technology and the Law of Naval Warfare, New York, p. 75
17. For a more detailed consultation: https://eur-lex.europa.eu/legal-content/EN/TXT/PDF/?uri=CELEX:52018DC0237 consulted the 12th of February 2025.
18. Trevithick J, The Navy is building a network of drone Submarines and sensors Buoys in the Artictc, theDrive, https://www.twz.com/36821/the-navy-is-building-a-network-of-drone-submarines-and-sensor-buoys-in-the-arcticconsulted l'11 febbraio 2025.
19. https://gjia.georgetown.edu/2024/07/12/war-artificial-intelligence-and-the-future-of-conflict/ accessed 17 February 2025.
20. Öztürk K, (2023) The Significance of Unmanned Aerial Vehicles (UAVs) in Strategic Contexts, pp. 75 ss.
21. Northrop Grumman Corp. Press Release, "Northrop Grumman's Next-Generation Fire Scout UAV on Track," June 23rd, 2005. FY2009–2034 Unmanned Systems Integrated Roadmap, p. 66.
22. For a reading on statistical data see the following link https://unctad.org/system/files/official-document/rmt2022_en.pdf accessed 12 February 2025.
23. See: US Department of the Navy (2016), Secretary of the navy instruction 5030.8C. General Guidance for the classification of naval vessels and battle force ship counting procedures.
24. About the definition of the ship under the main Conventions, read: https://cimsec.org/unmanned-maritime-systems-and-warships-interpretations-under-the-law-of-the-sea/ accessed 17 February 2025.

25. Li W, Hsu C Y, Tedesco M, (2024) Advancing Arctic Sea Remote Sensing with AI and Deep Learning Opportunities and Challenges, Remote Sensing, Basel, pp. 2–30.
26. About the definition of Maritime Domain Awareness, read: https://www.imo.org/en/OurWork/Security/Pages/Maritime-Domain-Awareness.aspx accessed on 17 February 2025.
27. About Geospatial Intelligence, read https://www.satcen.europa.eu/what-we-do/geospatial_intelligence accessed on 17 February 2025.
28. Simiòn D, Postolache F, Fleaca B, Fleaca E, (2014) AI Driven Predictive Maintenance in Modern Maritime Transport. Enhancing Operational Efficiency and Reliability, pp. 2–19.
29. The MUNIN project is a research project, co-funded by the European Commission under the Seventh Framework Programme. MUNIN aims to develop an autonomous ship, i.e. a vessel driven mainly by automated decision-making systems on board but controlled by a remote operator at a control station on land. See https://cordis.europa.eu/article/id/169600-towards-unmanned-cargo-vessels-for-more-sustainable-maritime-transport/it accessed the 10 of October 2024
30. This European funding instrument is intended to stimulate investment in artificial intelligence by involving different production sectors, for a consultation see: https://www.unioneeuropea.it/intelligenza-artificiale-nuovo-strumento-di-finanziamento-dellue-da-150-milioni-di-euro/
31. On the lesser environmental impact and the advantages of using unmanned ships in terms of sustainability, see: Liu D (2018) Autonomous Vessel Technology, Safety, and Ocean Impacts, in The Future of Ocean Governance and Capacity Development, Essays in Honour of Elysabeth Mann Borghese, Berlin, p. 494.
32. This includes the design of autonomous ships for scientific research purposes and for sea rescue in difficult and life-threatening situations. On sea rescue, see Severoni C (2020), Preliminary remarks on the legal regulation of the unmanned vessels, in Riv. dir. Econ., Trasp. Amb., (XVIII) pp. 67–75, where he writes: "However, the London Convention on Savage on the one hand does not provide for compensation from the salved persons (no remuneration is due from persons whose lives are saved: article 16, first paragraph of the London Convention), on the other it provides, among the ordinary criteria of definition of the reward, in multis, also the skill and the efforts of the salvors in salving vessel, other property or life" (article 13, letter e), a circumstance that may also apply to the salvage with unmanned ships, where skills and efforts are not those of the commander and of the crew inside the ship, but those of a pilot who operates outside the vessel, or of the ship-owner who has put instruments and rescue software on boar the autonomous vessel". By the same author: (2018) Soccorso e mezzi di trasporto autonomi, in Dir. trasp., I, pp. 27–66. For a historical reconstruction of the duty to rescue people lost at sea see: Le Clere J (1954) L'assistance aux navires et le sauvetage des épaves, (R. Pichon et R. Durand-Auzias, edited by); Righetti G, (1994) Trattato di Diritto Marittimo, Milan, III, p. 421; Rizzo P(1996) La nuova disciplina internazionale del soccorso in acqua e il codice della navigazione, Naples; Vincenzini E, (1985) *P*rofili internazionali del soccorso in mare, Milan. See also: Ardito S - Lazarevs D - Vasiliniuc B - Vukic Z– Massabayashi K- Caccia M, (2012) Cooperative Autonomous Robotic Towing system: definition of requirements and operating scenarios, IFAC Proceedings Volumes, vol. 45, 27, pp. 262–267. According to the authors, we will see a new concept of rescue in terms of optimising risks and reducing environmental impact:« The project CART (Cooperative Autonomous Robotic Towing system) proposes a new concept for salvage operations of distressed ships at sea based on the development of robotised un- manned marine platforms able to (semi-) automatically execute the high-risk operation of linking the emergency towing system of distressed ships to towing vessels. The CART device will be able to optimize the operations for safeguarding the environment, helping to prevent oil pollution at sea, and minimizing the risk for human lives».
33. See https://www.fog.it/convenzioni/inglese/visby-1979.htm#NR1 accessed on 10 of February 2025, In doctrine see: Lopez De Gonzalo M, (2020) Le navi autonome e le Hague-Visby Rules (spunti per un'interpretazione evolutiva, in *T*emas De Direito dos Transportes, M Januario Da Costa Gomes (edited by) Lisbon, pp. 609–619.
34. Article 136 of the navigation code in force since 21 April 1942, which literally states the following: "Navi e galleggianti 1. Per nave si intende qualsiasi costruzione destinata al trasporto

per acqua, anche a scopo di rimorchio, di pesca, di diporto, o ad altro scopo. 2. Le navi si distinguono in maggiori e minori. Sono maggiori le navi alturiere; sono minori le navi costiere, quelle del servizio marittimo dei porti e le navi addette alla navigazione interna. 3. Le disposizioni che riguardano le navi si applicano, in quanto non sia diversamente disposto, anche ai galleggianti mobili adibiti a qualsiasi servizio attinente alla navigazione o al traffico in acque marittime o interne.

35. In doctrine, see: Severoni C, (2018) Prime osservazioni in tema di responsabilità derivante da urto con navi senza equipaggio, in Dir. trasp., I, pp. 67–98, The author observes how, according to the Italian legal system, in order to trace the ship's liability in the event of a collision, it is necessary to investigate the wilful misconduct or fault of the liable party, while in the case of autonomous navigation, the absence of a pilot would give rise to many questions as to how to allow the search for possible fault in the conduct of the ship itself, given that guidance would be entrusted to sensors and through a virtual exchange of data. On the different degrees of autonomy of the vessel and the correlative responsibilities, see: Serdy A, Tsimplis M, Veal R, (2016) Liability for Operation in Unmanned Maritime Vehicles with Differing Levels of Autonomy (Southampton Institute of Maritime Law, p. 212.
36. On this point, read: "Personification of Vessels."*Harvard Law Review,* (1964) vol. 77, (n. 6), 1122–32. *JSTOR*, https://doi.org/10.2307/1339065. Last consultation 14 January 2025; G. L. C. "Maritime Law: Personality of Ship: Immunity of Government Property." *Michigan Law Review*, vol. 20, no. 5, 1922, 533–35. *JSTOR*, 10.2307 /1277641. Last consultation 14 January 2025.
37. For a critical reading in doctrine, see: Carey L, (2023), Contractual and Tortius Maritime Liability Regimes and the Introduction of Autonomous Vessels, in NUS Law Working Paper 2023/012 NUS Centre for Maritime Law Working Paper, 3, p. 29, in particular where the author points out that although it is essential for the shipowner to take out insurance for everything connected with navigation, this does not upset the balance of liability attribution between the shipowner and the injured third party known to date. In fact, however objective the autonomous shipowner's liability is in the event of an accident, there is always the possibility for him to prove that the damaging episode occurred due to the fault or negligence of the damaged party itself. The link consulted on 17 February 2025: https://papers.ssrn.com/sol3/papers.cfm?abstract_id=4403620, De Gioia Carabellese P, (2021) Unmanned vehicles e rischi legali ed assicurativi. Una visuale dal Regno Unito della disciplina della responsabilità dei veicoli senza guidatori in Diritto e Politica dei Trasporti, I, pp.1 ss.
38. Boi G M (2017) Navi-drone»: primi interrogativi in tema di disciplina giuridica, in Rivista di Diritto della Navigazione, p. 175 ss.; P Zampella (2019) Navi autonome e navi pilotate da remoto, spunti per una riflessione, in Diritto dei Trasporti, pp. 583 ss.; Lobianco R (2021), Navi senza equipaggio e profili di responsabilità, in Responsabilità Civile e Previdenza, 3, p. 756.
39. Committee on Legal Affairs, Draft Report with recommendations to the Commission on Civil Law Rules on Robotics (2015/2103(INL)), 31 of May 2016. The Draft Report is the first Community text that has been published in relation to modern robotics, thanks to which the regulation of robotics is to all intents and purposes placed high on the agenda of the European legislator, who will have to find ad hoc regulatory solutions within a maximum of fifteen years. On this subject, read: Zornoza A, Laukyte M, (2016) Robotica e diritto: riflessioni critiche sull'ultima iniziativa di regolamentazione in Europa, in 'Contratto e impr./Europa', 2, pp. 808 ss.
40. On the possibility of considering robots as an autonomous legal entity, read: Caroccia F, (2020) Soggettività giuridica dei robot? in Diritto e intelligenza artificiale: profili generali, soggetti, contratti, responsabilità civile, diritto bancario e finanziario, processo civile, Pisa, pp. 213-249.
41. Read: Bertolini A (2013) Robots as Product: The Case for a Realistic Analysis of Robotic Applications and Liability Rules, in Law, Innovation and Technology, 5(2), pp. 233 ss; Capilli G, (2020) I criteri di interpretazione della responsabilità, in *Diritto e Intelligenza Artificiale*, Alpa G(edited by), pp. 457 ss.

42. For a full consultation of the European Parliament Resolution of 16 February 2017 with recommendations to the Commission concerning civil law rules on robotics: https://eur-lex.europa.eu/legal-content/IT/TXT/PDF/?uri=CELEX:52017IP0051&from=IT consulted on 20 January 2025. In doctrine, for a critical reading of the European institutions' work to date, read: Ebers M – Hoch VRS – Rosenkranz F – Ruschemeier H – Steinrotter B, (2021) The European Commission's Proposal for an Artificial Intelligence Act – A Critical Assement by Members of the Robotics and AI Law Society (RAILS), in Multidisc. Sciet. Journ., 4, p. 589 ss.
43. Cerini D (2018), Dal Decreto Smart Road in avanti ridisegnare responsabilità e soluzioni assicurative, in Dann e resp., n. 4, p. 401, where the author points out that it is difficult to imagine a 'supine translation of the current product damage rules to the smart infrastructure sector'. More generally: M Kim - T-H Joung - B Jeong, HS Park, (2020) Autonomous shipping and its impact on regulations, technologies, and industries. Journal of International Maritime Safety, Environmental Affairs, and Shipping. 4 (n. 2, vol. 4) pp. 17–25; Duan W, GU J, Wen M, Zhang G, Ji Y, Mumtaz S, (2020), Emerging technologies, in*IEEE Netw.*,34, 5, pp. 283–289.
44. Ball M (2022) The Metaverse. And How It Will Revolutionize Everything, Liveright, New York, p. 29, which referred to a "persistent and interconnected network of 3D virtual worlds that will eventually serve as the gateway to most online experiences, and also underpin much of the physical world".
45. Brown S (2022) What Second Life and Roblox can teach us about the Metaverse, in MIT Management, of the 19 of July 2022, on the follow website last access the 24 of January 2024 https://mitsloan.mit.edu/ideasmade-to-matter/what-second-life-and-roblox-can-teach-usabout-metaverse; Yadav A (2022) The Metaverse:What is the IP Reality of the Virtual World, in NFTs, Cryptocurrency & the Metaverse, pp. 30 ss.
46. Lazzeroni L (2022) Metaverso, in AA.VV., Lavoro e tecnologie. Dizionario del diritto del lavoro che cambia, Turin, p. 162; Martone M, (2022) Prime riflessioni su lavoro e metaverso, in Arg. Dir. Lav., 6, pp. 1131 ss.; Costantino F (2023) Il lavoro nel metaverso, in Cassano G, Sforza G (edited by), Metaverso, Pisa, pp. 381 ss.
47. Balletti B, (1980) Rilevanza e tipicità del contratto di arruolamento, Naples, p. 172; Branca G, (1967) Contratto di arruolamento e lavoro subordinato, Milan, p. 104; Bianchi D'urso F (1988) La Corte costituzionale sui limiti al potere di licenziamento nel lavoro marittimo, in Rivista italiana del diritto del lavoro, p. 46: «la formulazione dell'art. 1 cod.nav. esclude qualsiasi relazione diretta — e dunque qualsiasi confronto in termini di compatibilità o meno — tra diritto della navigazione e diritto civile, il quale soccorre in via sussidiaria unicamente in presenza di vere e proprie lacune»
48. Borrelli S, Brino V, Faleri C, Lazzeroni L, Tebano L, Zappalà L (2022) Lavoro e tecnologie. Dizionario di diritto del lavoro che cambia, Turin, p. 162
49. Moschetti CM (1983), Pirateria in Enc. Dir. XXX, p. 873 ss.; Annati M, Caffio F, (2009) Pirati di ieri e di oggi *in* Supp. Riv. Mar., pp. 15 ss.; Gosse P (1992), Storia della pirateria, Florence, pp. 17 ss.
50. De Stefano V, Aloisi A, Countouris N, (2022) The Metaverse is a labour issue, in Social Europe, available on https://www.socialeurope.eu/the-metaverse-is-a-labour-issue consulted the 13 of February 2025 where authors emphasise: «The aim is to create a more interactive and collaborative work experience for remote workers. While this may sound like a good thing, a first concern is that such a combination will add to the stress of being subject to ever-more invasive and relentless forms of algorithmic surveillance, already experienced by remote workers, the sometimes toxic and oppressive dynamic of the office»
51. On this subject, read the study of the Busan port (South Korea) experiment, where through the introduction of the metaverse to facilitate and streamline all port logistics operations: Sunghyun S, Kim D, Park K, Bae H, (2024) Artificial Intelligence – based Smarts Port Logistics Metaverse for Enhancing Productivity, Enviroment and Safety in Port Logistics: A Case Stedy of Busan Port, https://doi.org/10.48550/arXiv.2409.10519.

Cultural Heritage in the Metaverse and the Impact of AI: Amplifying Creativity and Legal Protection in the Virtual World

Fabio Zambardino

Abstract This article explores the intricate legal framework surrounding cultural heritage within the expanding domain of the metaverse. It begins with a succinct overview of the metaverse concept before delving into its legal ramifications, particularly concerning the protection and ownership of digital assets that symbolize cultural heritage. The discussion contrasts traditional methods of cultural preservation with the digitization process, raising significant inquiries about authenticity and ownership rights. Additionally, the ephemeral nature of digital artworks within the metaverse introduces unprecedented challenges for legal frameworks, due to the intrinsic fluidity and replicability of virtual assets. Ultimately, the article assesses current legal mechanisms designed to protect cultural property and evaluates their relevance in the virtual realm, underscoring the necessity for adaptive legal frameworks that harmonize the safeguarding of cultural heritage with the innovative landscape of the metaverse.

Keywords Metaverse · Cultural heritage · Law · Art digitization

1 Introduction

In recent years, the appreciation of art has expanded beyond the traditional confines of galleries and museums, embracing the immense visual and experiential potential offered by modern technology. Digital applications have proven their ability to draw the public closer to culture in its most refined sense, transcending mere observation of art in its physical space by augmenting it with information, virtual reconstructions, and insights into its time of creation or site of discovery [1].

This technological mediation not only enriches the viewer's understanding but also offers the possibility of engaging with artworks or artifacts that are no longer

F. Zambardino (✉)
Università Degli Studi Della Campania "Luigi Vanvitelli", Caserta, Italy
e-mail: fabio.zambardino@unicampania.it

New Vision University, Tbilisi, Georgia

A. Borroni et al. (eds.), *Metaverses: Reshaping Law, Economy, and Society*, PoliTO Springer Series, https://doi.org/10.1007/978-3-032-08664-8_14

physically intact, or that have been lost to time, creating immersive experiences that rival, and in some ways surpass, the real. Through digitization, cataloguing, reproduction, and recovery, technology has made cultural heritage—especially submerged or inaccessible treasures—more comprehensible, accessible, and engaging to a global audience [2].

At the pinnacle of this convergence of art and technology lies the metaverse [3]: a digital realm where the boundaries between the real and virtual blur, and where users can interact, create, and participate in a world that is at once familiar and fantastical [4].

This immersive realm enables individuals to shape their digital identities, engage in gatherings and events, and establish virtual economies, reshaping human interaction [5]. By bridging the physical and digital worlds, it redefines social, economic, and cultural landscapes [6].

Within this digital expanse, a profound opportunity emerges: the chance to breathe new life into cultural narratives, to rediscover marginalized histories, and to engage with our shared heritage in ways previously unimaginable [7]. The metaverse, with its unprecedented capacity for interactivity and immersion, offers a platform for individuals and communities to explore identities, traditions, and historical legacies unfettered by the limitations of time and space. Furthermore, the integration of AI within this space enhances artistic creation, providing new tools for collaboration, innovation, and representation, thus igniting a renaissance of creative expression that melds human ingenuity with technological prowess [8].

Venturing into this brave new world, one is confronted with equally significant challenges. The transition of cultural heritage from the physical to the virtual realm raises complex legal and ethical considerations. Safeguarding the integrity of these digital representations becomes paramount, requiring the establishment of protections that ensure cultural legacies are preserved and respected within this virtual space [9].

As this intricate terrain is navigated, the development of robust legal frameworks and ethical guidelines emerges as an imperative, securing the protection of cultural assets in the metaverse. In this way, the vibrant expressions of the past are shielded from obscurity, allowed instead to flourish, enriched by the technological innovations of the present and future.

This moment is not one of mere technical advancement, but a pivotal point at which the collective history of humanity must be stewarded with care, foresight, and reverence.

2 Enhancing Cultural Representation in the Metaverse

The metaverse, a burgeoning digital universe, offers unprecedented potential for interconnection between people and experiences, especially through the interoperability of its virtual space [10].

This interconnected environment cultivates the perfect setting for the development and growth of emerging technologies like blockchain, cryptocurrencies, NFTs, etc. [11].

These innovations thrive in immersive realities, largely free from the constraints of traditional state authority.

In this rapidly evolving digital landscape, the law must retain its stabilizing function, ensuring protections for these emerging technologies despite the inherent instability of innovations such as blockchain and cryptocurrencies. The contractualization of market relations, especially in relation to NFTs, intellectual property, and extra-contractual obligations, introduces fresh legal complexities [12]. Furthermore, the interconnected nature of the metaverse [13], where daily activities increasingly merge with virtual interactions, entails an unprecedented collection of data [14]—from consumption patterns to biometric information—thus complicating traditional notions of privacy. The very concept of privacy, understood as the right to be left alone, becomes tenuous as users' personal data are continually harvested and exploited within this digital space [15].

Moreover, the corporations that control these virtual worlds wield immense power over users' data, predicting and shaping behaviors with unprecedented precision, effectively supplanting the regulatory role traditionally held by public authorities. This shift raises profound concerns regarding data portability, the delicate balance between property and user rights, and the safeguarding of personal data from both corporate exploitation and state interference [16].

Issues of data management, liability, and cybersecurity surface, as does the ethical dilemma of commodifying personal information—a practice that risks undermining individual autonomy, identity, and dignity [17]. In this regard, the metaverse demands robust security measures that both mitigate these risks and promote innovation [18]. Transparency, accountability, and an acute assessment of cyber risk are paramount in preserving the integrity of this digital ecosystem [19].

Yet, amidst these complexities, the metaverse also emerges as a beacon of cultural possibility, offering unparalleled opportunities for the amplification and preservation of cultural heritage. Within its boundless digital landscapes, users can embark on immersive journeys through meticulously recreated cultural sites, traditions, and practices from across the globe [20].

Augmented reality (AR) and virtual reality (VR) technologies breathe new life into human heritage, enabling users to engage with cultural artifacts, rituals, and narratives in ways that provoke reflection and foster deeper understanding [21].

However, this digital amplification of cultural heritage must be approached with great care. The metaverse offers the promise of preserving and revitalizing cultural traditions, ensuring that they remain vibrant and accessible for future generations. Yet, it also imposes a profound responsibility to portray these heritages with authenticity and respect [22]. Collaborations with the custodians of these traditions are essential to ensuring that the digital representation of cultural heritage honors its richness and complexity while promoting mutual understanding in an increasingly digital world [20].

This interplay between technological innovation and cultural preservation, however, demands vigilance. The metaverse, while offering immense opportunities for education and enlightenment, must not compromise the dignity, autonomy, and diversity of the cultures it seeks to portray [23].

In the context of cultural heritage law, this issue has been a matter of considerable debate, particularly in the international arena, as the indiscriminate extension of the notion of cultural property [24] has broadened the scope of legal protections [25]. In Italy, this complex subject is governed by Legislative Decree No. 42/2004, the "*Codice dei beni culturali e del paesaggio*". According to Articles 2, 10, and 11 of this decree, cultural property includes both immovable and movable objects of artistic, historical, archaeological, ethno-anthropological, archival, and bibliographical significance, as well as those recognized by law as having civilizational value [26].

This definition, however, reveals an inherent tension. The designation of cultural property is contingent upon a legislative act that formally recognizes its cultural significance, making it difficult to assess an object's cultural value without the prior intervention of the legislature [27]. Thus, cultural property is, to a large extent, created by the legislator, within a framework where the principle of legal typicality prevails. In effect, any object that bears witness to civilization may only be legally classified as cultural property once it has been formally recognized as such by the legislature [28].

This tension is exacerbated when cultural property is transposed into the digital realm of the metaverse. Here, the traditional processes of preservation, enjoyment, and circulation of cultural goods are upended. The digital reproduction of artworks, for instance, presents an opportunity to reconcile the often-competing interests of preservation and dissemination [29]. Through digitization and virtualization, technology transcends the limitations of the physical world, offering innovative ways to preserve cultural heritage while enhancing public access and enjoyment [30].

Digitization, in particular, serves as a vital tool for ensuring broader access to cultural resources, whether housed in museums, libraries, or archives [31]. The increased availability of cultural goods in the metaverse enhances their expressive potential, guaranteeing the widest possible usability of cultural heritage. However, this broadened access also brings into focus legal issues regarding the reproduction and exploitation of images of works of art, particularly those in the public domain [32].

Italian law provides little clarity regarding the so-called "freedom of panorama", which pertains to the use and reproduction of publicly visible works of art or architecture [33]. While works in the public domain should theoretically be freely usable by the community once copyright protections have expired, museums and other institutions often impose restrictions on the reproduction and use of such works, claiming ownership rights or exclusive rights to their images [34].

These practices highlight a tension between the rights of cultural institutions and the public's right to access and enjoy cultural heritage. In the metaverse, this issue becomes even more pronounced, as the digital reproduction of works must navigate the delicate balance between intellectual property protections and the community's right to access shared cultural goods [35].

Thus, the metaverse emerges as both a platform of boundless technological innovation and a crucible for cultural and legal challenges. It invites users to envision a future where cultural heritage is not relegated to the shadows of history, but rather thrives as a dynamic part of our collective human experience. Yet, as charting this new digital frontier, one must remain ever mindful of the legal and ethical responsibilities that accompany the preservation and celebration of shared cultural legacy.

3 Opportunities and Challenges for Creative Expression with AI

In the rapidly evolving landscape of the digital age, artificial intelligence (AI) emerges not merely as a tool but as a transformative force reshaping the essence of cultural heritage and creative expression. This technological advancement fosters a symbiotic relationship between human creativity and machine intelligence, creating fertile ground for artistic exploration and cultural dialogue [36].

The opportunities and potential offered by artificial intelligence are numerous and far-reaching. AI has the capacity to enhance the efficiency of systems within an archaeological park, automate repetitive tasks, and bolster decision-making processes in times of emergency. It can transform the experience of visitors to a museum, archive, or library [37]. Through the analysis of high-resolution images of paintings, frescoes, and artworks, artificial intelligence can detect signs of deterioration, guiding experts in the strategic planning of targeted restoration interventions [38].

Moreover, AI enables the automatic translation of historical (and non-historical) texts into various languages, facilitating global accessibility. It allows for the creation of virtual models of archaeological sites or historical centers that have suffered damage, preserving their essence in digital form [39].

In addition, artificial intelligence can support complex archival research, uncovering hidden correlations, patterns, trends, and stylistic influences, revealing new insights into the works themselves, the artists, and the contexts in which they were created. The use of AI, thus, opens doors to profound discoveries, enabling a deeper understanding of cultural heritage while safeguarding it for future generations [40].

Of particular interest are the applications of artificial intelligence in integrated security systems, designed to enhance both efficiency and effectiveness through the analysis of data streams generated by various IoT devices [41]. These include surveillance cameras, intrusion detection systems, access control mechanisms, and sensors for environmental monitoring. For instance, AI can process data from biological sensors, such as gas or radiation detectors, to swiftly identify the presence of biological or environmental threats [42].

In this context, artificial intelligence further offers the ability to analyze the preferences and interests of visitors to cultural sites, comparing them with the behavior of past tourists. Through this analysis, AI can provide personalized recommendations, guidance, and tailored suggestions, thereby enhancing the accessibility and enjoyment of the experience for all users [43].

As the boundaries of creativity are continually redefined, AI presents unprecedented opportunities for artists to engage with a multitude of cultural motifs, allowing for the reinterpretation and reimagining of these elements in ways that resonate profoundly with contemporary audiences [44].

The advent of AI-driven tools facilitates a renaissance of cultural expression, empowering creators to synthesize varied influences and meld disparate cultural elements into cohesive and original works. This confluence of technology and artistry engenders a rich tapestry of artistic expression that reflects the complexities of modern human experience [45].

As artists employ AI to explore new creative horizons, they are afforded the unique ability to traverse cultural boundaries, drawing from a vast array of traditions, histories, and practices. This broad engagement not only enriches individual artistic outputs but also cultivates a deeper understanding of the diverse cultural narratives that shape society [46].

Moreover, the iterative nature of AI invites a novel process of experimentation wherein creators engage in an ongoing dialogue with the technology [47]. This interaction fosters a dynamic environment for continuous feedback and exploration, allowing artists to refine their visions and challenge established norms [48].

However, as the integration of AI in artistic practices unfolds, it becomes imperative to navigate this terrain with discernment and ethical mindfulness [49]. While the potential for innovation is significant, it is essential to remain vigilant regarding the cultural contexts from which inspirations are drawn. The elevation and celebration of the voices of marginalized communities are critical to ensuring that their narratives and artistic expressions are not co-opted or diluted amid the rapid pace of technological advancements [50].

Furthermore, a critical examination of cultural appropriation is paramount in this discourse. The ease of access to diverse cultural elements through AI could inadvertently lead to the commodification of cultural heritage. Artists and technologists alike must engage in ethical practices that honor the origins of cultural expressions, fostering collaborations that prioritize authentic representation and mutual respect [51].

This commitment to ethical engagement is essential for ensuring that the flourishing of creativity within the metaverse does not come at the expense of the very cultures that inspire it. The potential risks of cultural appropriation and misrepresentation necessitate a thoughtful approach to the use of AI in creative contexts, underscoring the importance of ethical considerations in artistic practice [52].

In light of these challenges, it becomes clear that the intersection of artificial intelligence and cultural heritage presents a remarkable opportunity to enrich the landscape of human creativity. Embracing these technological advancements invites a vision of the future where art serves not only as a reflection of individual expression but also as a powerful conduit for cultural dialogue and understanding [53]. This endeavor calls for the cultivation of an inclusive and respectful creative landscape that honors the rich diversity of shared human experience, ensuring that the stories, traditions, and identities of all people are not merely protected but celebrated [54].

Ultimately, the establishment of AI-enhanced creative practices within the realm of cultural heritage represents a crucial step toward preserving and promoting these legacies in a manner that aligns with values of inclusivity and respect [55]. As society navigates this uncharted digital frontier, the responsibility lies in crafting a future that honors the complexity and richness of cultural heritage, safeguarding it while paving the way for innovative expressions that reflect the diversity of human experiences. This requires an ongoing commitment to dialogue among stakeholders—including artists, technologists, cultural practitioners, and policymakers—ensuring that all voices are heard and that the resulting frameworks for artistic expression are equitable and just [56].

In this quest, the metaverse can emerge as a testament to collective heritage—a digital space where cultural expressions flourish, are celebrated, and contribute to the ongoing dialogue that enriches the tapestry of human creativity [57]. By harnessing the potential of AI to deepen cultural engagement, artists can forge new pathways for expression that resonate with audiences across the globe [58].

In doing so, they can contribute to a more vibrant cultural landscape, one that acknowledges and honors the multiplicity of human experience while embracing the potential for innovation and transformation. As the intersection of AI and cultural heritage continues to unfold, it remains vital to uphold the values of ethical engagement, inclusivity, and respect, ensuring that the creative endeavors of the present serve to illuminate and enrich the legacies of the past [59].

Considering as stated so far, artificial intelligence is undoubtedly powerful and adaptable, but yet it remains unable to replicate certain uniquely human capabilities [60]. There are, in fact, several inherent limitations in AI's performance, particularly within the realm of professional tasks: AI cannot (as of yet) create, conceptualize, or manage complex strategic planning in the same way a human can; it lacks the ability to feel or engage with empathy and compassion, which are essential elements of human interaction [61]. Moreover, it is improbable that people will universally opt for interactions with robots over human-to-human engagement [62].

Regrettably, the rapid pace at which technological, business, and digital transformations are reshaping the way we work also introduces new domains of risk—risks to which traditional businesses were not previously exposed. Among these areas of concern are:

- security, as systems become more interconnected and vulnerable to cyber threats;
- the safeguarding of privacy, especially in terms of the personal data involved;
- the responsible and transparent use of the data collected, ensuring that it is not exploited or misused;
- the ethical and moral use of these new technologies, ensuring they serve the greater good;

- digital ethics as a whole, a broad field that demands constant attention and adaptation to evolving technologies [63].

It is crucial to note that security, privacy, and digital ethics cannot merely be retrofitted into a project after its inception; they must be woven into the very fabric of the development process from the outset. Failure to do so risks exposing businesses and institutions to unintended consequences, potentially undermining trust and integrity in the systems they create [64].

One of the current drawbacks of AI is that it requires access to vast amounts of data, often including personal and sensitive information. This reliance raises serious concerns about privacy and data security, particularly when dealing with highly valuable or culturally significant information, such as works of art or historical documents [65]. The potential for misuse or breach of this data necessitates the highest standards of data protection, as any lapse could compromise not only individual privacy but the cultural heritage itself [66].

To mitigate these risks, it is essential to adopt an ethical approach when applying AI to cultural heritage. This requires a deliberate focus on ensuring that the use of AI respects the integrity of both the data and the cultural artifacts it interacts with [67].

Furthermore, promoting awareness and understanding of AI among cultural professionals and the broader public is of paramount importance. This fosters a more responsible and informed application of this technology, encouraging its use in ways that enhance rather than diminish cultural preservation and engagement [68].

3.1 The Legal and Ethical Dimensions of Human-AI Co-Creation

"Combinational creativity" is an intrinsic trait of human cognition, yet it remains one of the most formidable challenges for artificial intelligence to emulate. While AI endeavors to replicate the intricate processes of idea formation that unfold within the human mind—mirroring the architecture and dynamics of the nervous system—it grapples with the profound subtleties of tacit knowledge. This implicit understanding, shaped by context, encompasses a myriad of human nuances, expressions, customs, and habits that elude conventional binary computational logic. However, advancements in expert systems have introduced "fuzzy logic", enabling AI to engage with ambiguity, vagueness, and multi-interpretational assertions in a manner that aspires to approximate human-like reasoning [69].

The advent of AI as a collaborative partner in artistic and cultural creation introduces profound questions regarding authorship, ownership, and the broader legal and ethical implications of co-created works [70]. Unlike traditional tools, AI systems possess the capacity to contribute creatively and autonomously, often generating content that transcends the original input provided by human collaborators. This dynamic introduces complexities that existing intellectual property (IP) frameworks, designed with human ingenuity and agency in mind, struggle to address adequately [71].

Thus, the intersection of AI and human creativity necessitates a re-evaluation of legal definitions, the redistribution of rights, and the ethical principles guiding the creative process.

At the core of the legal debate is the question of authorship: who is the rightful creator when AI and humans collaborate? Traditional copyright law operates on the premise of a single, conscious author—typically a human being—whose originality and creative effort confer ownership rights [72]. However, the involvement of AI challenges this model, especially when its contributions are substantial and autonomous. Should the AI system be recognized as a co-author, or does the credit rest solely with the human who provided the initial input or guidance? Legal systems worldwide have yet to reach a consensus, and approaches vary across jurisdictions [73].

Further complicating the issue is the role of AI developers. The algorithms and datasets that enable AI to generate artistic works are themselves the products of human ingenuity, often involving significant investment and technical expertise. Should developers or companies that create these systems hold partial ownership over the resulting works? This question becomes particularly pertinent when AI-generated outputs draw on vast datasets, including copyrighted materials, raising concerns about derivative works and fair use. Establishing clear guidelines around data provenance and usage rights is crucial to prevent potential legal conflicts and ensure that creators' original contributions are respected [74].

Beyond the legal dimensions, human-AI co-creation raises critical ethical questions about the values embedded in AI-driven works [75]. One key concern is the potential for the commodification of cultural heritage [76]. AI systems often draw from extensive databases of historical and cultural artifacts, enabling creators to produce works that blend or reinterpret elements from diverse traditions. While this offers exciting possibilities for innovation, it also risks cultural appropriation, particularly when the output of the AI commodifies or distorts elements of marginalized cultures without proper attribution or respect for their original context. To address this, ethical frameworks must prioritize transparency and inclusivity, ensuring that the voices and contributions of all cultural communities are honored [77].

Moreover, the collaborative process must safeguard human agency. While AI can augment human creativity, there is a risk that the technology could overshadow or diminish the role of the human artist [78]. Maintaining clear boundaries and ensuring that human creators retain control over the final output is essential to preserve the integrity and authenticity of artistic expression. This requires not only technical safeguards but also a cultural shift that emphasizes the importance of human oversight and ethical responsibility.

As the landscape of human-AI collaboration evolves, establishing adaptive and equitable frameworks will be essential. Policymakers, legal experts, technologists, and cultural stakeholders must engage in ongoing dialogue to address these challenges.

The goal is to create a balanced ecosystem where AI serves as a catalyst for innovation without undermining the rights, agency, or cultural integrity of human creators. By fostering a multidisciplinary approach that integrates legal, ethical, and cultural considerations, society can harness the transformative potential of human-AI co-creation while safeguarding the values that underpin artistic and cultural heritage.

3.2 The Problem of Ownership and Exercise of Rights

Generative art, in its manifold expressions, gives rise to works that can undoubtedly be deemed 'original', imbued with 'creativity', and, in numerous instances, marked by 'novelty'. As it is well established, the legal framework governing authorship necessitates 'originality' as a prerequisite for an intellectual creation to receive protection. This constitutes the foundational threshold for creativity [79], which may be further expanded through legal provisions that recognize, first, the safeguarding of works derived from pre-existing creations (Art. 2.3, Berne Convention for the Protection of Literary and Artistic Works, 9 September 1886), and second, the protection afforded to collections of works by various authors (Art. 2.5, Berne Convention) [80].

The concept of creativity admits both subjective and objective interpretations [81]: subjectively, an 'original' creation manifests the distinctive imprint of its creator; objectively, it must exhibit a certain degree of 'novelty'. Given that no author creates *ex nihilo*, despite what certain copyright statutes may implicitly suggest, but rather draws upon a shared cultural heritage and pre-existing works, this reality gives rise to the risk of minor alterations being made to prior creations and subsequently misrepresented as wholly 'original', when, in truth, they may constitute little more than the inadvertent appropriation of another's intellectual labor [82].

While both interpretations of the originality requirement present inherent challenges, national and European legislators have largely favored the subjective approach. This preference is grounded in the principle that the law extends protection to collections and compilations of pre-existing works, wherein the author's distinct contribution lies in their 'selection' and 'arrangement' of the materials [83]. In such cases, no manifest objective novelty is present, yet the work is nonetheless recognized as a 'new work'. Thus, it is argued that the legislator merely demands a 'minimum threshold of creative effort', which is reflected in the deliberate curation of content and the intellectual structuring, ordering, or presentation of it—constituting what is termed 'the author's own intellectual creation' [84]. These acts of selection and organization bear the personal imprint of the creator and thereby warrant legal protection. However, this creative input must extend beyond what is commonplace, routine, or self-evident. A mere chronological or alphabetical arrangement, or an arbitrary compilation of others' works devoid of meaningful coherence, would fail to constitute a 'new work' and would thus remain outside the scope of protection [80].

In this respect, the decisive factor in satisfying the originality requirement lies in the creative process itself, rather than in the final outcome. Indeed, when contrasted with the 'classical' model still enshrined in copyright legislation in Europe and the United States—where the author is perceived as an isolated figure, conjuring original works from pure inspiration [85]—advancements in artificial intelligence, digital technologies, and the Internet suggest a far more fluid and interactive paradigm. Within this evolving framework, the author is no longer an isolated creator but an active participant in an ongoing dialogue with the public, collaborators, and technological systems. The model of authorship shaped by the 'network of networks' thus shifts the emphasis from the individual creator to the creative process itself. Technology facilitates a state of perpetual evolution, wherein the act of creation is never conclusively finished but remains in constant flux—subject to refinement, transformation, and continuous enhancement [86].

When an algorithm generates a work, the creative process it undergoes may closely resemble, or even surpass, the mechanisms employed by the human mind in transformational and exploratory creativity, raising the question of whether such algorithmically produced outputs merit copyright protection. In such instances, the author's 'personal imprint' is so effectively emulated—if not exceeded—by the algorithm that it challenges traditional notions of authorship [87]. By contrast, works grounded in natural creativity, an area where AI remains far from replicating human ingenuity, would continue to fall outside the realm of copyright protection [88].

Thus, only those works produced by a human author, provided they exhibit at least a minimal degree of creative effort, would unquestionably qualify for protection. Conversely, where an algorithmic process replicates the human creative process with near fidelity, the question of protection hinges upon whether the algorithm itself can be deemed to possess an 'imaginative' faculty. This is more commonly observed in transformational creativity and evolutionary art than in exploratory or merely generative art, and least of all in cases where creativity is purely combinatorial—a process in which algorithms adhere strictly to predefined instructions, operating mechanically with little to no learning capability or capacity for independent innovation [89].

A fundamental dilemma thus emerges: under existing copyright paradigms, the term work is intrinsically linked to human authorship, excluding creations brought into existence by machines or even animals, regardless of their apparent 'creativity'. This raises the question of whether algorithmically generated outputs should be classified under an alternative conceptual category distinct from traditional 'works'.

If, as previously stated, copyright law is predicated upon the notion of authorship by a natural person [90], then any creation stemming from the spontaneous activity of an animal [91], an autonomous machine, or an algorithm falls outside the purview of intellectual property protection and, by default, enters the public domain [92]. However, where the presence of a human author remains fundamental, copyright law does, in certain circumstances, permit a presumption of authorship by a legal person. Under French, Italian, and Spanish legal frameworks, such a presumption arises in cases of so-called joint works [93], wherein authorship may be attributed either to a natural or legal person, or in the case of computer programs, where authorship may be vested in a collective entity or even an institution [94].

A parallel approach is found in the United States under the 'work made for hire' doctrine [95], whereby the employer of the individual responsible for the creative work is deemed its author. This distinction between the de facto and the *de jure* author is crucial: while in the first case the author is the individual who produces the work, in the second one, the author—usually the employer—holds the rights of exploitation [96].

Thus, whether through legal fiction or presumption, copyright law recognizes exceptions to the principle that only a natural person, as de facto author, may hold exclusive rights over a work. From this premise, one might infer that protection could extend to works generated by algorithms, provided the creative process emulates human cognition. Computer scientists argue that this occurs predominantly in instances of exploratory and transformational creativity, where a legal fiction may be invoked to assign authorship rights. In such cases, the de jure author might be the individual or entity commissioning the algorithm or repurposing an existing one for creative ends. This entity would, in turn, hold the moral and economic rights associated with the work, mirroring the structure applicable to human-authored works. The de facto author, in this scenario, would be the machine or algorithm itself [97].

Certain common-law jurisdictions have already codified a framework for this interpretation [98], recognizing computer-generated works as those produced by a computational system in circumstances where no human author can be identified. Under these statutes, the individual or entity responsible for making the necessary arrangements for the creation of such works is deemed the legal author [99].

If authorship cannot, by legal definition, be ascribed to an expert system, then—given the foundational premise of copyright—the autonomous output of an algorithm, absent human involvement, would default to the public domain. Alternatively, a *sui generis* right could be introduced to ensure compensation for those who invest substantial human and financial resources in the development of expert systems or intelligent agents. Under such a scheme, any individual or entity acquiring such a system for creative purposes would be required to remunerate its developer [100].

The implementation of such compensatory measures could be managed through automated electronic systems, mitigating the risk of discouraging investment in technological innovation.

This approach would foster continued advancements in AI-driven creativity while safeguarding incentives for both technological progress and artistic development [101].

4 Legal Frameworks for Protecting Cultural Assets: Balancing Innovation and Preservation

The advent of the metaverse marks a transformative era in the digital landscape, where the question of legal protection for cultural assets becomes a matter of paramount importance. Within this virtual space, users can buy, sell, and exchange goods, fostering an innovative economy that parallels real-world markets. Yet, the legal implications of such transactions, particularly regarding virtual property rights, remain complex, especially when ownership of a work is divided between its author and another party [102].

In such cases, the reproduction of a work requires the agreement of both the author and the proprietary owner, a scenario most commonly encountered with museum collections, many of which, particularly in Italy, have entered the public domain following the expiration of copyright. Despite this, state ownership and the absence of copyright constraints have not always guaranteed free reproducibility. On the contrary, public legislation governing cultural heritage, traditionally structured around a proprietary model, has often imposed exclusive control over the reproduction and use of such works [103].

As early as 1965, Italian law granted the administration the authority to regulate the lucrative use of cultural assets, a provision reaffirmed in subsequent legislation [104].

These regulations enabled the imposition of fees for the reproduction of cultural goods, reserving gratuitousness only for personal or academic purposes [105]. However, with the emergence of the metaverse, cultural assets can now be disseminated and appreciated independently of their physical form, transforming how such works are experienced [106].

This shift was codified in 2014 with Law No. 106, which amended Article 108 of the Cultural Heritage Code, "liberating" the reproduction and dissemination of cultural assets [107].

The law permitted the free reproduction of these assets—whether online, on social media, or through other digital means—without prior authorization, provided the use was non-commercial. Despite these advancements, however, the reform could have been more far-reaching, as the reproduction of cultural goods remains, in certain cases, contingent upon prior authorization and the payment of a fee [108].

The evolution of these legal frameworks, therefore, calls for a reexamination of copyright laws, intellectual property rights, and the mechanisms by which cultural heritage is preserved and protected in this unprecedented digital landscape [109]. As we venture deeper into the metaverse, the rapid evolution and fluidity of the digital realm present unique challenges for safeguarding intellectual property and cultural heritage. In this space, where boundaries blur and creations can be replicated with ease, the need for robust legal frameworks becomes ever more pressing. Existing legal paradigms often fail to account for the distinctive nature of digital creations, necessitating a comprehensive reconsideration of how such assets are regulated [110].

Beyond mere legal protections, there is also a need to address the ethical dimensions of digital representation and the potential for cultural misappropriation in this new virtual domain [111]. Legal safeguards must extend beyond ownership, encompassing principles of ethical engagement that respect the origins and significance of cultural expressions. This holistic approach to legal frameworks will not only protect intellectual property but will also nurture a respectful and equitable digital ecosystem where cultural heritage is revered rather than commodified [112].

Ultimately, the establishment of effective legal frameworks in the metaverse represents a vital step toward ensuring that cultural heritage is preserved and promoted in a manner aligned with values of inclusivity and respect. As this digital frontier expands, those tasked with crafting its legal architecture bears the profound responsibility of safeguarding the legacies of the past while paving the way for the cultural innovations of the future. In doing so, the metaverse can be envisioned as a testament to our collective heritage—a virtual space where the stories, traditions, and identities of all peoples are not only protected but celebrated [113].

5 Conclusive Remarks

In conclusion, the intersection of cultural heritage, the metaverse, and artificial intelligence weaves a compelling narrative rich with both opportunity and responsibility, inviting profound reflection on the collective role in shaping the future of our cultural landscapes. As humanity navigates this uncharted territory, it becomes imperative to cultivate a discourse that values not only creativity and innovation but also inclusivity and the diligent protection of our cultural legacies. The metaverse, as an expansive digital canvas, offers remarkable potential for the diverse stories of humanity to flourish, yet it simultaneously demands engagement with these narratives characterized by the utmost respect and ethical consideration.

Embracing the transformative potential of technology unlocks new dimensions of artistic expression and cultural engagement, fostering connections that transcend geographical and temporal boundaries [114].

The ability of AI to synthesize and reinterpret cultural motifs empowers creators to transcend traditional limitations and craft works that resonate deeply with contemporary audiences. However, as we harness these innovative tools, we must remain vigilant guardians of the sanctity of cultural heritage. This vigilance is crucial to ensure that our creative endeavors do not succumb to the pitfalls of commodification and appropriation, which threaten to dilute the significance of cultural expressions [26].

It is our collective responsibility to honor the voices of marginalized communities, amplifying their stories and ensuring that their contributions to the rich tapestry of human experience are recognized and celebrated. In a world increasingly mediated by digital technologies, the risk of cultural misrepresentation and appropriation looms large, making ethical engagement not merely a choice but a necessity. By prioritizing authenticity and respect in creative practices, stakeholders can forge pathways that elevate and empower those historically underrepresented in cultural narratives.

Moreover, as the trajectory into this digital realm unfolds, there arises an urgent need for advocacy regarding legal frameworks that protect cultural assets while simultaneously promoting collaboration and creativity. Such frameworks should not only safeguard the rights of creators but also cultivate an environment where cultural heritage can thrive in all its forms, evolving alongside the innovations characteristic of the digital age. The establishment of equitable legal structures can enable artists to engage with their cultural heritage responsibly, ensuring that artistic expression remains rooted in respect for its origins [115].

In envisioning the future, the aspiration must be to create a metaverse where our collective history transcends the confines of mere relics of the past and emerges as a vibrant, dynamic presence that continually resonates within the consciousness of all who inhabit it. This digital tapestry ought to reflect the complexities and richness of our shared human experience, celebrating diversity and fostering mutual understanding among disparate cultures. By providing a platform for diverse narratives to intersect, the metaverse can become a space for genuine cultural dialogue, promoting empathy and connection across differing backgrounds.

Ultimately, as stewards of both culture and technology, humanity is presented with a unique opportunity to shape a future that harmonizes the advancements of the digital age with the timeless essence of our cultural heritage. In this endeavor, it is vital to proceed with a spirit of collaboration, innovation, and reverence, ensuring that the stories we tell and the legacies we uphold are imbued with meaning and purpose. In this pursuit, we can not only preserve the past but also illuminate the path forward, crafting a narrative that resonates with the hearts and minds of future generations [26].

Looking forward, the possibilities are expansive. The confluence of AI, cultural heritage, and the metaverse presents opportunities for creating immersive experiences that engage audiences in novel ways. Virtual and augmented reality technologies can provide interactive platforms through which individuals can experience cultural heritage firsthand, fostering deeper appreciation and understanding. Furthermore, these technologies may allow for the digitization and preservation of endangered cultural practices, ensuring their survival for future generations.

As stakeholders embark on this transformative journey, ongoing dialogue among artists, technologists, cultural custodians, and policymakers will be essential. Such collaboration can lead to innovative solutions that bridge the gaps between tradition and modernity, allowing cultural heritage to flourish in a digital landscape characterized by continuous change. By nurturing a creative ecosystem that respects and honors the richness of our shared past, society can craft a metaverse that not only entertains but also educates, inspiring a deeper engagement with the multifaceted dimensions of human culture.

In closing, the intersection of cultural heritage, the metaverse, and artificial intelligence invites an optimistic vision for the future—one where the stories of all peoples are woven together in a vibrant, interconnected digital tapestry. It is within this shared narrative that humanity can find strength, resilience, and a renewed commitment to the values that underpin our collective existence, ensuring that as we move forward, we do so with a profound sense of purpose and a commitment to honoring the legacies that have shaped us.

References

1. Be permitted a reference to Zambardino F (2024) Preserving identities in the "Digital Realm": safeguarding cultural heritage in the Metaverse. Aedon 1, p. 71 ff.

2. De Meo R (2021) Quale disciplina per accesso e uso delle immagini digitalizzate di opere d'arte in pubblico dominio?. In Cortese B. (ed.) Il Diritto dei Beni Culturali, p. 189. Roma Tre Press, Rome. On the matter of audience diversification, as a direct consequence of the multifaceted purposes that the museum institution is able to realise through technological fruition, Guerzoni G, Stabile S. (2013) I contenuti museali digitali. In Negri-Clementi G, Stabile S (eds.), Il diritto dell'arte, II, La circolazione delle opere d'arte, p. 217, Skira, Milan. The authors, in particular, highlight how digitisation of works of art can increase cultural activities of education, teaching, conservation and access to heritage.
3. For a conceptual analysis of the metaverse see Perboli G, Borroni A, Pratali S, Vandoni C (2024) Metaverso e Web 3.0. In Borroni A (ed.), Le nuove frontiere dei beni culturali: il metaverso. Percorsi tra diritto, cultura e tecnologia, p. 39 ff, Tipografia Gutenberg, Salerno; Zanenga P, Perboli G (2024) Metadominio: un nuovo modello economico per il Metaverso. In Borroni A (ed.), Le nuove frontiere dei beni culturali: il metaverso. Percorsi tra diritto, cultura e tecnologia, *op.cit.*, p. 267 ff.
4. Zambardino F (2024) La rappresentazione dei beni culturali nel metaverso: alla ricerca di un equilibrio tra creatività e tutela giuridica. In Borroni A (ed.), Le nuove frontiere dei beni culturali: il metaverso. Percorsi tra diritto, cultura e tecnologia, *op.cit.*, p. 53 ff.
5. See for an in-depth analysis, Sarzana D'Ippolito F (2022) Il diritto del metaverso, p. 3 ff, Giappichelli, Turin; Moro Visconti R, Cesaretti A (2022) Il metaverso tra realtà digitale e aumentata: innovazione tecnologica e catena del valore. In Dir. Internet, fasc. 3, p. 627–634.
6. Zambardino F (2024) Preserving identities in the "Digital Realm": safeguarding cultural heritage in the Metaverse. *op.cit.*, pp. 71–72. See generally on the opportunities of participation deriving from new technologies Piccinali M (2023) Il patrimonio artistico culturale e la promozione virtuale globale. In Piccinali M, Puccio A, Vasta S (eds.), Il metaverso. Modelli giuridici e operativi, p. 103, Giuffrè, Milan; Colombo ME (2020) Musei e cultura digitale, 2020, Giuffrè, Milan; De Meo R (2019) La riproduzione digitale delle opere museali fra valorizzazione culturale ed economica. Dir. Infor.; Sandulli MA (2019) Codice dei beni culturali e del paesaggio, Giuffrè, Milan; Sbarbaro E (2016) Codice dei beni culturali e diritto d'autore: recenti evoluzioni nella valorizzazione e nella fruizione del patrimonio culturale. Riv. Dir. Ind.; Stanzione G (2019) Beni culturali, realtà aumentata e nuove tecnologie dell'informazione: profili giuridici. Comp. Dir. Civ.
7. Zambardino F (2024) La rappresentazione dei beni culturali nel metaverso: alla ricerca di un equilibrio tra creatività e tutela giuridica. *op.cit.*, pp. 53–54.
8. See on this point Frosini TE (2022) L'orizzonte giuridico dell'intelligenza artificiale. Dir. Infor., p. 7 ff.
9. De Marco F (2024) Le nuove frontiere dei Beni Culturali. La tecnologia del metaverso. In Borroni A (ed.), Le nuove frontiere dei beni culturali: il metaverso. Percorsi tra diritto, cultura e tecnologia. *op.cit.*, p. 77 ff.
10. Floridi L (2022) Metaverse, a Matter of Experience. Philosophy & Technology, p. 4 ff, although, as the author himself explains, it is possible that this choice is still uneconomical today, because every company operating in its metaverse tries to attract as many users as possible to its location, to the exclusion of others.
11. Zambardino F (2024) Preserving identities in the "Digital Realm": safeguarding cultural heritage in the Metaverse. *op.cit.*, p. 73. For a general overview of these new technology see Zambardino F (2024) The Applications of the Blockchain Technology. A Comparative Law Study. ESI, Naples; Poncibò C (2020) Il Diritto Comparato e la Blockchain. In Memorie del Dipartimento di Giurisprudenza dell'Università di Torino, ESI, Naples; Borroni A (2019) Blockchain: Uses and Potential Value. In Borroni A (ed.), Legal Perspective on Blockchain. Theory, Outcomes, and Outlooks, p. 1–55, ESI, Naples; Borroni A (2019) Bitcoin e blockchain: un'analisi comparatistica dalla nascita alla potenziale regolamentazione, pp. 275–314, IANUS, Siena. Be also permitted a reference to Zambardino F (2022) La blockchain e la protezione dei dati personali: una tecnologia privacy compliant by design?. In EJPLT, p. 136–152; Zambardino F (2019) Blockchain and Regulatory Issues. In Borroni

A (ed.), Legal Perspective on Blockchain. Theory, Outcomes, and Outlooks. *op.cit.*, pp. 87–107. For a overview on cryptocurrencies, see Pernice C (2020) Criptovalute, tra legislazione vigente e diritto vivente. In Gimigliano G, Il diritto alla prova dei cryptoasset, IANUS, Siena; Vardi N (2015) "Criptovalute" e Dintorni: alcune considerazioni sulla natura giuridica dei bitcoin. Dir. Inf.

12. Perboli G, Borroni A, Pratali S, Vandoni C (2024) Metaverso e Web 3.0. *op.cit.*, p. 39 ff. On the subject of contractualisation through digital platforms in the strict sense of the term see Alpa G (2022) Sul potere contrattuale delle piattaforme digitali. Contr. Impr., n. 3; Pardolesi R (2021) Piattaforme digitali, poteri privati e concorrenza. Dir. Pubb., n. 3; Aceto di Capriglia S (2019) Contrattazione algoritmica. Problemi di profilazione e prospettive operazionali. L'esperienza «pilota» statunitense Federalismi.it; Resta G (2018) Digital Platforms and the Law: Contested Issues. MediaLaws, n. 1; Sica S, Parisi AG (2011) La tutela del consumatore nel contratto online. In Gambino AM (ed.), Rimedi e tecniche di protezione del consumatore, Giappichelli, Turin.
13. See generally Riccio GM (2023) Metaverso, logiche proprietarie e poteri privati. In Cassano G, Scorza G (2023) Metaverso, p. 85 ff, Pacini Editore, Rome. Of the same author Riccio GM (2023) Il metaverso e la necessità di superare i dogmi proprietari. Dir. Internet, pp. 223–243
14. Sarzana D'Ippolito F (2022) Il diritto del metaverso, *op.cit.*, p. 38 ff.
15. For a general discussion of the subject, please refer to Stanzione P (2022) I "poteri privati" delle piattaforme e le nuove frontiere della privacy, Giappichelli, Turin.
16. Zambardino F (2024) Preserving identities in the "Digital Realm": safeguarding cultural heritage in the Metaverse. *op.cit.*, pp. 74–75.
17. Cerrina Feroni FG (2022) Il metaverso tra problemi epistemologici, etici e giuridici. MediaLaws, p. 29 ff.
18. On these issues, see, generally Aceto di Capriglia S (2018) Gli illeciti on line e le nuove frontiere della responsabilità civile nell'era digitale. Federalismi.it.
19. Zambardino F (2024) Preserving identities in the "Digital Realm": safeguarding cultural heritage in the Metaverse. *op.cit.*, p. 75.
20. Zambardino F (2024) La rappresentazione dei beni culturali nel metaverso: alla ricerca di un equilibrio tra creatività e tutela giuridica. *op.cit.*, pp. 54–55.
21. De Marco F (2024) Le nuove frontiere dei Beni Culturali. La tecnologia del metaverso. *op.cit.*, p. 78 ff. These immersive experiences allow diverse cultural stories to flourish, providing a powerful platform for the representation of marginalized voices and underrepresented cultures.
22. Mastrolia P (2024) La tutela penale dei Beni Culturali nel metaverso. Profili de iure condito e prospettive de iure condendo. In Borroni A (ed.), Le nuove frontiere dei beni culturali: il metaverso. Percorsi tra diritto, cultura e tecnologia. *op.cit.*, p. 91 ff.
23. *Ibid*.
24. On the definitional problem, without any claim to completeness, Miranda A (2004) Introduzione alla "tutela privatistica dei beni ambientali e culturali". In Miranda A (ed.), La Tutela Privatistica dei Beni Ambientali e Culturali tra Prescrizione e Cooperazione. Modello inglese e prospettive italiane, pp. 1–20, CEDAM, Padua; Demuro GP (2002) Una proposta di riforma dei reati contro i beni culturali. Riv. It. Dir. Proc. Pen., n. 4, pp. 1358–1369; Pitruzzella G (2000) La nozione di bene culturale (artt. 1, 2, 3 e 4 d.lg. 490/1999). Aedon, 2011; Sciullo G (2010) I beni. In Barbati C, Cammelli M, Sciullo G (a cura di), Diritto e gestione dei beni culturali, pp. 21–52, Zanichelli, Bologna; Alpa G *et al.*, I beni culturali nel diritto - Problemi e prospettive, ESI, Naples.
25. Di Bonito F (2021) Il problema definitorio del «bene culturale» tra 'reale' e 'dichiarato': consapevolezza della culturalità e ricadute penalistiche. In Cortese B (ed.), Il Diritto dei Beni Culturali, p. 51, Roma Tre Press, Rome.
26. Zambardino F (2024) Preserving identities in the "Digital Realm": safeguarding cultural heritage in the Metaverse. *op.cit.*, p. 76.
27. As already stated in consolidated doctrine, with reference to Cerulli Irelli V (1988) Beni culturali, diritti collettivi e proprietà pubblica. In Scritti in onore di M. S. Giannini, vol. 1, pp. 135–178, Giuffrè, Milan.

28. Di Bonito F (2021) Il problema definitorio del «bene culturale» tra 'reale' e 'dichiarato': consapevolezza della culturalità e ricadute penalistiche. *op.cit.*, p. 62.
29. De Meo R (2019) La riproduzione digitale delle opere museali fra valorizzazione culturale ed economica. *op.cit.*, p. 190. See also Negri-Clementi G, Stabile S (eds.), Il diritto dell'arte, II, La circolazione delle opere d'arte, cit., pag. 228 ff. The authors note how digitization of collections in the analog world has created new opportunities for exchange and creative reuse, enabling the community to explore and interact with cultural heritage in innovative ways. Cultural institutions are thus transformed from custodians of analog collections into providers of digital services.
30. Zambardino F (2024) Preserving identities in the "Digital Realm": safeguarding cultural heritage in the Metaverse. *op.cit.*, pp. 76–77.
31. For an analysis on the subject of copyright protection with reference to the web from a comparative perspective see Pollicino O (2013) Tutela del diritto d'autore e protezione della libertà di espressione in chiave comparata. Quale equilibrio su Internet?. In Pizzetti F (ed.), Il caso del diritto d'autore, I diritti nella "rete" della rete, p. 97 ff, Giappichelli, Turin.
32. On the matter, Resta G (2011) L'immagine dei beni. In Resta G (ed.), Diritti esclusivi e nuovi beni immateriali, pp. 568–569, Giappichelli, Turin.
33. Magnani P (2020) Profili di tutela del diritto d'autore nella creazione di cataloghi digitali del patrimonio culturale: la protezione della banca dati e la protezione dei contenuti. Aedon, n. 3:7
34. Resta G (2011) L'immagine dei beni. *op.cit.*, pp. 557–558.
35. Magnani P (2020) Profili di tutela del diritto d'autore nella creazione di cataloghi digitali del patrimonio culturale: la protezione della banca dati e la protezione dei contenuti. *op.cit.*, p. 8 ff. See also more generally Ricolfi M (2001) Il diritto d'autore. In Abriani N, Cottino G, Ricolfi M, Diritto industriale. In Cottino G (ed.), Trattato di Diritto Industriale, p. 458 ff, CEDAM, Padua; Aliprandi S (2017) Vincoli alla riproduzione dei beni culturali oltre la proprietà intellettuale. Arch. Calc., p. 93 ff.
36. Pansoni S et al. (2023) Artificial Intelligence and Cultural Heritage: design and assessment of an ethical framework. In The International Archives of the Photogrammetry, Remote Sensing and Spatial Information Sciences, Volume XLVIII-M-2, p. 1149 ff. For a general overview of the regulation of Artificial Intelligence see Casonato C, Fasan M, Penasa S (eds) (2022) Diritto e Intelligenza Artificiale. DPCE Online, n. 1; Finocchiaro G (2022) La regolazione dell'intelligenza artificiale. Riv. Trim. Dir. Pubbl., n. 4; Pajno A, Bassini M, De Gregorio G et al. (2019) AI: profili giuridici Intelligenza Artificiale: criticità emergenti e sfide per il giurista. BioLaw Journal – Rivista di BioDiritto, n. 3; Resta G (2022) Cosa c'è di "europeo" nella proposta di regolamento UE sull'intelligenza artificiale?. Dir. Infor., n. 2.
37. See for a general overview of the impact of artificial intelligence on the experience of visitors Caramiaux B (2023) AI with Museums and Cultural Heritage. AI in Museums, Transcript Verlag, pp. 117–130, De Gruyter, Berlin.
38. On this argument, see amplius Mishra M, Lourenço PB, Artificial intelligence-assisted visual inspection for cultural heritage: State-of-the-art review. J. Cult. Heritage, pp. 536–550.
39. Wang N, Zhao X, Wang L, Zou Z (2019) Novel system for rapid investigation and damage detection in cultural heritage conservation based on deep learning. J. Infrastruct. Syst. Chatbots, equipped with AI, can provide a wealth of information, respond to inquiries, and even offer personalized virtual tours, enriching the cultural experience for visitors both near and far.
40. Pansoni S et al. (2023) Artificial Intelligence and Cultural Heritage: design and assessment of an ethical framework. *op.cit.*, p. 1150. This can inform conservation and preservation efforts and aid in decision-making processes. Moreover, AI can assist in the creation of digital models of cultural heritage, allowing for accurate reconstructions and simulations. This combination of geomatics and AI also enables the development of intelligent systems for the interactive and immersive exploration of digital cultural heritage, enhancing the user experience and promoting cultural awareness. On the argument see Vasic I et al. (2022) Virtualization and vice versa: A new procedural model of the reverse virtualization for the user behavior tracking

in the virtual museums. In Extended Reality: First International Conference, XR Salento 2022, Lecce, 2022, Proceedings, Part II, pp. 329–340, Springer, Heidelberg.

41. Mishra M, Lourenço PB, Ramana GV (2022), Structural health monitoring of civil engineering structures by using the internet of things: a review. J. Build. Eng., p. 103954 ff.
42. Mishra M, Lourenço PB, Artificial intelligence-assisted visual inspection for cultural heritage: State-of-the-art review. *op.cit*., p. 541–542. Drones, too, can contribute to this network of security, with AI generating alarms when critical thresholds are breached, identifying suspicious or abnormal behaviors – such as visitors entering restricted areas or leaving suspicious objects – and promptly triggering mitigation measures to prevent potential harm to both people and property. Generally, on this point Ciampa E, De Vito L, Pecce MR, Practical issues on the use of drones for construction inspections. In Journal of Physics: Conference Series, vol. 1249, IOP Publishing, Bristol.
43. Caramiaux B (2023) AI with Museums and Cultural Heritage. AI in Museums, Transcript Verlag. *op.cit*., p. 121 ff. This not only improves the individual visitor's journey but also contributes to an enriched, more engaging overall experience, expanding the cultural offering to a broader audience.
44. Abgaz Y, Rocha Souza R, Methuku J, Koch G, Dorn A (2021) A methodology for semantic enrichment of cultural heritage images using artificial intelligence technologies. J. Imaging, p. 121 ff.
45. Pansoni S et al. (2023) Artificial Intelligence and Cultural Heritage: design and assessment of an ethical framework. *op.cit*., pp. 1150–1151.
46. On this point, see amplius Trier ØD, Reksten J, Løseth K (2021) Automated mapping of cultural heritage in Norway from airborne lidar data using faster R-CNN. Int'l J. Applied Earth Observation and Geoinformation, p. 95 ff; Argyrou A, Agapiou A (2021) A Review of Artificial Intelligence and Remote Sensing for Archaeological Research. Remote Sensing, p. 6000 ff.
47. On which see amplius Arzberger A, Lupetti MR, Giaccardi E (2024) Reflexive Data Curation: Opportunities and Challenges for Embracing Uncertainty in Human-AI Collaboration. ACM Trans. Comput. Hum. Interact., p. 1 ff.
48. Heikkilä M (2023) Machine Learning in Interaction: Tool, Material, Culture. MIT Tech. Rev. According to the author, through this collaborative relationship, AI serves not only as a facilitator of creativity but also as a catalyst for innovation, pushing the boundaries of what is considered possible in artistic endeavors.
49. On the ethical issues, see Abulibdeh A (2024) Navigating the confluence of artificial intelligence and education for sustainable development in the era of industry 4.0: Challenges, opportunities, and ethical dimensions. J. Clean. Prod.; Gupta H et al. (2021) Industry 4.0, cleaner production and circular economy: an integrative framework for evaluating ethical and sustainable business performance of manufacturing organizations. J. Clean. Prod.
50. Pansoni S et al. (2023) Artificial Intelligence and Cultural Heritage: design and assessment of an ethical framework. *op.cit*., pp. 1152–1153. In this regard, the creative use of AI must be approached with a profound sense of responsibility, recognizing the historical legacies and cultural significances that inform the very motifs being employed.
51. On the commodification of cultural heritage, see generally Craith MN, Cultural Heritages: Process, Power, Commodification. In Kockel U, Craith MN (eds.), Cultural Heritages as Reflexive Traditions, pp. 1–18, Springer, London.
52. Macneill P (2014) Ethics and the Arts, pp. 1–8, Springer, London.
53. Bordoni L, Mele F, Sorgente A (eds) (2016) Artificial Intelligence for Cultural Heritage. Cambridge Scholars Publishing, Cambridge
54. Mulholland P et al. (2016) Modelling Museum Narratives to Support Visitor Interpretation. In Bordoni L, Mele F, Sorgente A (eds.) Artificial Intelligence for Cultural Heritage. *op.cit.*, pp. 3–24.
55. Macneill P (2014) Ethics and the Arts. *op.cit*., pp. 1–8.
56. *Ibid*.

57. Zambardino F (2024) La rappresentazione dei beni culturali nel metaverso: alla ricerca di un equilibrio tra creatività e tutela giuridica. *op.cit.*, pp. 57–58.
58. Chi J (2024) The Evolutionary Impact of Artificial Intelligence on Contemporary Artistic Practices. Comm. Hum. Res., pp. 6–11.
59. *Ibid.*
60. Shai George A (2024) Artificial Intelligence and the Future of Work: Job Shifting Not Job Loss. Partners Universal Innovative Research Publication (PUIRP), p. 17 ff.
61. Alawamleh M (2024) Examining the limitations of AI in business and the need for human insights using Interpretive Structural Modelling. Journal of Open Innovation: Technology, Market, and Complexity, vol. 10, Issue 3.
62. *Ibid.*
63. Zambardino F (2024) Preserving identities in the "Digital Realm": safeguarding cultural heritage in the Metaverse. *op.cit.*, pp. 75–76.
64. Pascuzzi G, Izzo U, Macilotti M (eds.) (2013) Comparative Issues in the Governance of Research Biobanks. Property, Privacy, Intellectual Property, and the Role of Technology. Springer, Heidelberg-New York-Dordrecht-London.
65. See generally Resta G (2012) Le fotografie delle catacombe e la proprietà intellettuale. Dir. Infor. p. 845 ff.
66. *Ibid.*
67. Pansoni S et al. (2023) Artificial Intelligence and Cultural Heritage: design and assessment of an ethical framework. *op.cit.*, p. 1150 ff.
68. *Ibid.*
69. Navas S (2020) Creativity of Algorithms and Copyright Law. Cambridge University Press, p. 223.
70. Deserti A, Real M, Schmittinger F (2022) Co-creation for Responsible Research and Innovation. Experimenting with Design Methods and Tools. Spinger, New York.
71. Resta G (2012) Le fotografie delle catacombe e la proprietà intellettuale. *op.cit.*, p. 845.
72. Söderberg J (2017) Comparing two cases of outlaw innovation: file sharing and legal highs. In Godin B, Vinck D (eds.) Critical Studies of Innovation. Alternative Approaches to the Pro-Innovation Bias, pp. 115–136, Edward Elgar Publishing, Cheltenham.
73. Deserti A, Real M, Schmittinger F (2022) Co-creation for Responsible Research and Innovation. Experimenting with Design Methods and Tools. *op.cit*. For instance, in some regions, such as the European Union, copyright is only granted to works that reflect human creativity, explicitly excluding AI-generated content. In contrast, discussions in other countries, including the United States, continue to explore whether the role of AI in creation merits its own form of legal recognition.
74. Söderberg J (2017) Comparing two cases of outlaw innovation: file sharing and legal highs. op.cit., pp. 118–120.
75. On which Rodrigues R (2020) Legal and human rights issues of AI: Gaps, challenges and vulnerabilities. J. Resp. Econ.
76. Siliutina I, Tytar O, Barbash M, Petrenko N, Yepyk L (2024) Cultural preservation and digital heritage: challenges and opportunities. Amazonia Investiga, p. 264.
77. *Ibid.*
78. Eapen TT, Finkenstadt DJ, Folk J, Venkataswamy L (2023) How Generative AI Can Augment Human Creativity: Use it to promote divergent thinking. Harv. Bus. Rev., pp. 56–64.
79. Yu R (2017) The Machine Author: What Level of Copyright Protection Is Appropriate for Fully Independent Computer-Generated Works? U. Pa. L. Rev., p. 1241 ff.
80. Navas S (2020) Creativity of Algorithms and Copyright Law. *op.cit.*, p. 227.
81. Perry M, Margoni T (2010) From Music Tracks to Google Maps: Who Owns Computer-Generated Works? Comp. L. & Sec. Rev.
82. Rahmantian A (2011) Copyright and Creativity. The Making of Property Rights in Creative Works. Edward Elgar Publ.
83. Art. 2.5, Berne Convention.

84. Margoni T (2016) The Harmonisation of EU Copyright Law: The Originality Standard. Available at SSRN: <https://ssrn.com/abstract=2802327>. Date of access: March 2025.
85. Grimmelmann J (2016) Copyright for Literate Robots. Iowa L. Rev., p. 657. The author, in particular, states that «Copyright's ideal of romantic readership involves humans writing for other humans [...] Copyright ignores robots».
86. Navas S (2020) Creativity of Algorithms and Copyright Law. *op.cit.*, p. 228.
87. On this point, Navas S (2016) Creation and Witticism in the User-Generated Online Digital Content. In Actas de Derecho Industrial, pp. 403–415.
88. Eapen TT, Finkenstadt DJ, Folk J, Venkataswamy L (2023) How Generative AI Can Augment Human Creativity: Use it to promote divergent thinking. *op.cit.*, p. 59 ff.
89. Navas S (2020) Creativity of Algorithms and Copyright Law. *op.cit.*, pp. 228–229.
90. See on this point § 2(2) of the *Urheberrechtsgesetz* in Germany, that considers that only works consisting of '*persönliche geistige Schöpfungen*' can be considered objects of protection; Art L 111–1 of the French *Code de la propriété intellectuelle*, that alludes to '*ouvrages de l'esprit*', which implies that these are created by man. In the same sense, see the wording of s 9 UK Copyright, *Design and Patent Act* (1988).
91. On which see Guadamuz A (2016) *The monkey selfie: copyright lessons for originality in photographs and internet jurisdiction.* Internet Pol'y Rev., p. 1 ff.
92. Art. 18 Berne Convention.
93. In France, see Art L 113–2 *Code la propriété intellectuelle*; in Italy Art 7 *Legge di protezione del diritto d'autore e di altri diritti conessi al suo esercizio* (23 April 1941); in Spain Art 8 *Ley de propiedad intelectual.*
94. Art 2.1 Directive 2009/24/EC of the European Parliament and Council, 23 April 2009, on the legal protection of computer programs. Furthermore, when authorship occurs within an employment relationship, the rights in the resulting program typically vest in the employing entity.
95. § 201(b) U.S. Copyright Act.
96. Lee E (2012) Digital Originality. Vanderbilt J. Ent. & Tech. L., p. 919 ff.
97. Samuelson P (1985) Allocating Ownership Rights in Computer-Generated Works. U. Pitt. L. R., p. 1224 ff.
98. In particular, the reference is to the UK *Copyright, Designs and Patents Act* (1988), § 9(3), along with corresponding provisions in the New Zealand *Copyright Act* (1994), paragraphs 2 and 5; the Irish *Copyright and Related Rights Act* (2000), Part I, § 2 and Chapter 2, § 21; and the South African *Copyright Act* (1978), No. 98.
99. Navas S (2020) Creativity of Algorithms and Copyright Law. *op.cit.*, pp. 231–232. However, the precise scope of these provisions remains open to interpretation: the phrase necessary arrangements does not unambiguously confirm the eligibility of fully autonomous, human-independent creations for protection. The nature of these arrangements – whether they require direct human intervention or may be executed by an expert system – remains unresolved, raising fundamental questions about whether an algorithm alone can be deemed the guiding agent in the creative process. See also Schafer B, Komuves D, Niebla Zatarin JM, Diver L (2015) A Fourth Law of Robotics? Copyright and the Law and Ethics of Machine Co-production. Artif. Intell. L., pp. 217–240.
100. Navas S (2020) Creativity of Algorithms and Copyright Law. *op.cit.*, p. 232. A precedent for this approach can be found in the legal framework governing databases, where protection mechanisms operate independently of traditional authorship rights. Directive 96/9/EC of the European Parliament and Council, 11 March 1996, on the legal protection of databases (OJL 77, 27.3.1996, 20–28). See also McCutcheon J (2013) Curing the Authorless Void: Protecting Computer-generated Works Following ICETV and Phone Directories. Melbourne U. L. Rev., p. 46 ff, who argues against this solution and in favour of recognising, at least in Australian law, work created by an algorithm as original and attributing the author's rights to an individual or organisation.
101. Samuelson P (1985) Allocating Ownership Rights in Computer-Generated Works. *op.cit.*, p. 1224 ff.

102. Forte P (2019) Il bene culturale pubblico digitalizzato: prime note per uno studio giuridico. Pers. Amm., p. 283. See also Resta G (2011) L'immagine dei beni. *op.cit.*, p. 585. The author bases his conclusion on a comparative jurisprudential analysis observing that "the image of movable or immovable property that is freely visible is not subject to any restriction on reproduction and exploitation by third parties, with the exception of any protection deriving from the right to privacy or intellectual property rights that may insist on the property. Resta G (2012) Le fotografie delle catacombe e la proprietà intellettuale. *op.cit.*, p. 845.
103. De Meo R (2021) Quale disciplina per accesso e uso delle immagini digitalizzate di opere d'arte in pubblico dominio?. *op.cit.*, p. 194.
104. Modolo M (2018) Promozione del pubblico dominio e riuso dell'immagine del bene culturale. Arch. Calc., p. 76, reconstructs the parliamentary affair of law no. 340/1965, which led to the choice to make the authorisation regime prevail over the free reproducibility regime: in that circumstance, the alleged reasons of physical protection of the property prevailed, but above all, and this was even more interesting, the intention to limit, through the inhibiting power of authorisations, the multiplication of photographs.
105. These articles expressly establish the limits within which cultural goods may be reproduced. On this point, widely, Galli C (2021), Tutela e valorizzazione dei beni culturali pubblici e privati attraverso la proprietà intellettuale. Dir. Ind.
106. Morbidelli G (2014) Il valore immateriale dei beni culturali. In Bartolini A, Brunelli D, Caforio F (eds.), I beni immateriali tra regole privatistiche e pubblicistiche, p. 171 ff, Jovene, Naples; Barile S (2012) Verso una nuova ipotesi di rappresentazione del concetto di bene culturale. In Golinelli G (ed.), Patrimonio culturale e creazione di valore. Verso nuovi percorsi, p. 71 ff, CEDAM, Padua; Manfredi G (2017) La tutela proprietaria dell'immateriale economico nei beni culturali. Dir. Econ., p. 29 ff; Forte P (2019) Il bene culturale pubblico digitalizzato: prime note per uno studio giuridico. *op.cit.*, pp. 245–301.
107. De Meo R (2021) Quale disciplina per accesso e uso delle immagini digitalizzate di opere d'arte in pubblico dominio?. *op.cit.*, p. 198.
108. Aliprandi S (2017) Vincoli alla riproduzione dei beni culturali oltre la proprietà intellettuale. *op.cit.*, p. 93 ff.
109. Zambardino F (2024) Preserving identities in the "Digital Realm": safeguarding cultural heritage in the Metaverse. *op.cit.*, p. 77.
110. On this point, an extensive examination has been carried out by Bartolini A (2014) L'immaterialità dei beni culturali. Aedon. The crafting of these legal frameworks requires an inclusive and collaborative dialogue among policymakers, legal scholars, cultural practitioners, and technologists. Such collaboration is essential in establishing legislation that not only protects but also promotes the flourishing of cultural heritage within the metaverse, ensuring that creators and custodians retain their rights over their works. Moreover, it is crucial that these discussions include marginalized voices to ensure that laws and regulations reflect the diverse perspectives and needs of all communities.
111. For a general overview, please refer to F. Zambardino F (2024) La rappresentazione dei beni culturali nel metaverso: alla ricerca di un equilibrio tra creatività e tutela giuridica. *op.cit.*, pp. 58–59.
112. On this point, for a comparative analysis Caso R (2018) Il diritto d'autore accademico nel tempo dei numeri e delle metriche. Trento Law and Technology Research Group Research Paper n. 36; Pascuzzi G (2017) Il fascino discreto degli indicatori: quale impatto sull'università?. Foro It., p. 2549 ff. But also, among the first on the topic, Pardolesi R, Motti C (1990) «L'idea è mia!»: lusinghe e misfatti dell'economicis of information. Dir. Infor., p. 345 ff; Pardolesi R (1988) «Software» di base e diritto d'autore: una tutela criptobrevettuale?. Foro It., p. 3133 ff.
113. Zambardino F (2024) Preserving identities in the "Digital Realm": safeguarding cultural heritage in the Metaverse. *op.cit.*, pp. 78–79.
114. On this point, an extensive examination has been carried out by Bartolini A (2014) L'immaterialità dei beni culturali. Aedon, p. 1.

115. On this point, for a comparative analysis Caso R (2018) Il diritto d'autore accademico nel tempo dei numeri e delle metriche. Trento Law and Technology Research Group Research Paper n. 36; Pascuzzi G (2017) Il fascino discreto degli indicatori: quale impatto sull'università?. Foro It., p. 2549 ff. But also, among the first on the topic, Pardolesi R, Motti C (1999) "L'idea è mia!": lusinghe e misfatti dell'economicis of information. Dir. Infor., p. 345 ff; Pardolesi R (1988) "Software" di base e diritto d'autore: una tutela criptobrevettuale?. Foro It., p. 3133 ff.

Contracts in the Metaverse: Between Virtual Avatars, Instant Clicks, and New Legal Frontiers

Andrea Russo

Abstract The emergence of the Metaverse has introduced new forms of digital interaction where economic exchanges and agreements are concluded through avatars, clicks, and virtual interfaces. This chapter explores how traditional contract principles adapt to virtual environments, where users negotiate, buy, and sell digital goods and services. It examines the challenges of consent, identity verification, and enforceability when agreements are made within immersive platforms. The discussion highlights real-world examples of disputes arising from virtual transactions and analyzes how legal frameworks may need to evolve to balance user protection with the need for innovation and seamless digital experiences. The chapter advocates for simple, clear contractual models suitable for the Metaverse's dynamic nature, to avoid overregulation while safeguarding fundamental rights.

Keywords Metaverse · Contract · Law · Digitization

1 Introduction: New Deals in Virtual Worlds

The progressive dematerialization of human activities, catalyzed by the exponential growth of digital technologies, has culminated in the rise of the Metaverse: an intricate amalgamation of virtual environments where users interact, transact, and construct complex social and economic architectures [1].

This paradigm shift marks a profound transformation not only in the way individuals communicate and engage with one another, but also in the underlying structures that support interpersonal and commercial relationships across digital domains. What was once confined to isolated digital interactions has now evolved into expansive, immersive ecosystems that mirror, and at times transcend, real-world dynamics.

Unlike traditional online platforms, the Metaverse offers a persistent, immersive reality wherein individuals, represented by avatars, undertake actions that can give

A. Russo (✉)
Università Degli Studi Della Campania "Luigi Vanvitelli", Caserta, Italy
e-mail: andrea.russo@unicampania.it

A. Borroni et al. (eds.), *Metaverses: Reshaping Law, Economy, and Society*, PoliTO Springer Series, https://doi.org/10.1007/978-3-032-08664-8_15

rise to legally relevant consequences. The digital persona, no longer a mere intermediary, becomes a conduit for agency and intention, embodying the user's will and capacity to act within a sophisticated, often gamified, environment. This evolution prompts the reconsideration of foundational legal concepts and raises profound challenges for conventional legal theory, particularly with regard to the identification of parties, the expression of consent, and the enforceability of obligations [2].

he redefinition of these core elements becomes all the more pressing as interactions within virtual spaces increasingly approximate, or even substitute, real-life exchanges.

In this novel dimension, contracts are often concluded through the mere activation of a virtual command, the click of a mouse, or the movement of a digital proxy—gestures that, while seemingly ephemeral, may embody the intentionality and agreement necessary to generate binding legal effects [3]. The apparent informality and immediacy of such acts belie their potential to fulfill the substantive requirements of contractual formation. Consequently, the ontological shift from tangible signatures and verbal negotiations to intangible, avatar-mediated interactions necessitates a re-examination of the criteria by which agreements are recognized, validated, and enforced under existing legal frameworks [4].

This challenge is compounded by the fluidity and spontaneity inherent in many Metaverse environments, where the boundaries between play and commerce, between fiction and juridical relevance, are often indistinct.

Moreover, the Metaverse blurs the traditional demarcations between public and private spaces, between commercial and social spheres, thereby creating hybrid contexts in which the spontaneous emergence of contractual relationships is both facilitated and obscured [5]. Within these overlapping domains, the architecture of virtual platforms often enables rapid, informal exchanges that may possess legally binding implications, even when such outcomes are neither intended nor foreseen by the users involved. The informal nature of these exchanges, combined with the relative anonymity of users and the global, decentralized structure of many virtual platforms, complicates the application of jurisdictional norms, consumer protection laws, and dispute resolution mechanisms [6].

These complexities necessitate a careful reconsideration of existing legal tools and their capacity to adapt to contexts where normative expectations are continuously being reshaped.

Considering the premises, this chapter endeavors to explore the complexities inherent in the formation and regulation of contracts within the Metaverse, emphasizing the need to preserve the essential principles of private law—such as autonomy of will, good faith, and the protection of weaker parties—while adapting their practical application to the specificities of virtual environments. These values, which constitute the bedrock of contractual justice in the offline world, must be recalibrated to address the distinct modalities of interaction that characterize digital realities, without diluting their normative force.

It is posited that legal systems must develop interpretative and normative tools capable of addressing the *sui generis* nature of Metaverse transactions without stifling the innovative potential of these digital ecosystems. Through a careful analysis of practical cases, doctrinal considerations, and prospective regulatory approaches, the discussion aims to contribute to the ongoing dialogue concerning the evolution of contract law in the era of virtual realities.

2 Forming Agreements Through Avatars

The formation of contracts within the Metaverse necessitates a re-examination of traditional legal doctrines, particularly concerning the manifestation of consent [7]. The act of agreeing, which in physical contexts often involves explicit verbal declarations or handwritten signatures, acquires a novel form within virtual environments, where users engage through digital proxies—avatars—interacting via interfaces that mediate reality and agency in complex ways. In these spaces, agreements are frequently concluded through simplified yet legally consequential mechanisms such as clicking buttons, selecting digital assets, or navigating virtual pathways designed by platform developers. These actions, while seemingly trivial or game-like in nature, may carry significant legal implications when assessed through the principle of functional equivalence, which recognizes electronic actions as valid expressions of intent under certain conditions [8].

This conceptual bridge between digital conduct and legal effect serves as a cornerstone in the effort to transpose established legal norms into the evolving architecture of the Metaverse.

The jurisprudence surrounding electronic contracts provides a foundational framework for understanding these dynamics and offers valuable analogies for the regulation of avatar-mediated agreements. Notably, in *Feldman v. Google Inc.*, the court upheld the enforceability of a clickwrap agreement, emphasizing that the user had a reasonable opportunity to review the terms before assenting by clicking "I agree" [9].

This decision underscores a broader principle: that the enforceability of digital agreements hinges not only on the act of clicking but on the fairness and clarity of the conditions under which consent is obtained. Such precedents illustrate that consent in digital contexts must be informed, voluntary, and manifest through a deliberate act of acceptance, even when traditional forms of contracting are absent.

However, the application of these principles to the Metaverse introduces additional complexities that challenge their direct transposition. The immersive and often gamified nature of virtual environments can obscure the user's understanding of the legal consequences of their actions [10]. In these visually rich and experientially dynamic settings, users may engage with content that blends entertainment and transaction, blurring the line between play and legal engagement. For instance, the act of purchasing a virtual item or entering a digital space may not be perceived as entering into a binding contract, especially when traditional indicators of contractual intent, such as written agreements or formal negotiations, are absent [11].

The architecture of these platforms may foster user behaviors that, while legally significant, are undertaken with minimal awareness of their juridical weight.

Moreover, the issue of identity in the Metaverse further complicates contract formation. Avatars, which serve as the outward representation of users in virtual spaces, may not accurately reflect the individuals behind them—either in appearance, intent, or legal capacity—raising questions about the attribution of actions and the validity of consent. This ambiguity is exacerbated by the possibility of multiple users sharing a single avatar or, conversely, a single user operating under multiple digital identities. As discussed in the literature, the potential for avatars to act autonomously or be controlled by artificial intelligence agents adds another layer of complexity to determining intent and capacity [12].

These factors necessitate a refined understanding of agency and representation in digital environments, where the visual or behavioral cues associated with intent may be highly stylized, ambiguous, or even automated.

The distinction between clickwrap and browsewrap agreements also becomes particularly pertinent in this context. Clickwrap agreements, which require users to affirmatively assent to terms by clicking a button, are generally enforceable when the terms are presented clearly and the action evidences a conscious decision to accept them. In contrast, browsewrap agreements, where terms are accessible via a hyperlink and do not require explicit consent, have been met with judicial skepticism due to concerns about adequate notice and assent [13].

In the immersive environments of the Metaverse, where attention is often diverted by visual stimuli and interactive narratives, ensuring that legal terms are both visible and understandable becomes an even greater challenge. The subtle presentation of terms may go unnoticed, and the user may unwittingly find themselves bound by obligations to which they did not knowingly agree.

Given these challenges, it is imperative for developers and platform operators in the Metaverse to implement mechanisms that ensure users are adequately informed of contractual terms and that their consent is clearly manifested. This may involve designing user interfaces that prominently display terms and require affirmative actions—such as scrolling through contract clauses, acknowledging key provisions, or confirming understanding—thereby aligning with established legal standards for contract formation. Moreover, adopting layered disclosures and real-time prompts tailored to user behavior can enhance awareness without unduly interrupting the immersive experience.

Therefore, as the Metaverse continues to evolve in complexity and scale, legal frameworks must adapt to address the unique aspects of virtual interactions. Ensuring that consent is informed, deliberate, and unequivocally expressed is not merely a formal requirement, but a substantive safeguard essential to upholding the integrity of contractual relationships in these digital spaces. The legitimacy and sustainability of economic exchanges in the Metaverse depend on the trust that such agreements are both fair and enforceable in light of the distinctive nature of virtual consent.

3 The Challenge of Identity

T he question of identity lies at the heart of legal personhood and contractual enforceability. In classical private law, the determination of the contracting parties is not a mere formal requisite but a precondition for the existence and validity of the legal relationship itself. Contracts are binding because they emanate from the will of legal subjects—natural or juridical persons—capable of bearing rights and obligations [14]. This foundational premise enables legal systems to allocate liability, enforce obligations, and uphold predictability in private dealings. Within the Metaverse, however, the traditional foundations of this framework are profoundly destabilized, as the emergence of virtual personae and disembodied interactions introduces radical indeterminacy into the identification of contracting parties. Digital environments characterized by persistent pseudonymity and user-generated representations introduce significant ambiguity as to the true identity of the individual or entity behind a given contractual act [15].

Avatars—graphical proxies manipulated by users—serve as the primary vectors of interaction in virtual worlds. These digital embodiments operate as visible stand-ins for users, facilitating both expressive and transactional functions. Yet their ontological detachment from the corporeal subject raises substantial difficulties in ascertaining whether the individual controlling the avatar possesses the requisite legal capacity to contract. In the absence of embodied presence or verified credentials, one cannot assume that the party behind the avatar is of legal age, mentally competent, or even a natural person. Indeed, the emergence of algorithmic agents, autonomous bots, and generative AI-controlled avatars further complicates the matter by blurring the line between human and machine agency [16].

Such hybrid actors challenge classical assumptions of voluntariness, consent, and accountability, which are all predicated on human intentionality and discernment.

As the European Parliament has noted, "the increasing presence of intelligent autonomous systems in daily life raises questions about legal accountability and the attribution of actions" [17].

This concern becomes particularly acute in the context of contract law, where the ability to assign responsibility for obligations, misrepresentations, or breaches depends on the capacity to attribute acts to legal persons. The absence of an identifiable natural or juridical subject behind the avatar or agent fundamentally threatens the enforceability of virtual agreements and disrupts the coherent functioning of private law norms.

Furthermore, the use of decentralized architectures—particularly blockchain-based platforms—eschews traditional identity verification in favor of cryptographic authentication [18]. A user's public key or wallet address may suffice to authorize a transaction, but it does not disclose the legal identity of the signatory. Such identifiers operate with mathematical certainty but social opacity. While such systems offer undeniable advantages in terms of security and user autonomy, they deprive counterparties—and potentially courts—of the ability to establish a concrete link between virtual behavior and legal responsibility [19].

As a result, trust becomes displaced from interpersonal recognition to systemic architecture, creating a transactional environment that is secure but potentially unaccountable.

The issue is compounded when smart contracts—self-executing code deployed on decentralized networks—are involved, as these transactions are often irreversible and may be triggered by pseudonymous parties acting anonymously [20].

Once a smart contract is deployed and conditions are met, execution proceeds automatically, irrespective of whether the initiating party can later be identified or legally pursued. This characteristic, while technologically efficient, introduces a latent vulnerability in legal terms, particularly where fraud, coercion, or incapacity may later be alleged.

The problem of identity in the Metaverse is not merely academic; it has practical and doctrinal repercussions. First, it undermines the traditional model of liability, which presupposes that wrongful or non-performing parties can be identified and held accountable. Without reliable identity mechanisms, enforcement becomes impracticable, and the legal system's deterrent function is diminished [21]. Second, it compromises contractual formation itself: one cannot enter into a legally binding agreement with an entity that cannot be identified or verified *ex post* [22]. Third, it exposes users—particularly consumers—to elevated risks of fraud, impersonation, and manipulation, with little recourse to remedies [23].

In such an environment, the asymmetry of information and power is exacerbated, and traditional safeguards designed to protect vulnerable parties may prove ineffective.

Comparative legal systems have long recognized the centrality of identity in contractual matters. For example, under Italian civil law, Articles 1325 and 1427 of the Civil Code require both identifiable parties and valid consent for a contract to be enforceable, with error *in persona* constituting grounds for annulment [24].

Similarly, under common law principles, the doctrine of mistaken identity may render an agreement void *ab initio* if the identity of the contracting party was essential to the formation of consent [25]. These doctrines, although developed in a pre-digital age, continue to reflect enduring concerns that remain salient in virtual contexts [26].

Their continued relevance underscores the fact that, despite technological innovation, the human dimension of contracting cannot be entirely abstracted from legal analysis.

Proposals to address these challenges vary in scope and ambition. Some advocate for mandatory Know Your Customer (KYC) procedures for high-value transactions in the Metaverse, drawing inspiration from existing anti-money laundering frameworks [27], as developed by the World Wide Web Consortium (W3C), which allow for verifiable, user-controlled identities anchored in blockchain infrastructure. Such solutions may provide a compromise between anonymity and accountability, enabling users to prove specific attributes (e.g., age, nationality, accreditation) without fully disclosing their legal identity [22].

These selective disclosure mechanisms aim to reconcile the privacy-enhancing features of decentralized systems with the minimum identifiability requirements of legal enforceability.

However, the implementation of such frameworks must navigate a delicate balance between privacy, user autonomy, and legal enforceability [28]. Excessive demands for identification may deter user engagement and undermine the very openness and creativity that make the Metaverse attractive [29]. Conversely, insufficient mechanisms risk fostering an unregulated domain where legal certainty is eroded, and unlawful behavior is incentivized [30].

The regulatory challenge, therefore, is not merely to impose identity checks but to design a coherent legal infrastructure that accommodates layered, context-sensitive approaches to identity—without flattening the pluralism and fluidity that characterize virtual presence.

De iure condendo, the challenge of identity in the Metaverse is emblematic of the broader tension between technological innovation and legal order. As virtual environments evolve from playful experimentation to fully-fledged economic and social ecosystems, legal systems must develop adaptive tools capable of reconciling pseudonymity with legal accountability. Such tools must remain faithful to the foundational principles of private law while embracing the plural ontologies of digital identity. Failure to do so would not only impair the enforceability of virtual contracts but may ultimately hinder the legitimacy and trustworthiness of the Metaverse as a space for meaningful human interaction.

4 The Need for Simple and Effective Rules

As the Metaverse consolidates its role not only as a locus of digital leisure but as a site of economic and contractual significance, the imperative of legal regulation becomes increasingly inescapable. What was once an experimental frontier for technologists and gamers has now evolved into a complex arena of social, cultural, and commercial activity. Yet, such regulation must be guided by a principle of proportionality: the legal architecture surrounding virtual transactions must be sufficiently robust to offer predictability and protection, while remaining sufficiently agile to preserve the fluidity, inclusivity, and spontaneity that characterize virtual ecosystems. Overregulation—particularly through the imposition of rigid, formalistic requirements inherited from traditional contract law—risks stifling innovation and alienating the very communities that have fostered the Metaverse's rapid growth. Legal regimes must be carefully calibrated to avoid chilling creativity, user participation, and platform experimentation, which are essential to the vitality of these digital environments.

It is axiomatic in comparative contract law that formal requirements should not serve as barriers to the expression of will, but as instruments of clarity and evidentiary certainty. In this regard, form should serve function, not obstruct it. Indeed, modern legal systems have progressively embraced a functionalist approach to contractual form. As articulated by Atiyah, "the law of contract is best seen as a set of rules designed to facilitate and secure exchanges, rather than to punish informality or to police morality" [31]. This conception of contract law, rooted in pragmatism and responsiveness, is particularly relevant in virtual spaces where interactions are often informal, rapid, and layered with social meaning. In this spirit, the regulation of contractual relations in the Metaverse should privilege substance over form, intent over technicality, and accessibility over legalistic precision [22].

The challenge lies in crafting rules that safeguard fairness and legal certainty without imposing disproportionate burdens on casual or non-expert users.

The average user in the Metaverse is often not a legal expert, nor is he or she always operating in a commercial or professional capacity. Rather, many participants are ordinary individuals engaging in a mix of recreational, expressive, and transactional behaviors. Transactions—ranging from the acquisition of digital clothing for avatars to the rental of virtual meeting spaces—occur at the intersection of social interaction and commercial conduct [32]. This hybrid nature of virtual exchanges complicates their legal qualification, but also underscores the need for regulatory sensitivity to user experience and context. Therefore, regulatory frameworks must recognize that many users are not repeat players in the legal system and may lack both the knowledge and resources to navigate intricate contractual mechanisms. This calls for the development of "lightweight" contractual models: simplified templates, intuitive terms and conditions, and visual or interactive aids that facilitate informed consent without overwhelming users with legalese [33].

Such models should lower cognitive and procedural barriers, enabling users to understand their rights and obligations with minimal effort or legal training.

Digital platforms and developers bear a particular responsibility in this context. The design of user interfaces (UI) and user experiences (UX) plays a pivotal role in shaping legal awareness and decision-making. In virtual worlds, the point at which a user clicks, accepts, or enters a space may be the moment a contract is formed—yet the visibility and comprehensibility of terms at that moment are often lacking. Legal design thinking—an emerging interdisciplinary field—advocates for embedding legal information in user-centered ways, enhancing comprehension and participation through visuals, plain language, and contextual cues. When well implemented, such techniques do not undermine legal enforceability but rather strengthen it, by ensuring that consent is truly informed and voluntary [32].

This approach reframes the presentation of legal content not as an obstacle course but as a communicative act between the platform and the user.

Moreover, the law should accommodate modularity and self-executing mechanisms suited to digital environments. The use of smart contracts, for instance, offers potential for clarity, speed, and reliability [34]. These coded agreements can automate performance and reduce transaction costs, especially in high-volume or low-value exchanges. However, their deployment must be accompanied by human-readable summaries and fallback mechanisms for dispute resolution, to prevent opacity and safeguard equity in the case of malfunction or misunderstanding [35]. Codification cannot be a substitute for interpretation; the contract must remain not merely a line of code, but a mutual understanding anchored in reciprocal expectations [34].

Legal enforceability depends not only on mechanical execution but on interpretive coherence and the possibility of remedial intervention when things go wrong.

Simplification must not be mistaken for deregulation. Rather, it should be understood as a normative strategy aimed at enhancing transparency, accessibility, and fairness in a context where traditional safeguards (such as legal counsel, paper trails, or face-to-face negotiations) are absent or impractical. In this regard, it is instructive to recall the European Commission's approach to consumer protection in digital contracts, which emphasizes pre-contractual clarity and digital fairness while avoiding excessive formalism [36].

This balanced model recognizes the importance of clear information and meaningful consent, especially when asymmetries of knowledge and power are pronounced, as is often the case in virtual settings.

Therefore, the effective governance of contracts in the Metaverse does not require the transposition of classical formalism into digital domains. Instead, it demands the construction of adaptive, intelligible, and interoperable legal instruments that respect the distinct nature of virtual interaction while upholding essential legal values. Such instruments should support a layered legal ecology: one that accommodates informal peer-to-peer transactions, platform-mediated agreements, and high-stakes commercial contracts alike, each according to its own risk profile and communicative needs.

Only by embracing regulatory simplicity with legal rigour can the law support the sustainable evolution of virtual economies and ensure that participation in the Metaverse remains open, fair, and meaningful. In doing so, legal systems reaffirm their role not as impediments to innovation, but as enablers of trust, inclusion, and reciprocal accountability in an increasingly dematerialized world.

5 Conclusive Remarks

The juridical challenges posed by the Metaverse are not merely circumstantial anomalies to be remedied through incremental doctrinal adjustments; rather, they constitute the outward signs of a more profound epistemological shift. Contract law—long anchored in the materiality of written documents, physical presence, and state-sanctioned enforcement—now finds itself confronted with environments that are simultaneously persistent and fluid, communal yet anonymous, global yet fragmented. In this new topography of human interaction, the foundational coordinates of contract theory—identity, consent, form, and enforceability—require not only adaptation but conceptual recalibration.

The preceding chapters have mapped some of the core tensions at the heart of this emerging legal landscape. The dematerialization of parties through avatars, the gamification of interface-driven interactions, the substitution of natural persons by algorithmic agents, and the decentralization of transactional authority all signal a rupture with the ontological premises of classical private law. The consequences are far-reaching. The absence of stable identities undermines the attribution of responsibility; the ambiguity of gestures complicates the recognition of consent; the irreversibility of smart contracts subverts traditional remedies; and the opacity of code erodes legal intelligibility. These are not technical nuisances—they are structural fractures that force us to revisit the normative scaffolding upon which contract law has historically relied.

Yet this moment of juridical rupture is not to be feared—it is to be seized. The Metaverse, in its protean and still-gestating form, offers an unprecedented opportunity to imagine a contract law that is not bound by the sedimented habits of paper-based legalism but is responsive to the social and technological affordances of digital life. Such a re-imagination must be guided by a dialectical sensibility—one that resists both technophilic utopianism and nostalgic conservatism. Law must neither fetishize innovation nor recoil from it; it must inhabit the interstitial space between continuity and transformation, mediating between the imperative to preserve legal certainty and the necessity to accommodate new forms of agency and expression.

In this light, the persistence of core contractual principles—such as autonomy of will, pacta sunt servanda, and the protection of the weaker party—must be reaffirmed not despite the rise of virtuality, but because of it. These principles embody the ethical and functional rationale of contract law: to facilitate voluntary cooperation among legal subjects, to structure expectations, and to provide mechanisms of trust and recourse. But their implementation in virtual spaces demands an act of juridical translation. Consent must be recognized not solely through signatures or verbal assent, but through gestures, interactions, and user journeys. Identity must be anchored not necessarily in civil registries, but in cryptographically verifiable, privacy-preserving credentials. Form must serve clarity and intelligibility, not ritual. And enforcement must combine code-based execution with human oversight and procedural fairness.

This reconfiguration of contract law cannot be left to private ordering alone. While platform design, smart contract coding, and decentralized governance models have a pivotal role to play, they cannot substitute the legitimizing function of public law. Regulation must provide a normative horizon within which innovation unfolds—a set of guiding principles that prevent abuse, ensure accessibility, and protect the human dignity of all participants, regardless of technical literacy or bargaining power. This regulatory framework must be modular, experimental, and iterative. It must accommodate the rapid evolution of virtual environments while ensuring a baseline of legal intelligibility and procedural justice. The use of sandbox regimes, soft law instruments, and collaborative governance initiatives may offer a viable way to balance agility with accountability.

Moreover, the challenge is not simply one of enforcement, but of imagination. Legal systems must cultivate a new jurisprudence—attuned to the symbolic and material grammars of digital interaction, capable of articulating what it means to "agree," to "contract", and to be "bound" in a space where spatiality is simulated, time is elastic, and agents may be non-human. This requires interdisciplinary insight: from computer science, to semiotics, to behavioral economics, to philosophy of law. It also requires humility. The law cannot control the Metaverse. But it can aspire to co-evolve with it—providing normative guidance without ossifying the dynamism that gives virtual spaces their generative potential.

Ultimately, the Metaverse confronts us with a choice. We may cling to inherited forms, struggling to fit the new within the contours of the old, at the risk of irrelevance.

Or we may take seriously the task of conceptual renewal, crafting legal instruments that speak the language of the digital without surrendering the ethical commitments of the legal. To do so is not to abandon legal tradition, but to fulfill its deeper vocation: to serve as a medium through which freedom, responsibility, and reciprocity are made possible—even in worlds not made of atoms, but of code.

A contract law fit for the Metaverse must be, above all, a law that listens—to users, to technologies, to values. It must be capable of abstraction without alienation, of enforcement without rigidity, and of interpretation without anachronism. It must be technical and humanist, experimental and principled, agile and fair. Only such a law will be worthy of the new realities it seeks to govern—and only such a law will ensure that, in the age of avatars and algorithms, the contract remains not a relic, but a promise.

References

1. For a conceptual analysis of the metaverse see Perboli G, Borroni A, Pratali S, Vandoni C (2024) Metaverso e Web 3.0. In Borroni A (ed.), Le nuove frontiere dei beni culturali: il metaverso. Percorsi tra diritto, cultura e tecnologia, p. 39 ff, Tipografia Gutenberg, Salerno; Zanenga P, Perboli G (2024) Metadominio: un nuovo modello economico per il Metaverso. In Borroni A (ed.), Le nuove frontiere dei beni culturali: il metaverso. Percorsi tra diritto, cultura e tecnologia, op.cit., p. 267 ff.
2. Janssen A (2024) Contract law in the metaverse. In: Di Matteo LA, Cannarsa M (eds) Research Handbook on the Metaverse and Law. Edwar Elgar Online, pp 279–291
3. Garon JM (2022) Legal Implications of a Ubiquitous Metaverse and a Web3 Future. Marq. L. Rev., p. 163 ff.
4. Sitnikov MS (2023) Law and metaverse: selected issues in theory and practice. Digit. Law J., pp. 51–71.
5. De Elizalde F (2024) Consumer protection in the metaverse. In Di Matteo LA, Cannarsa M (eds.), Research Handbook on the Metaverse and Law, op.cit., pp. 98–117.
6. Perboli G, Borroni A, Pratali S, Vandoni C (2024) Metaverso e Web 3.0. In Borroni A (ed.), Le nuove frontiere dei beni culturali: il metaverso. Percorsi tra diritto, cultura e tecnologia, op.cit., p. 39 ff.
7. Finocchiaro G (2018) Il contratto nell'era dell'intelligenza artificiale. Riv. Trim. Dir. Proc. Civ., p. 441 ff.
8. Werbach K, Cornell N (2017) Contracts Ex Machina. Duke L. J., p. 318 ff.
9. Feldman v. Google, Inc. United States District Court 513 F. Supp. 2d 229 (E.D. Pa. 2007).
10. Cheong BC (2022) Avatars in the metaverse: potential legal issues and remedies. Int. Cybersecur. Law Rev., pp. 467–494.
11. On this point, generally Zambardino F (2024) Preserving identities in the "Digital Realm": safeguarding cultural heritage in the Metaverse. Aedon 1, p. 71 ff.
12. See amplius on the topic of identity and data protection Tortorano P (2024) La protezione dei dati personali nel metaverso. In Borroni A (ed.), Le nuove frontiere dei beni culturali: il metaverso. Percorsi tra diritto, cultura e tecnologia, op.cit., pp. 149–158; Borroni A, Zambardino F (2024) The data protection in the metaverse. Revista Româna de Drept al Afacerilor - Romanian Business Law Review, pp. 17–32, Wolters Kluwer, no. 3.
13. Pereira de Lima CR (2023) Adhesion Electronic Contracts ("shrink-wrap" and "click-wrap") and the terms of use ("browse-wrap"). Brazilian J. L. Techn. & Innov., p. 108 ff.
14. Clarizia R (2018) Informatica e contratto: la identificazione dei contraenti. Dialoghi con Guido Alpa. Roma Trepress, Rome, pp 93–100

15. On this point see, Rodotà S (2018) Quattro paradigmi per l'identità. In Vivere la democrazia, p. 20 ff, Laterza, Bari; Alpa G (2017) L'identità digitale e la tutela della persona. Spunti di riflessione. Contr. Impr., p. 725 ff; Resta G (2007) Identità personale e identità digitale. Dir. Infor.
16. Finocchiaro G (2018) Il contratto nell'era dell'intelligenza artificiale. Riv. Trim. Dir. Proc. Civ., pp. 441–460.
17. European Parliament, Report with recommendations to the Commission on Civil Law Rules on Robotics, 2017/2103(INL). In doctrine see Oriti S (2025) Brevi note sulla risoluzione del Parlamento europeo del 16 febbraio 2017 concernente le norme di diritto civile sulla robotica. In Ratio Iuris, no. 4. See also in terms of civil liability Zambardino F (2025) Artificial Intelligence and law: hypothesis of civil liability. In Opinio Juris in Comparatione, no. 2, pp. 1–37.
18. For an in-depth analysis of the cryptographic authentication see amplius Zambardino F (2024) The applications of the blockchain technology. A comparative law study. Edizioni Scientifiche Italiane, Naples.
19. Ibid.
20. Cuccurru P (2017) Blockchain ed automazione contrattuale. Riflessioni sugli smart contracts. Nuova Giur. Comm.; De Caria R (2019) The Legal Meaning of Smart Contracts. Eur. Rev. Priv. L., n. 6; Martone I (2022) Gli smart contracts. Fenomenologia e funzioni, in Faro S, Lettieri S, Perlingieri C, Driritto Scienza Tecnologia, Edizioni Scientifiche Italiane, Naples.
21. Zambardino F (2025) Artificial Intelligence and law: hypothesis of civil liability. op.cit., pp. 1–37.
22. Finocchiaro G (2018) Il contratto nell'era dell'intelligenza artificiale. op.cit., pp. 441–460.
23. On these risks Zambardino F (2024) The applications of the blockchain technology. A comparative law study. op.cit., p. 29 ff.
24. Bianca CM (2019) Il Contratto. In Diritto Civile, Giuffrè, Milan; Zeno-Zencovich V (1991) Il contenuto del contratto. In Alpa G, Bessone M (eds.), I contratti in generale, in Giur. sist. dir. civ. e comm. fondata da W. Bigiavi, III, Giappichelli, Turin.
25. See Cundy v. Lindsay (1878) 3 App Cas 459. In doctrine see Thomas SR (2008) Mistake of Identity: A Comparative Analysis. In Lloyd's Maritime and Commercial Law Quarterly, pp. 188–215.
26. Mckendrick E (2022) Contract law. Oxford University Press, Oxford; Beatson J, Burrows A, Cartwright J (2020) Anson's Law of Contract. Oxford University Press, Oxford; Adams JN, Brownsword R (2004) Understanding Contract Law. Sweet & Maxwell, London; Atiyah PS (1995) An Introduction to the Law of Contract. Oxford University Press, Oxford. From the Italian perspective see Criscuoli G (2001) Il contratto nel diritto inglese. Cedam, Padua.
27. See, e.g., Directive (EU) 2015/849 on the prevention of the use of the financial system for the purposes of money laundering or terrorist financing). Others promote the integration of Decentralized Identifiers (DIDs).
28. See generally Zambardino F (2022) La blockchain e la protezione dei dati personali: una tecnologia privacy compliant by design? EJPLT, no. 2, pp. 136–152.
29. Perboli G, Borroni A, Bruni ME (2025) Developing Metaverse Use Cases: The LETO Methodology. In Borroni A, Perboli G, Piccinelli GM (eds.) Metaverses: Reshaping Law, Economy, and Society. Springer.
30. Ibid.
31. Atiyah PS (1979) The Rise and Fall of Freedom of Contract. Oxford University Press, Oxford
32. Perboli G, Borroni A, Bruni ME (2025) Developing Metaverse Use Cases: The LETO Methodology. In Borroni A, Perboli G, Piccinelli GM (eds.) Metaverses: Reshaping Law, Economy, and Society. op.cit.
33. On which see amplius Owe O, Fazeldehkordi E (2022) A lightweight approach to smart contracts supporting safety, security, and privacy. J. Log. Algebraic Methods Program.
34. Zambardino F (2024) The applications of the blockchain technology. A comparative law study. op.cit., p. 39 ff.
35. Indeed, one of the elements related to the use of smart contracts would be the substantial elimination of judicial intervention, given the impossibility of disputes arising between the

parties due to their particular operating mechanism. Amato C (2020), La 'computerizzazione' del contratto (Smart, data oriented, computable e self-driving contracts. Una panoramica). Eur. dir. priv., fasc. 4, p. 1266.
36. See Directive (EU) 2019/770 on contracts for the supply of digital content and services.

Metaverse and Tourism: Current Applications and Future Perspectives

Melissa Capelli and Giulia Roasio

Abstract Tourism is an economic sector firmly anchored in the territory, not reliant on producing physical commodities but rather on providing diverse services that facilitate a unique and unrepeatable experience for the traveller within the destination. Consequently, the concept of tourism within the metaverse understood as a virtual realm facilitated by advanced technologies such as augmented reality (AR), virtual reality (VR), and mixed reality (MR), may appear to be a rather implausible proposition. However, it should be noted that the advent of the pandemic has given rise to new forms of 'travel' that have gained increasing popularity, enabling tourists to 'visit' places from the comfort of their homes. Consequently, new technologies have precipitated the emergence of new services and novel forms of cyber-tourism, thereby signifying that tourism competition is no longer confined to a physical realm. It is incumbent upon operators and destinations to adapt to this emergent paradigm. This contribution examines the interactions between tourism and the metaverse to illuminate their potential and criticalities.

Keywords Metaverse tourism · Cyber-tourism · Smart destinations · Web 3.0 · Virtual tourism

1 Introduction

Tourism is a universal language that transcends barriers of age, gender, and geography. From young adventurers to seasoned travellers, people across the globe share a passion for exploration and cultural discovery. The desire to learn about new destinations, experiences, and stories fuels an ever-growing interest in travel.

M. Capelli (✉) · G. Roasio
University of Turin, Turin, Italy
e-mail: melissa.capelli@unito.it

G. Roasio
e-mail: giulia.roasio@unito.it

A. Borroni et al. (eds.), *Metaverses: Reshaping Law, Economy, and Society*, PoliTO Springer Series, https://doi.org/10.1007/978-3-032-08664-8_16

The global tourism industry reflects this trend. With a market size of USD 1.8 trillion in 2023 [1], tourism has evolved from a simple leisure activity to a global conversation. Travel trends are now shaped by shifting consumer preferences, technological innovations, and sustainable practices. Travellers focus not only on itineraries but also on cultural immersion and connecting with like-minded explorers. Visual storytelling through travel photography and videos has become integral as people increasingly explore destinations with their eyes first.

Over the past decade, tourism has emerged as a dominant content driver across television, OTT platforms, and social media. Whether exploring hidden gems on Instagram or following travel critiques, audiences are captivated by travel content.

The COVID-19 pandemic amplified interest in virtual travel experiences. With global lockdowns halting physical travel, people turned to virtual tours, online travel guides, and immersive VR experiences to satisfy their wanderlust [2]. Metaverse travel platform extend these innovations, creating a virtual realm where travel enthusiasts can interact with blockchain technologies, explore virtual destinations, and build a global community centered around tourism.

While travellers "*can't explore physically*" in the digital world, the Metaverse provides new channels of engagement that blend digital and physical experiences. Tourism companies must adapt to ensure consistency in their brand presence across these dimensions. For example, airlines and travel agencies can leverage AI-powered tools to create personalised itineraries, reduce planning time, and enhance customer engagement. Hospitality businesses can utilise AI to design tailored recommendations and enhance guest experiences. "*For instance, Leven, a hospitality brand, is creating an engaging and entertaining virtual hotel in the metaverse where people from all over the world can interact. Likewise, Facebook, now known as Meta, has commenced the development of metaverse concerts and metaverse tours. The BCB group launched a metaverse city project, meticulously designing renowned attractions such as the Pyramids of Giza, Niagara Falls, and the Great Wall of China, aiming to attract customers and encourage their engagement with travel and tourism products*" [3]. Social media insights can also guide travel brands in identifying emerging trends and preferences.

However, technological investments in this space require financial resources and expertise, which could pose barriers to adoption. Data ownership is another key factor—brands may struggle to deliver personalised experiences and engagement without actionable insights.

A renewed focus on health and wellness is increasingly influencing travel preferences. Wellness tourism, eco-friendly accommodations, and mindfulness retreats are gaining attraction. Mental health considerations have also become prominent, driving interest in destinations that promote relaxation and rejuvenation. Likewise, sustainable tourism is on the rise, with a growing percentage of consumers avoiding businesses perceived as socially or environmentally irresponsible. Transparency in supply chains and eco-certifications are becoming essential differentiators.

Yet, sustainability may face challenges in an inflationary environment. Rising costs may deter consumers from paying premiums for eco-conscious options, and tourism-related companies should adapt to changing consumption patterns.

Metaverse platforms aim to deliver a unique and immersive experience for travellers while acting as a powerful marketing and engagement tool for the global travel and hospitality industry. Additionally, the metaverse holds significant potential to promote education on sustainable and mindful tourism practices. Establishing a clear, sustainability-focused governance framework could enhance its impact, ensuring long-term benefits for the industry and the environment.

This chapter aims to analyse, through a holistic perspective, tourism interactions with the metaverse and its current and future applications. This perspective facilitates an analysis of the problem in terms of economic impact, trying to identify possible new business models (Sect. 2) and in terms of governance and lack of regulation, focusing on critical issues (Sect. 3). Section 4 analyses the main current applications underlying benefits and limits of metaverse tourism evolution, while Sect. 5 focuses on sustainability issues, strategies and opportunities detecting strategies and opportunities.

Section 6 offers concluding remarks and highlights further challenges warranting the tourism sector in the metaverse era.

2 Tourism in the Metaverse: New Business Models and Economic Impact

"*Tourism is a social, cultural and economic phenomenon which entails the movement of people to countries or places outside their usual environment for personal or business/professional purposes. These people are called visitors (which may be either tourists or excursionists; residents or non-residents), and tourism has to do with their activities, some of which involve tourism expenditure*" [4].

As the preceding definition indicates, the tourism sector is characterised by multiple facets and various forms of "*tourism*" (the delineation of which is often indistinct). It consists of various services that uniquely allow visitors to experience the destination. The practice of tourism has not remained static over time but has evolved, adapting to different eras and the changing needs of tourists. From essentially elite tourism, we have moved on to mass tourism, primarily experienced in seaside resorts during August, up to the most recent evolutions, linked to the spread of new technologies, which have modified its characteristics (both on the demand and supply side). The tourism industry has experienced considerable evolution in recent years, with changes in tourist behaviour resulting in shifts in demand. Contemporary trends indicate that 67% of individuals embarking on a journey undertake research on the Internet, with two-thirds of trip reservations being facilitated online. Furthermore, it has become imperative for contemporary tourists, who are often connected digitally even while on holiday, to express their experiences through feedback, comments and photographs. "*These considerations are of absolute importance for the tourism industry in general, considering the evolutions brought to the sector by the recent dynamics of new types of tourism (smart/digital tourism). The last twenty years have witnessed a radical change in the tourism sector, both quantitatively, with constant growth in the number of travellers, and qualitatively: tourists, thanks to innovation in the transport system and the consolidation of low-cost flights, are increasingly global; in their holidays they seek the local aspect, the quality of services and the authenticity of experiences, but above all, they are increasingly young and digitally interconnected and make extensive use of technology to organise their holidays*" [5]. This paradigm shift has consequently enabled consumers to assume an active role in co-creating their experiences. In this context, technological advances have facilitated the personalisation of tourist services to align with the needs and preferences of individual tourists. Concurrently, real-time information has enhanced decision-making processes [6]. This phenomenon has come to the fore in the aftermath of the pandemic. Data collated by the European Commission reveals that in 2019, the proportion of Italians who booked travel online was only 39%. Moreover, internet services at destinations were of inferior quality, and the utilisation of e-commerce was well below the European average.

The tourist offer must, therefore, change and adapt to the new characteristics of the tourist. Suppose the traditional tourist experience begins long before arriving at the destination (when planning the trip) today. In that case, this statement adds value with the possibilities of "visiting" the destination through the metaverse and augmented reality and checking in online or via mobile devices.

The proliferation and diffusion of ICT in urban infrastructures has increased interest in smart cities, whose ultimate goal is to improve the quality of services provided to citizens and, consequently, their quality of life [7]. But the evolution of tourism has not stopped there: from smart cities, we have moved on to smart destinations, which not only revolutionise the concept of tourism offers but also represent new ecosystems [8, 9] "*based on an innovative, accessible tourist space consolidated on a cutting-edge technological infrastructure that guarantees the sustainable development of the territory, smart destinations facilitate tourists' interaction and integration within the environment and improve visitors' experience with destinations as well as residents' quality of life*" [10]. Territories cannot ignore the role of the intangible elements that drive technological and digital innovation. The links between the various players in the smart destination are thus multiplying. While this leads to greater integration of the tourism product, from a purely IT point of view, each player introduces new vulnerabilities for the entire chain and, in turn, for the ICT product created by the chain itself. The development of technology, therefore, affects not only consumers and tour operators but also destinations, increasing their competitiveness. Technology and tourism are closely linked, not only in terms of new services that can be offered locally but also in the sector, which can explore new avenues and potentials thanks to the metaverse. As mentioned, even though tourism is strongly linked to the territory, many destinations are already "visited" virtually: competition is a game played on a ground that is no longer just physical. For example, thanks to virtual reality (VR) and augmented reality (AR), tourists can take a virtual 'tour' of a city, a museum or a natural landscape. This can help travellers make more informed decisions about where to go and what to see, giving them a 'taste' before they book their trip. This is just one application of the metaverse in the tourism sector, but there may be many others. Before assessing the different applications of the metaverse within the tourism sector, it seems useful to briefly digress what the metaverse is and outline its evolution. The notion of the metaverse was first theorised in the 1992 science fiction text Snow Crash. "*In the world of Snow Crash, every citizen is a user, connected to the metaverse through their avatars. All users explore the metaverse to engage in social interactions with friends, family, or colleagues, find amusement or relaxation in digital recreational spaces, and even practice their professions or manage their businesses and earn real-world income. Within the metaverse, entities (known as agents) exist that offer a range of services and functions. These agents access real-world databases and process data with unimaginable speed, enabling users to obtain information on any subject or receive guidance on any aspect of their lives*" [11]. As a computer-generated universe, the metaverse concept has been defined through various concepts, including lifelogging, collective space in virtuality, embodied Internet, and Omniverse. The primary utilisation of early VR technology was in military training, flight simulation, medicine, and other professional sectors, owing to its substantial cost and hardware constraints. The advancement of hardware has led to the emergence of commercial VR devices, such as the Oculus Rift and the HTC VIVE, which have catalysed the dissemination of VR technology into private households. Consequently, VR technology applications have gradually expanded into diverse fields, including the retail, energy, education, and tourism sectors. [12]. "*In tourism, the metaverse is seen as a platform through which greater experiences can be offered to travellers before, during, and after their travels while also facilitating interactions in the virtual world*" [13].

But this is not the only definition of the metaverse. Tsai defines the metaverse as "*an interoperated persistent network of shared virtual environments, engendering perception of holistic presence through avatar embodiment as well as psychological and physiological immersion*" [14]. This definition fits well with the functions the metaverse can perform in the field under consideration. The possible uses of the metaverse in tourism will be better outlined later in the discussion. At this stage, we merely point out that the metaverse can influence the various stages of the tourist experience, which can be categorised as follows [15]:

(i) Pre-trip: the phase preceding travel, characterised by meticulous planning and preparation, is represented by this stage. During this period, the metaverse makes virtual access to information about tourist attractions, transportation options, accommodation, dining establishments, and shopping centres possible. Consequently, travellers are empowered to formulate itineraries, considering budgetary constraints and tour logistics. The execution of the plan entails the acquisition of necessary transportation, such as bus, train, or flight tickets, as well as the reservation of accommodation, rental of vehicles, and other relevant services;
(ii) On-site: in this phase, visitors arrive at their destination and can conduct virtual research using the metaverse to locate subjects of interest. They are then able to disseminate their findings via social media in the form of photographs, narratives, or video recordings;

(iii) Post-trip: in this phase, the trip has been completed, and the tourist has returned home. The activities undertaken during this phase include anticipating the subsequent trip and disseminating travel experiences via social media platforms. This dissemination may take the form of photographs, videos, or written accounts detailing the experiences both en route and at the destination. The content uploaded to social media can be viewed on a screen in two dimensions or experienced in 3D via a virtual reality headset.

However, it is important to remember that not everything that appears beneficial is necessarily so. The metaverse is not without its limitations, which must be given due consideration: lack of commonality, continuity, and global standards; problems of accessibility, inclusivity, and global connectedness; low levels of social acceptability; and last but not least, the gap between the latest technologies and the requirements of the metaverse [16].

3 Metaverse and Governance: Lack of Regulation and Critical Issues

The metaverse already plays, and will increasingly play in the future, a crucial role in people's lives, serving as a transversal and pervasive element across every sphere of daily life. However, precisely what the metaverse is and its relationship with Artificial Intelligence, Blockchain, Virtual Reality, and Augmented Reality remains an unresolved question.

Numerous definitions have been proposed, often resulting from interdisciplinary approaches. While they differ in detail, they all highlight certain key characteristics: (i) digitalisation of assets, (ii) application-based scenarios, (iii) economic scope, (iv) industrial universality, (v) integrated innovation, and (vi) platform ecosystem [17]. In this context, the metaverse can be described as "*a crossroads between 'meta' (meaning transcendence) and 'universe,' depicting a persistent and immersive three-dimensional online environment (decentralised) where users, represented by avatars, can participate socially and economically in a creative and collaborative manner within virtual spaces detached from the physical world*" [18].

Nevertheless, there is no single metaverse but multiple metaverses, as numerous as the number of investors and developers creating them, each designed for vastly different purposes [19]. In the context of tourism, this implies the potential for an unlimited number of virtual worlds specifically tailored to offer unique meta-tourism experiences. These metaverses could enable users to explore cultural heritage sites, engage in virtual travel experiences, and participate in immersive environments that replicate or even transcend real-world tourism. For instance, users might virtually walk through ancient ruins, participate in local festivals recreated digitally, or interact with avatars representing local communities.

Given the plethora of technological applications already available in this sector, issues surrounding the regulation of the metaverse emerge as a priority, along with the development of innovative governance models to address the unique challenges and opportunities presented. Governance in this context must balance the facilitation of technological innovation with the imperative of protecting users and real-world society while ensuring legal certainty and sustainability. Furthermore, the question arises as to whether regulation is necessary, to what extent, and how regulation might be implemented. This regulatory issue is closely intertwined with the socio-philosophical debate on how the development of the metaverse could transform individuals, society, governance, and even our ways of thinking.

Since the Industrial Revolution, an asymmetrical, bidirectional relationship has characterised law and technology. While technology profoundly impacts the legal system, the latter has only a limited influence on technological evolution.

While significant technological advancements have brought about substantial socio-cultural changes and can be regarded as primary drivers of legal transformations [20], the law is traditionally a reactive discipline.

The dissemination of digital technologies enabled people worldwide to access unimaginable volumes of data with the advent of the World Wide Web (Web 1.0) and subsequently decentralised knowledge production with the emergence of Web 2.0. Web 3.0 and the metaverse development will further extend decentralisation to data storage [21].

Nevertheless, relegating the law to a mere descriptive science of reality would mean denying its very essence and capacity to generate norms of conduct to regulate social interactions. Indeed, the technological advancements of recent centuries have been facilitated by a reliable legal framework, within which innovation has benefited from the positive effects generated by clear and predictable rules of conduct. It is, therefore, crucial to establish a robust legal framework capable of supporting and guiding technological progress, especially in sectors such as tourism, where the economic and cultural stakes of the metaverse are particularly pronounced.

First and foremost, it is essential to identify the governance model most suited to fostering the technological evolution of the metaverse while simultaneously protecting users and real-world society. The governance models conceivable within the rule of law include regulation, co-regulation, self-regulation, and non-regulation. Typically, within the rule of law, any strategy of regulation or co-regulation necessitates a certain degree of centralisation in the production of norms. The two models differ in terms of the transparency of the legislative procedure. In both scenarios, large technology companies and, more broadly, developers of new technologies participate in the legislative process. However, in the case of regulation, this occurs through lobbying. In contrast, co-regulation occurs within an institutionalised co-governance framework, ensuring greater transparency and more efficient resource allocation. Conversely, self-regulation entrusts the creation of minimum normative standards to the developers or service providers, with the risk of creating regulatory atomisation and fragmentation [22]. Finally, the extreme alternative of non-regulation poses challenges for developers and service providers and constitutes a legal certainty risk.

Establishing legal norms within the Metaverse could begin with projecting the laws of physical society onto electronic social relations.

This strategy appears to have been chosen by the European Union, which, with the Digital Services Package, including the Digital Service Act [23] and Digital Markets Act [24], aims to regulate online platforms according to the principle "*what is illegal offline must also be illegal online*". The ambition of this regulation package is vast, aiming at ensuring the protection of fundamental rights for users, limiting the spread of illegal goods, content, and services, protecting consumer rights, and promoting competition within the digital single market. However, given the global nature of the Metaverse, the mere application of national laws would not ensure effective regulation of the system. The decentralisation of the virtual world in the Web 3.0 era introduces new and diverse legal challenges, chief among them being determining which jurisdiction applies. One of the first hurdles legal systems will face in the age of the metaverse will be to decide whether a "traditional" concept of jurisdiction, tied to nation-states and international institutions, will persist or whether this will evolve to accommodate a new e-jurisdiction [25]. Such a framework would regulate interactions developed in the metaverse while potentially departing definitively from local, national, or international legal paradigms [26].

In tourism, this challenge becomes particularly relevant, as determining jurisdiction is crucial for resolving disputes over intellectual property, liability for virtual travel experiences, and consumer protection standards for services offered within the metaverse. Furthermore, creating a unified e-jurisdiction model could provide clarity and coherence in addressing cross-border issues inherent in virtual tourism ecosystems.

From this perspective, ad hoc anti-monopoly and antitrust laws should be developed to prevent the dangerous centralisation of metaverse development [27]. Developing a robust legal framework in this regard necessitates rethinking the principles governing the protection of intellectual property rights. While the current framework does provide certainty to technological developers by safeguarding their inventions against unauthorised use by third parties, thereby incentivising technological advancement, it also hinders generalised access to technological innovations. This, in turn, slows potential further developments and increases the public's access cost.

Moreover, envisioning each metaverse as a multidimensional and immersive online environment with decentralised jurisdiction, where users, developers, and administrators are represented by avatars, an additional critical issue concerns the interplay between the metaverse and legal personality. In defining a legal system to regulate the metaverse—or the metaverses—it is essential to determine which entities should be granted legal personality. Indeed, not only human beings, with their data and right to privacy, may interact with the metaverse through avatars, but also legal entities and even emerging technologies such as Artificial Intelligence may create avatars and engage in virtual social relations.

In tourism contexts, this raises questions about the legal recognition of virtual tour operators, AI-guided tourism avatars, and the protection of user data in these virtual interactions. Balancing the interests of users, service providers, and developers requires nuanced legal approaches to ensure trust and transparency in virtual tourism transactions.

Connected to this issue is the question of whether legal situations within the metaverse can be attributed by treating avatars as persons within the metaverse itself, or whether avatars—or a series of avatars—must necessarily be linked to a physical or, potentially, legal person. Furthermore, if a correlation between a person and an avatar is required, what rights and obligations can be transferred from the person to the avatar and vice versa? To what extent will the avatar's existence be tied to that of the person? What legal effects produced in the metaverse may have tangible consequences in the real world? What are the conditions for attributing civil and criminal liability for actions occurring in the metaverse? Will the principle that "*what is illegal offline is also illegal online*" prevail? [28].

Finally, other critical issues for achieving a systematic legal framework for the metaverse include identifying new intangible assets and property rights [29] and the socio-economic links between the metaverse and the real world, establishing the framework for sovereignty, defining mechanisms for the protection of human rights, and designing an institutional system aimed at ensuring long-term sustainability. Within meta-tourism, such a framework must account for the dynamic interplay of cultural preservation, economic innovation, and user engagement, ensuring that the tourism industry can thrive in virtual spaces while adhering to ethical and legal standards.

4 Current Applications: Benefits and Limits of Metaverse in the Tourism Sector

As indicated earlier, the potential of the metaverse in tourism has been previously delineated, and it has been demonstrated how it can impact the different stages of the tourism experience. In this section, a concerted attempt will be made to highlight the various practical applications of these new technologies to the sector, highlighting their potential benefits and limitations.

A wide range of solutions is represented, including the Internet of Things (IoT), cloud computing, artificial intelligence, mobile devices and applications, big data, Wi-Fi, virtual reality, augmented reality, chatbots, wearable devices, QR codes, near-field communication (NFC), radio frequency identification (RFID), social networks and beacons [10]. Other instruments are digital twins that predict tourist trajectories and blockchain, which has the potential to completely change the tourism industry by providing innovative solutions to its most pressing needs. Thanks to these technologies, we are also seeing some innovations in the field of tourism, such as introducing new distribution channels and product/service use. The first examples of NFT deployment in travel can be found in Marriott and Alpitour [30]. Another potential application of blockchain technology within the tourism hospitality industry is manifold. These include establishing loyalty programmes and online booking services and assuring the reliability and traceability of customer testimonials on review sites. Furthermore, utilising blockchain technology within the metaverse, an innovative and immersive virtual universe technology gaining traction within the tourism industry is instrumental in safeguarding data.

Another application of metarverse, can be the Virtual Destination Creation. Some tourism boards and brands have already started to build entirely virtual destinations in the metaverse. These destinations may include reproductions of iconic places such as squares, monuments and beaches that tourists can explore without leaving home. Such experiences offer a unique opportunity to visit places inaccessible in physical reality, such as underwater worlds or imaginary lands, as well as historical or futuristic places. These could stimulate interest in new types of tourism. Destinations in the metaverse offer unique experiences, such as tours of cities from the past or imaginary worlds, which could inspire tourists to explore similar real places, driving market expansion into new types of tourism, such as cultural or experiential tourism. The utilisation of virtual destinations, digital twins, or the reproduction of a singular attraction has been demonstrated to enhance the tourist experience and encourage socialisation. Virtual reality (VR) technology simulates a virtual environment, immersing users to the extent that they perceive themselves as present in the simulation. To achieve this, the software and hardware must be compatible. Advances in computer hardware and computing capability have resulted in the development of versatile devices for VR environments, ranging from single head-mounted displays to several complementary devices. These include the EyePhone (a head-mounted display), the Audio-Sphere (a sound system), and the DataGlove and DataSuit (for measuring movements). In the metaverse, tourists can not only explore places but also interact with other people from all over the world. Through virtual events such as concerts, festivals and exhibitions, travellers can participate in shared experiences without physically travelling. This social aspect is a great strength, allowing people to feel part of a global community without geographical limitations. For instance, people could participate in activities such as skydiving, scuba diving, hiking or virtual safaris, which might otherwise be difficult or too expensive to undertake in reality. These immersive experiences can attract a wide range of users, stimulating interest in physical tourism and generating new forms of digital adventure-based tourism.

The metaverse can also be integrated with traditional tourism. For example, hotels could offer check-in experiences or virtual tours of facilities in the metaverse. At the same time, travel agencies could provide interactive platforms where customers can "try out" different tourism experiences before booking. The metaverse could also act as a link between the physical supply of destinations and the digital demand. Virtual reality also allows businesses to showcase products through immersive advertisements, giving consumers a clear expectation of offerings [31].

The metaverse has the potential to make tourism more accessible. People with disabilities or living in remote regions could participate in tourism experiences that are not usually accessible. In addition, virtual tourism allows anyone to explore distant or exotic destinations without high costs or logistical difficulties. The Metaverse offers opportunities for individuals unable to physically travel to explore and tour diverse landscapes and experiences virtually. Although replicating every aspect of the physical world in the virtual realm is impossible, the Metaverse enables real-time experiential visits that entice travellers and complement physical tourism [32].

The applications of the metaverse in cultural and museological contexts are more widely recognised. "*Technological contribution, in this sector, can be useful for the activities of cataloguing, reproducing, recovering, digitising and making works available to the public, thus making the cultural heritage (starting from the submerged one) more evident, comprehensible and usable to visitors*" [33].

The Metaverse can aid in preserving and conserving heritage and cultural attractions endangered by urbanisation, terrorism, climate change, natural disasters, and other threats. Unlike physical sites, digital assets are harder to damage or manipulate, ensuring long-term conservation. Moreover, the Metaverse can enhance interest in physical sites, generating financial resources to support preservation efforts, expanding the labour force, and fostering the creation of new tourism-related products and services. This growth would promote employment opportunities in site maintenance and hospitality-related industries. Additionally, blockchain technology and non-fungible tokens (NFTs) can be employed to preserve historical sites, facilitating transactions through digital currencies. This approach appeals to digital enthusiasts and helps maintain the value of these culturally significant sites [2].

The advent of the Internet has profoundly impacted the generation and dissemination of knowledge. The development of information and communication technology (ICT) has facilitated the transformation of content from analogue to digital formats, leading to the dematerialisation of media. This has resulted in a significant paradigm shift in storing and circulating knowledge. The progression from analogue to digital formats, coupled with the subsequent evolution of ICT, has played a crucial role in this transformation. The advent of the Internet and the technical solutions that have facilitated its use over recent years have determined and will continue to choose a profound paradigm shift. This shift involves restoring visibility and depth to the subjective dimension of each individual. The Internet and technical solutions have made their use increasingly easy and ubiquitous, including wireless mobile access, broadband, smartphones and the latest wearable devices [34].

Notwithstanding this potential, the relationship between museums and the Internet in Italy has frequently been problematic. The online presence of museums in Italy is often limited in quality and quantity, resulting in fragmented, static content that lacks continuity. The museums' approach to digital communication is predominantly unidirectional, perpetuating the problem. This is evidenced by the underutilisation of interactive tools, such as social media, typically employed to foster dynamic engagement with their respective audiences. Moreover, when these tools are considered, their implementation is often characterised by an absence of a comprehensive strategy, a stark contrast to utilising digital tools in other contexts, where systematic planning at the system level ensures their effective integration.

In light of the above, the most apparent advantage of the metaverse lies in that Virtual reality enables tourism businesses to offer a "try before you buy" experience, allowing users to explore a destination in advance. A pivotal element in this regard pertains to formulating innovative business models. The metaverse has the potential to generate new business models in the tourism sector, such as the sale of 'digital souvenirs' or tourism experiences through NFTs (Non-Fungible Tokens).

Tour operators can create exclusive experiences in the metaverse that are only accessible to those who own a specific token or digital collection, thus adding a dimension of exclusivity and value that could attract tourists to participate in unique events or special tours, thereby expanding monetisation options. However, the impacts of such technologies are not only relegated to this aspect. Thanks to the metaverse, there is, in fact, an impact on the economy and monetisation. In other words, tourist destinations in the metaverse could become a new source of revenue for cities or tourism brands. These virtual experiences could be sold as 'tickets' for virtual tours or experience packages. Furthermore, the metaverse offers opportunities to create digital merchandise, such as virtual souvenirs, or to integrate virtual advertising for travel brands, hotels, restaurants, etc. It is essential to note the close association of this aspect with both the high level of customisation available and the opportunity for increased visibility and marketing. About the first point, in the metaverse, tourism experiences can be highly personalised: travellers could choose customised virtual itineraries, interact with personalised virtual guides, or participate in unique experiences based on their preferences, such as immersive activities or themes-based events. This type of customisation not only enhances the experience but also increases customer satisfaction, driving expansion through word-of-mouth and loyalty. About the secondary point, the metaverse offers destinations and operators a unique opportunity for experiential marketing. Creating immersive experiences in virtual reality can increase the visibility of a destination. For example, a virtual tour of a city or a particular attraction could attract potential tourists who had never considered visiting. Tour operators can use platforms in the metaverse to promote packages, events or special offers to a much wider audience beyond physical boundaries. The metaverse has the potential to transcend geographical and physical limits, thereby signifying an imminent expansion into the global marketplace. It breaks down geographical barriers, allowing people worldwide to access tourism experiences without the cost or hassle of physical travel. This means destinations can attract a global audience, even from countries or regions with limited physical tourism. A tourist who has never visited a physical destination can do so in the metaverse, stimulating interest in a real trip in the future. Linked to the fact that tourists visit locations without necessarily having to travel from home, there is the fact that thanks to the metaverse, more sustainable tourism can be developed (reducing emissions and without burdening the carrying capacity of destinations). However, these aspects will be discussed more in the next section. Last but not least, the metaverse can have an educational function. It can function as an educational tool, with virtual experiences providing information to tourists regarding the attractions, local culture and customs of a destination. Furthermore, simulations can be used to teach responsible tourism practices, thereby contributing to creating a more conscious tourist culture [35].

5 Strategies and Opportunities for Sustainable Development

When addressing the new challenges of virtual worlds, uncertainties often outweigh certainties. It is, therefore, essential to adopt a human-centred, rights-oriented approach that fosters innovation and promotes investment while ensuring a focus on human rights and the protection and empowerment of individuals.

Sustainability is a concept that has gained prominence across all fields globally; it means "*fulfilling our requirements without endangering the abilities of future generations to meet their own needs*" [36]. In a manner analogous to the experiences in other sectors of the economy, consumers within the tourism sector are becoming progressively more attentive and sensitive to ESG issues. In the contemporary context of tourism, there has been an observable shift in the behaviour of tourists. The modern tourist is increasingly active, participative and learning-oriented. This trend has been compounded by technological advancements and the development of intelligent destinations, which have facilitated greater autonomy in terms of action and the potential for collaborative design and creation between visitors and guests. The traditional tourist, characterised by routine, repetitive behaviour, is gradually replaced by a more dynamic and competent individual. This paradigm shift aligns with the tenets of Service-Dominant Logic, emphasising the concept of the "competent tourist" [37]. The sustainability of the tourism sector's development within the metaverse depends on the equitable balance of technological innovation, legal frameworks, and economic dimensions. The metaverse's potential to create immersive and interactive tourism experiences that transcend physical boundaries can only be realised by implementing strategies that ensure long-term economic, social, and environmental sustainability. From an operational point of view, metaverse technologies may even be able to serve as a substitute for travel, for example, when attractions are closed for renovations (such as the long-delayed Grand Egyptian Museum), unreachable because of unrest (Peru in early 2023, or Syria and Yemen during their civil wars), environmentally sensitive (such as Antarctica) or otherwise inaccessible, metaverse technologies such as VR can provide a substitute. Additionally, for would-be travellers with mobility impairments or other health issues, metaverse travel experiences may be an acceptable substitute [38].

As with most metaverse applications, creating a robust legal and regulatory framework is fundamental to developing sustainable tourism in this context. Such a framework must, among other things, promote accessibility and inclusivity in virtual tourism experiences. These experiences should be designed to be accessible to all, regardless of geographical, economic, or physical barriers.

This entails technological considerations, such as optimising platforms for lower bandwidths or less advanced devices and legal safeguards to prevent discrimination or exclusion. The risk of algorithmic discrimination is particularly concerning in Web 3.0 [26]. While the metaverse holds the potential to democratise access to tourism, cultural, and recreational experiences, it also risks exacerbating existing inequalities, raising concerns about its social sustainability. To promote a virtuous evolution of the metaverse in this regard, it is essential to subsidise infrastructure and digital literacy programs for underserved communities, thereby preventing the exclusion of the most vulnerable populations. The Metaverse raises several ethical and social risks, including privacy, security, consent, fairness, inclusivity, diversity, accessibility, and social cohesion [39, 40]. Thus, ethical guidelines for virtual interactions should be established, including measures to address harassment, discrimination, and other harmful behaviours. Such initiatives are fundamental to ensuring that the metaverse develops as an inclusive and equitable space that fosters positive and meaningful engagement among all users. Moreover, as metaverse tourism inherently involves collecting and processing vast amounts of personal data, compliance with high privacy and data protection standards—such as those outlined in the European General Data Protection Regulation [41] is essential for fostering user trust.

From an economic sustainability perspective, metaverse tourism requires establishing fair and transparent financial models. Virtual tourism must operate within an equitable system that avoids undue advantages for dominant players while promoting inclusive participation. Additionally, using blockchain technology to verify transactions and ownership can enhance transparency and reduce fraud, further supporting the economic sustainability of the metaverse tourism ecosystem.

From an environmental standpoint, virtual tourism offers the opportunity to reduce the ecological footprint associated with traditional tourism. By enabling users to explore virtual replicas of destinations, participate in events, and engage in other activities without physical travel, the metaverse can significantly mitigate the environmental impact of mass tourism. However, Metaverse could even hurt the environment, demanding a massive increase in energy use and posing significant ecological challenges due to carbon emissions from data centres. This risk should be mitigated by dematerialising consumption and shifting towards virtual goods [36].

Virtual tourism also provides a valuable opportunity to digitise and make available cultural artefacts, monuments, and heritage assets, thereby ensuring their global accessibility and contributing to their secure preservation. Collaborative governance models involving the participation of local communities, cultural institutions, and other stakeholders are essential to ensure that this is done in a manner that respects the intellectual property rights and cultural sovereignty of the communities that serve as custodians of these assets.

Initial context analyses reveal that sustainable tourism development within the metaverse requires an integrated approach that establishes a robust governance framework, fosters inclusivity, promotes fair economic practices, minimises environmental impact, and preserves cultural heritage.

A promising starting point for implementing sustainable practices in virtual tourism is the European Union's Digital Decade Policy Programme 2030 [42]. This program identifies key pillars of the strategy, including (i) incentivising public–private cooperation for the development of rights-oriented and cybersecurity-safe technologies, (ii) establishing a dynamic ecosystem that bridges the gap between research, innovation, and market implementation, (iii) promoting investment and financial support measures, and (iv) fostering inter-institutional and multi-stakeholder dialogue.

Establishing international dialogue that encourages cooperation and the exchange of best practices is essential to promoting a conscious, trustworthy, transparent, and accountable ecosystem. The European strategy can serve as a preliminary step towards a globally shared governance system that ensures the metaverse contributes positively to the global tourism sector while respecting the rights and needs of all stakeholders involved.

6 Conclusions

Recent statistics indicate that the VR market is valued at US$6 billion and is projected to expand at a compound annual growth rate of 25% over the forthcoming decade. The number of HMDs is estimated to increase from approximately 20 million units in 2020 to over 60 million by 2026. A 14% surge in VR start-ups has been observed within one year, with 75% of Forbes' "Most Valuable Brands" employing VR or AR technology. By 2030, VR will influence over 20 million jobs and contribute US$1.9 trillion to the global economy. These statistics indicate a rapid growth trajectory and widespread adoption of VR technology across various business and societal sectors, including healthcare, education, defence, entertainment, logistics, manufacturing, and tourism. [13]. A survey of marketers in the United States who are professionally aware of the metaverse identified the top three target markets for the metaverse: Millennials, Generation Z, and Generation X. It is estimated that within a decade, 700 million people worldwide will be participating in the metaverse and 15% of the digital economy will transfer to the metaverse [43].

As discussed in this chapter, the tourism industry is set to be significantly influenced by the proliferation of the metaverse in the forthcoming decades. The advent of virtual reality (VR) is poised to facilitate the development of virtual worlds, avatars, digital twins, and novel social networks. These technological advancements hold considerable potential for the tourism sector, enabling travellers to engage with real-world experiences before their physical presence, augment their on-site experiences, reminisce about past experiences post-travel, and, in select circumstances, even substitute for actual travel.

The relationships between metaverse applications and tourism can be categorised into two dimensions. The first dimension pertains to the impact of metaverse applications on tangible entities within the tourism sector. In contrast, the second dimension focuses on the influence of metaverse applications on abstract entities in the tourism sector. The resulting ecosystem for tourism is novel, encompassing stakeholders with unique business models. The provision of novel and creative tourist products within the metaverse environment, the revision of existing touristic products to provide different experiences, and the structuring of beta tourist worlds are the ways of creating an intense new touristic universe, a combination of meta tourism universes [43].

From the standpoint of the sector's operational dynamics, from a demand-side perspective, the metaverse is expected to impact the following aspects: how tourists select their vacation destination, the process of booking accommodation or other destinations to visit during their holiday, the selection of activities in which they choose to participate; the maintenance of anonymity in all interactions with tourist services (blockchain technology can ensure anonymity or pseudo-anonymity); the perception and enjoyment of ongoing experiences; the experience of resources and the phenomenon of word-of-mouth. Regarding the supply side, the planned changes concern the introduction of new marketing channels, exploring new brand potential, and formulating novel forms of destination governance. In consideration of supply and demand in the metaverse, the following changes are projected: disintermediation (the blockchain facilitates peer-to-peer communication, leading to disintermediation within the tourism industry); change in the way tourists and service providers interact through avatar-to-avatar conversation; increased reliability of online reviews and ratings; cultivation of strong ties between users and destinations or users of a particular brand between them, leading to community building; the possibility of using cryptocurrencies as well as the potential for both tourists and tourism service providers to generate additional income through their activities on the metaverse; increased feeling of trust and confidence between all parties; and finally, prevention of overbooking.

The metaverse confers a range of utilitarian benefits to its users. Service providers and developers in the metaverse face a primary challenge: providing a balanced array of advantages to ensure user engagement and retention. This can be achieved by integrating utilitarian benefits, such as virtual tours of renowned landmarks, gamified challenges that enhance hedonic benefits, and awarding exclusive badges or titles representing symbolic benefits. This comprehensive approach could enrich the user experience, increasing user stickiness to the platform. Moreover, this finding is particularly relevant to marketers, who can utilise this knowledge to formulate promotional strategies that emphasise these advantages. Marketing narratives can showcase the multifaceted nature of the metaverse tourism experience, encompassing practical and educational tours, engaging gaming aspects, and the prestige associated with acquiring exclusive rewards. The final practical implication derived from our research underscores the pivotal role of symbolic benefits in influencing continuance intention. Users of the metaverse seek more than functional or entertainment value; they also seek symbols that resonate with their identity or status.

Notwithstanding the plethora of advantages, it is imperative to acknowledge the critical concerns arising from the pervasive use of technology. The excessive utilisation of technological devices has the potential to attenuate the quality of travel experiences. This can engender impediments to the desired state of escapism and the ability to engage in a 'momentary mental absence' during tourism. Preserving memories through mobile devices may hinder tourists' capacity to fully engage with and remember the experiential nature of their travels. Moreover, excessive focus on connectivity and social media sharing can deleteriously affect tourists' enjoyment. Therefore, it is the tourist's responsibility to exercise discernment in their use of technology to avoid the potentially alienating effects of hyperconnectivity and the risk of missing out on enriching experiences. The potential negative consequences of over-exposure to technology are manifold. They may include phenomena such as cognitive overload and the emergence of new terms such as technostress and technological stress. Therefore, the need for digital detox and disconnection is becoming increasingly apparent [10]. Virtual environments can facilitate illegal and unethical conduct, including harassment and cybercrime. These practices can carry significant consequences for user safety and well-being. Documented cases have emerged of users on prominent metaverse platforms encountering sexual harassment or even assault, which can lead to considerable psychological distress and harm. Furthermore, the immersive nature of these environments can give rise to addictive behaviours, thereby causing a dissolution of the boundary between the physical and virtual spheres and exerting a negative impact on the well-being of users. A significant issue identified is privacy and data security arising from the convergence of virtual and real-world interactions. This has given rise to major concerns about user data collection, storage and use within virtual environments. Concerns about privacy and data security are of the utmost importance due to the substantial collection and processing of personal data, which can result in potential identity-related concerns and security risks. In the context of the metaverse, there have been instances of fraudulent activities related to sensitive personal data and digital wallet credentials. Furthermore, concerns regarding accessibility, particularly those related to the digital divide, may result in excluding individuals with limited resources or technological capabilities, thereby exacerbating pre-existing inequalities [11].

Considering the observations, it can be concluded that "*Although the metaverse offers numerous benefits to the hospitality and tourism industry, its adoption and implementation face many challenges and barriers*" [44]. It is important to note that the barriers to integrating technology in this field are numerous and complex. These barriers can be categorised into several different aspects, including security and privacy concerns, the integration and interaction of different technologies, the absence of recognised industry standards, ethical and legal issues, addictive tendencies, high financial costs, uncertainty regarding needs, underdeveloped infrastructure, limited form and content, and a lack of both understanding and awareness and professional skills. Regarding this last point, in a period of talent shortages in hospitality and tourism, the increasing utilisation of technology, including the metaverse, is pivotal for organisations seeking competitive advantage in their field. Despite the growing significance of metaverse implementation in enhancing customer experience in the tourism sector, the potential of such applications in facilitating learning and workforce development within this industry is frequently underestimated [45].

The comprehension of these impediments assists stakeholders in prognosticating the trajectory of metaverse adoption and its prospective impact on the industry. For instance, investment in infrastructure and standards can drive more widespread adoption, whilst comprehending ethical, legal, and cost-related issues can mitigate risks. Enhancing infrastructure and standards will enable the more seamless integration of the metaverse, resulting in more personalised and immersive travel experiences. This transformation will likely engender the strengthening of client relationships and the opening of new avenues for value creation, ultimately leading to the revolutionisation of the industry.

Tourism business managers must develop novel policies or modify existing ones to address the challenges associated with metaverse adoption. Such modifications should be undertaken to eliminate or reduce the impact of major barriers. To facilitate the adoption of the metaverse by the hospitality and tourism industry, governments must invest in metaverse research and development activities, provide more subsidies or introduce incentives, and encourage private enterprises to generate revenue.

* The work is the result of the joint reflection of the Authors. However, Sects. 2, 4, and 6 were written by M. Capelli, while Sects. 1, 3, and 5 by G. Roasio.

References

1. UNWTO International Tourist Arrivals hit 96% of pre-pandemic levels through July 2024, available at https://www.unwto.org/news/international-tourist-arrivals-hit-96-of-pre-pandemic-levels-through-july-2024, last accessed, as every link hereby referenced, 19/01/2024
2. Allam Z, Sharifi A, Bibri SE, Jones DS, Krogstie J (2022) The metaverse as a virtual form of smart cities: opportunities and challenges for environmental, economic, and social sustainability in urban futures. Smart Cities 5:771–801
3. Liu H, Park KS (2024) Exploring the impact of metaverse tourism experiences on actual visit intentions: an integrated model of presence, the technology acceptance model, and the theory of planned behavior. Int J Tour Res 26(1):e2616
4. UNWTO Glossary, available at https://www.unwto.org/glossary-tourism-terms, last accessed 2025/01/09
5. Mariotti G, Carrus S, Panai E, Martinez V, Camerada MV (2018) Smart destinations e competitività in ambito turistico. Il ruolo della cyber security. Geotema, Sup 59–77
6. Buhalis D, Amaranggana A (2015) Smart tourism destinations enhancing tourism experience through personalisation of services. In: Tussyadiah I, Inversini A (eds) Information and communication technologies in tourism. Springer, Cham
7. Khatoun R, Zeadally S (2017) Cybersecurity and Privacy Solutions in Smart Cities. IEEE Commun Mag 55(3):51–59
8. Boes K, Buhalis D, Inversini A (2015) Smart tourism destinations: ecosystems for tourism destination competitiveness. Int J Tour Cities 2(2):108–124
9. Gretzel U, Werthner H, Koo C, Lamsfus C (2015) Conceptual foundations for understanding smart tourism ecosystems. Comput Hum Behav 50:558–563
10. Sustacha I, Baños-Pino JF, Del Valle E (2023) The role of technology in enhancing the tourism experience in smart destinations: a meta-analysis. J Destin Mark Manag 30:1–12
11. Ioannidis S, Kontis AP (2023) Metaverse for tourists and tourism destinations. Inf Technol Tour 25:483–506
12. Li J, Hu P, Cui W, Huang T, Cheng S (2023) Deepsea: a meta-ocean prototype for undersea exploration. Inf Technol Tour 25:391–409
13. Baker J, Nam K, Dutt CS (2023) A user experience perspective on heritage tourism in the metaverse: Empirical evidence and design dilemmas for VR. Inf Technol Tour 25:265–306
14. Tsai S (2022) Investigating metaverse marketing for travel and tourism. J Vacat Mark 30:479–488
15. Behera RK, Bala PK, Rana NP (2024) Reaching new heights: investigating adoption factors shaping the moon landing of metaverse tourism. Inf Technol Tour 26:219–253
16. Ulubaş Hamurcu A (2022) The metaverse, online communities, and (real) urban space. Urbani Izziv 33(2):73–81
17. Song C, Shin SY, Shin KS (2023) Exploring the key characteristics and theoretical framework for research on the metaverse. Appl Sci 13(13):7628

18. Ritterbusch GD, Teichmann MR (2023) Defining the metaverse: a systematic literature review. IEEE Access 11:12368–12377
19. Ball M (2022) The Metaverse and how it will revolutionise everything. Liveright Publishing Corporation
20. Friedman LM (1969) Legal culture and social development. Law Soc Rev 4(1):29–44
21. Dwivedi YK, Baabdullah AM, Hughes L, Ribeiro-Navarrete S, Giannakis M, Al-Debei MM, Dennehy D, Metri B, Buhalisj D, Cheung CMK, Conboy K, Doyle R, Dubey R, Dutot V, Felix R, D.P. Goyal, Gustafsson A, Hinsch C, Jebabli I, Janssen M, Kim YG, Kim J, Koos S, Kreps D, Kshetri N, Kumar V, Ooi KB Papagiannidis S, Pappas IO, Polyviou A, Park SM, Pandey N, Queiroz MM, Raman R, Rauschnabel PA, Shirish A, Sigala M, Spanaki K, Wei-Han Tan G, Tiwari MK, Viglia G, Fosso Wamba S (2022) Metaverse beyond the hype: multidisciplinary perspectives on emerging challenges, opportunities, and agenda for research, practice and policy. Int J Inf Manag 66:1–55
22. Garon JM (2022) Legal implications of an ubiquitous metaverse and a Web3 future. Marquette Law Rev 106(1):163–242
23. Regulation (EU) 2022/2065 of the European Parliament and of the Council of 19 October 2022 on a Single Market for Digital Services and amending Directive 2000/31/EC
24. Regulation (EU) 2022/1925 of the European Parliament and of the Council of 14 September 2022 on contestable and fair markets in the digital sector and amending Directives (EU) 2019/1937 and (EU) 2020/1828
25. Kostenko O, Furashev V, Zhuravlov D, Dniprov O (2022) Genesis of legal regulation web and the model of the electronic jurisdiction of the metaverse. Bratislava Law Rev 6(2):21–36
26. Bengoetxea Caballero J, Cruz Balbuena RL (2024) Institutions of law in the metaverse. Oñati Socio-Legal Series 14(6):1531–1554
27. Mangada Real de Asúa E, Otter V, Tsukuda T, Vivenot B (2022) The metaverse: challenges and regulatory issues. Student paper, SciencesPo. https://www.sciencespo.fr/public/sites/sciencespo.fr.public/files/Metaverse-Group-report-final-draft-June-12-1.pdf
28. European Parliament, European Council, European Commission, Declaration of 23 Jan 2023 on the European Digital System, 2023/C 23/01
29. Murray MD (2023) Trademarks, NFTs, and the Law of the metaverse. Arizona Law J Emerg Technol 6:3
30. Alaio E (2022) I possibili sviluppi del metaverso. Amministrazione & Finanza 11:74–79
31. Roy A (2021) Doing business in the metaverse: opportunity or threat? https://www.xrtoday.com/virtual-reality/ng-business-in-the-metaverse-opportunity-or-threat/
32. Sunny B (2023) An analysis of future prospects of metaverse. In: El Khoury R, Alareeni B (eds) How the metaverse will reshape business and sustainability. Springer Nature, Singapore
33. Zambardino F (2024) Preserving identities in the "Digital Realm": safeguarding cultural heritage in the Metaverse. Aedon 1:71–80
34. Solima L (2026) Smart Museums. Sul prossimo avvento della Internet of Things e del dialogo tra gli oggetti nei luoghi della cultura. Sinergie Italian J Manag 34:263–283
35. Hassanein Taha M, Elhamid Eldalash SA, Zakaria HM (2023) The role of accounting disclosure of sustainable development activities using metaverse in the field of education and training. In: El Khoury R, Alareeni B (eds) How the metaverse will reshape business and sustainability. Springer Nature, Singapore
36. Shaju A, Opportunities and risks of the "Metaverse" for environmental sustainability. In: El Khoury R, Alareeni B (eds) How the metaverse will reshape business and sustainability. Springer Nature, Singapore
37. Lusch RF, Vargo SL (2009) Service-dominant logic—a guiding framework for inbound marketing. Mark Rev St Gallen 26(6):6–10
38. Rubio-Escuderos L, García-Andreu H, Michopoulou E, Buhalis D (2021) Perspectives on experiences of tourists with disabilities: implications for their daily lives and the tourist industry. Tour Recreat Res 49(2):1–15
39. Wang G, Badal A, Jia X, Maltz JS, Mueller K, Myers KJ, Niu C, Vannier M, Yan P, Yu Z, Zeng R (2022) Development of metaverse for intelligent healthcare. Nat Mach Intell 4(11):922–929

40. Zallio M, Clarkson PJ (2022) Designing the metaverse: a study on inclusion, diversity, equity, accessibility and safety for digital immersive environments. Telematics Inform 75:101909
41. Council of the European Union, General Data Protection Regulation (Regulation (EU) 2016/679)
42. Decision (EU) 2022/2481 of the European Parliament and of The Council of 14 December 2022 Establishing the Digital Decade Policy Programme 2030
43. Tas Gursoy I, Aktas E, Tecim V, Avsar Kurgun O (2023) Beta tourist world: a conceptual framework for organising an event in the metaverse. Inf Technol Tour 25:529–548
44. Chi M, Chen Y, Xu Y, Wu Y (2024) Modelling barriers to metaverse adoption in the hospitality and tourism industry. Inf Technol Tour 26:711–743
45. Dutta D, Srivastava Y, Singh E (2023) Metaverse in the tourism sector for talent management: a technology in practice lens. Inf Technol Tour 25:331–365

The Value and Exchange of Wealth in the Metaverse. The Juridical Frontier of a Law We Do not Own

Francesco Petrucciano

Abstract Metaverse is a challenging environment, offering users and observers a quite new perspective on the very concept of reality. Given its nature of a virtual yet permanent environment, where exchange of wealth exists and generates consequences that are relevant in a patrimonial sense, defining the juridical nature of the value that in such an environment is produced (or lost) is worth of attention. In Metaverse, literally a place "put after" the physical sphere, principles such as the monopoly of the State-issued currency or the territoriality of law are, maybe, to be understood under different lenses: in fact, the "diverseness" of this environment which lies "after" the normal ecosystem is the backbone of the whole discourse. This contribution explores the conceptual quality of value that is produced and interchanged in the Metaverse and reads the current legal environment defining it as a "frontier", being it at the same time the last existing regulating line of the sensible world and, perhaps, the first glance of a new legal field for private law. Shall we think about a "metalaw"?

Keywords Metaverse · Private law · Cyberspace · Value · Money

1 Introduction

1.1 A Matter of Space

The unexpectedly literary origin of the metaverse is to be found in Neal Stephenson's "Snow Crash" of 1992, a science fiction novel about the life at the times of Hiroaki Protagonist ("Hiro"), a hacker for the intelligence and a pizza delivery working for the Mafia in a 21st century fictional Los Angeles. Stephenson, a geographer by education, developed the term "Metaverse" from a mixed Greek Latin etymology.

F. Petrucciano (✉)
University of Rome Tor Vergata, Rome, Italy
e-mail: francesco.petrucciano@uniroma2.it

A. Borroni et al. (eds.), *Metaverses: Reshaping Law, Economy, and Society*, PoliTO Springer Series, https://doi.org/10.1007/978-3-032-08664-8_17

The item may be quite freely rendered into English as "over the world", "over the sensitive sphere", or—in a more intuitive way, "over the universe" [1].

The setting of the novel is not simply characterized by digital representation but starts on a (useless to say) futuristic environment: the world has lost any kind of State-based control, and the new order is made up by private corporations. The land of the United States has been extra-territorialized, subdivided into parcels where the rule of law has been substituted, as said, by privates. The juridical meaning of territoriality has therefore been destroyed, and the public power has been replaced by multinational corporations whose authority replaced the rule of law. Therefore, any kind of juridical relation is defined by private contracts, despite the fact that they are often concluded between unequal subjects. The same intelligence Hiro works for is the Central Intelligence Corporation, a name that suggests the complete privatization of all formerly public entities and powers: America is now parceled into franchises and the same capitalism has collapsed on itself, paving the way to an anarcho-capitalistic system [2].

Stephenson called this digital environment the "Metaverse", and imagines it as the representation of the evolution of the internet based on a "virtual reality" (an oxymoron, of course, to which the contemporary world has nevertheless got used to) [3].

There, an alternative world exists, where people are living their own social interaction in a shared virtual environment, where they are represented as "avatars", another cultured term coming from Sanskrit *avatāra*: in Indian religious confessions such as Hinduism and Brahmanism, an *avatāra* denotes the physical body a deity gets incarnated while descending on earth, therefore meaning a representation of the self, alternative to the natural one [4].

In "Snow Crash", this virtual reality on three dimensions shared on an optical fiber is accessible even by public terminals, actual gates through which the person can enter a different word and enjoy a much better life, walking through bars or shops, creating and exchanging wealth and generating new value. That creates, of course, a double environment in a person's existence. Stephenson's work seems to have been prophetical in the sense of predicting the existence of a tool, that kind of parallel society in another dimension, that would exist at our times, taking even the same name Stevenson gave it in his book [5].

The existence of these two intertwined parallel realities is exactly the sensitive point for the jurist, whose point of view is necessarily influenced and limited by either the concept of territoriality, or that of personality, of law. Are these two concepts enough for investigating the Metaverse legal environment (here, I mean the actual one, the one we are experiencing)? Or is the latter a brand-new legal space, maybe a "metalaw"? [6]

1.2 Where Things Happen

If metaverse as a name is born in a novel, in real world a kind of "proto-metaverse" has been existing the last three decades, fueled by the powerful evolutional push of videogaming experiences [7]: as naif as it may seem, videogaming has represented a milestone in the conceptual evolution of an alternative environment, suggesting a first admissibility of a space where actions made by a person do "exist", do create their effects "somewhere else" than reality and exclude personal responsibility [8], given they represent the first alternative to the material world - where laws are instead applicated. Moreover, except for the money that the user may be bound to pay in order to play, the videogame does not create an actual, permanent environment alternative to the sensible one where wealth and value are created: I would like to underline here that the importance lies all in the permanent/impermanent nature of wealth creation in the parallel virtual world [9]. The video game does (can) create an avatar, does let the player do and perfect acts that are generative of effects in that environment, does let others join this environment and suffer/benefit from the other players' moves while playing together and keeps everything in the interior of its systems. To paraphrase, "what happens in the videogame stays in the videogame". And gaming is usually related to connections between players operating in nets belonging to the same legal environments, if not in the same physical machine.

From another point of view, a video game may involve several players, but it does not recreate a social environment where personal relations are expanded up to the economic sphere (besides the money paid as a title to join in the game), nor where juridical instruments are used and create permanent consequences in one's juridical sphere [10].

In the metaverse, consumers and companies are already experimenting with a life built in parallel: from socializing to fitness (for how absurd that may sound), to exercise commerce, to virtual learning, and (most importantly) to executing any kind of life's activity [11], or even buying real estate properties "existing" in only in this "ultra" world.

I believe it is important to note, before tacking on the juridical part, how this phenomenon may be read and explained under the lights of anthropology, and in particular under proxemics, seen as the study of the organization of physical space and how human beings unconsciously relate to the world around themselves: the first works on this subject are due to Edward Hall, while Albert Moncusi pioneered studies on borders and national identities [12] developed a reflection on borders (and therefore on the space within it), be it a physical or a conceptual one. The latter mark attitudes, beliefs, cultures, the linguistics of power and the discursive fields of inclusion and exclusion [13]. This concept may be of an extreme interest for the jurist, given that they might (should?) define the subjective position of joining this other space, in the sense of being conscious of being in another juridical realm.

We may now wonder whether jumping from the video game to the metaverse may mean that jump into "something else" in a juridical sense, where the "meta" (over, ter, *ultra*) may mean, in a juridical sense, another realm, another jurisdiction.

How does all this “other” reality, that is there exactly such because it is not real, be subject to the legal concepts of value and wealth? How can law regulate it, and what is the actual jurisdiction?

2 The Frontier of the Territory

2.1 A Terrestrial Legal Framework

Both legal territorialism and personalism are the milestones in which the juridical experience has been developed in the Western world. The first conceptual challenge to this duality got manifest during the European wars of religion, when Hugo Grotius defined modern Jusnaturalism, re-taking Heraclitus’ elaboration of the shared origin of all human laws from the divine one and defining a new starting point of law: internationalism, based on *jus naturæ* [14].

Talking about front-line technology such as the metaverse starting from classical legal-philosophical sources may seem out of place, improper or at least out of fashion, but the reality seems to be exactly the opposite: understanding a new “verse”, that is a “new territory”, “over” the one we are living but where nevertheless the juridical discourse takes place, is exactly the same theorical battlefield where Grotius, and Heraclitus before him, fought their conceptual battles. Modern jurists shall be extremely interested in using such precedents to elaborate a (hopefully) new and truly innovative conceptual ground.

At the base of the need to define new concepts stands the fact that metaverse is something that was born overcoming the understanding capacity of its contemporary jurist, that is (exactly like for Grotius) something that started to exist that contemporary legal theory was unable to digest. Grotius was then forced to look for solutions that he found digging backward and not looking further the theorical limits of the legal thoughts of his times: his first and foremost intuition seems to be that of having understood that the legal evolution of its time was a concretization of the classical juridical thought, but that those specific results were not necessarily the only ones possible. Grotius had understood that new grounds need new capacities to walk on them, and his return to Heraclitus is as ingenuous as necessary: how to resolve a case where we do not see any applicable law?

His vision, that is - as said - the return to the primitive base, *natura*, is that positive law (*jus in civitate positum*) is inherently uncapable to adapt completely to natural law, given that it contains variable and accidental elements, which are different and subject to change in any place and in any time: *jura in civitate posita* are therefore not the law, but approximations, imperfect phenomena while the jus per se is the *norma*, natural, perfect and—especially—universal, common to all places and times.

As we all know, Grotius would then “create” international law [15], that new *jus gentium* that with the old one shared only the name. There would be no *praetor peregrinus* to decide between the possession of a ship found *in terra nullius* during

European laws of the XVI century but rather a legal system based on consensus, and treaties would loose their nature of bi-lateral instrument to define specific cases to embrace multilateralism and global value [16]. The shift of economic activities to the metaverse seems to propose the same intellectual challenge, stimulating jurists to find a new "platform" to develop another field of law.

2.2 *Plus Ultra*

The problem of insufficiency in understanding the metaverse thoroughly, in a juridical sense, is apparent even from the way that single States, or Organizations, or *suj generis* entities such as the European Union (EU) are facing the issue: the Legal Strategy of the European Commission of the European Union dedicated to the development of the Web 4.0 concept, *exempli gratia*, seems to be a proof of that. Despite the fact that "virtual worlds" is the term used to mean the scope of the application of the strategy, it most likely refers to the metaverse. And in fact, they are defined as "permanent immersive environments based on technologies, including 3D and mixed reality (XR), that allow the integration of physical and digital worlds in real time for various purposes such as design, simulation, collaboration, learning, communication, transactions and entertainment" [17]. This strategy is made up of four pillars: Empowering people and reinforcing skills, Business (supporting a European Web 4.0 industrial ecosystem) Government (supporting societal progress and virtual public services), Openness and global governance of Web 4.0 and virtual worlds (provide a fair and more equal governance). It is useless to underline how no perspective exists on the evolution of a juridical platform, how nothing makes this programmatic strategy a place for a new definition of the digital as a new space [18].

It is worth noting that this is not the first time that the EU is trying to face the issue of the metaverse, given that in 2022, the European Parliamentary Research Service published the results of the briefing titled "Metaverse: Opportunities, risks and policy implications", again considering the metaverse as a tool to use/by which being defended. This way, the EU does show an understanding of the metaverse as an ecosystem and not as a simple further technological innovation whose nature is simply more advanced than the previous existing tools, but ends up never getting the sense of taking that as an opportunity to build up a coherent, sustainable and shared new juridical platform.

3 The Frontier of the Value

3.1 Not Anymore a State-Issued Currency Approach

Now, let us pass on to the main object of this contribution, that is: value and wealth in the metaverse.

It is intuitive that a pivotal role in the building up of the metaverse consists in the integration of payment systems enabling seamless transactions [19]. As anticipated above, and understandable from the absence of a coherent legal infrastructure, the development of an integrated system finds several limitations whose nature is related to: lack of standardization of payment infrastructure, diversity of the existing virtual currencies, distribution of the latter through different platforms [11].

The current system seems therefore to be lacking solutions to manage assets coherently: the consumer is now able to use different metaverse platforms. That means, in practice, the possibility to choose different currencies built following different methods (blockchain) [20]. The metaverse is under the jurists' attention in relation to the nature and to the consequences of what is produced in what is called the Web 3.0, that is the digital environment of metaverse. It is a further development of virtual interactions that, as said, presents the characteristic of being a complete environment [21].

These technologies both enable global transactions, that is cross-border operations and decentralized exchange of value. It is worthy now to focus on what it actually means, especially in the field of digital property: tokens, also in their *species* of the Non-Fungible ones (NFTs), are the protagonist in this environment [22]. This does not mean that standard payment methods finished to exist in the Web 3.0. That shows of course the need for integration between old and new methods inside this world of digital models, which requires new payment infrastructures and the technological maturity of the metaverse.

Cryptocurrencies and NFTs are the tools enabling the "tokenization" of wealth, which is of course the structure of what can be defined as ownership in the metaverse [23]. These technologies are able to define an object (the token, in this case) as a virtual one and therefore to make them the object of an asset's ownership. Those tokens are therefore experiences to be found between what is tangible and what is not: this kind of ownership stands in the middle of the physical and the digital worlds.

Cryptocurrencies in real word are alternative means of payment, as they do not represent a currency in the strict sense of the term, but for sure they represent an effective form of money. In the metaverse their nature is no longer that of a mere speculative asset, as they sometimes may be in the real world, but they find their way as an ordinary component of the metaverse: as the key to materializing digital ownership, and therefore as the way to the tokenization of wealth [24].

This transformation of property rights into blockchain tokens is surely not only a new economic issue, but the sign of a new kind of economy: classical tools are now turn into objects of trade and management, true digital assets whose value is perceived as if they were tangible objects.

3.2 Metamoney

Another important point in the metaverse is the absence, in perspective, of any kind of banking system. Decentralization is in this sense the keyword for understanding how the offer of money, passing from a State-managed system to a completely private one, has the important effect of eliminating financial intermediaries [25].

Here again, the two tools of the virtual space and the virtual money demonstrate how interactive and reciprocally interdependent they are, facilitating that "direct peer-to-peer transactions" that is nothing else than the very sense of the existence of bitcoin, the first and most important of all these tools [26].

The application of all these technologies within the metaverse consists in the ecosystem that is still waiting to be considered as a brand-new juridical environment.

Cryptocurrencies do consist in a tool that is more structured than the already used "in-game currencies", created to support and play a specific game, because they exist outside those specific games they can be used for and because they support two-way exchange with actual currencies (*in primis* the USD) and given their interoperability. Once a client/user creates their own account on a metaverse platform (being Decentraland the biggest at the moment, and Sandbox the second, other three exists at the moment), they subsequently have the possibility to create a "crypto wallet" with which exchanging the *fiat* currency for the chosen cryptocurrency(es), to then use the token contained in the crypto wallet to conduct transactions [27].

As anticipated, cryptocurrencies are the most adapt, useful and surely spread method of payment in the metaverse, but this does not mean a monopoly of that category: traditional payment methods and direct use of State-issued currencies, and this may happen through two types of transactions: the first uses central processors (PayPal, mayor card networks), while the second uses virtual currency that can directly be purchased using *fiat* currency [28].

Operating cryptocurrencies still present the difficulty of the user/consumer not being able to conclude the operations for participating to the virtual world with just a one-click payments, and for sure the fact that the crypto issuers are private users does not help in the sense of creating a sense of the familiarity of traditional payment methods used (and trusted) by a larger part of the population. Combining the security of the cryptoassets with the ease of processes that is typical of traditional payment methods would make of the metaverse a tool more accessible to the market [29].

As the metaverse evolves, another way to obtain a coherent and maybe integrating modern blockchain-based payment systems with traditional banking systems becomes a key challenge, especially given that blockchain is at the end a computational technology which evolution has not ended, leaving therefore room to further evolutions. In the sense of both the payment agility and the possibility of a more harmonized juridical environment to discipline payment methods by crypto, the need arises to define a "common denominator" within the blockchain technology and the experienced, trusted paying infrastructure based on traditional banking systems [30].

4 Conclusions

Modern legislation on metaverses is not yet born: the world is still waiting to realize it is pregnant of a system whose unicity is given by the different ontology of its structure. While, as said, videogames have been a middle process between the real and the purely virtual, they did not present yet the unicity of creating an environment where activities have juridically relevant and permanent consequences (such as the purchase of virtual properties), and where life is not limited to a game or ancillary activities, but where a person feels and believes they are conducting a true "parallel life". Independently from the fact that a single Company has given itself the name of metaverse (with capital M), the phenomenon does not collapse in a single economic agent. At the opposite, there is a technology, a virtual space, a "ultra-world" that will be able to be used by different actors which shall be regulated. The classical tools of conflicts of laws/international private law are not sufficient to elaborate such a task. Especially, we need a different idea of juridical environment in which possible future metaverses may coexist. We likewise have to be ready to define instruments able to structure it coherently. Different ideas on how to build a system do exist, but they still sit at "phase zero" of its formation. Not to mention the fact that under some views, the metaverse is considered as tomorrow's thing, a future issue.

Despite the growing attention of some States or international entities to the metaverse, in the sense of utilizing their capacities to enhance their potential and consolidate their presence in what we can call the international leadership in digital development, the solutions for such a strategy seem to be unable to leave the environment of the policy of short-time strategies (often related to a decade-long programme). Effects of the metaverse definitions developed in doctrine and practice, despite not having been to call the jurists' ears and suggest that a new Grotius' era may be approaching. National solutions, based on a great fragmentation in both currency and contract issues, always requiring adjustments and mutual understandings based on old and already dated platforms, risk to delay with no reason the formation of a new "international" law, alien to the idea of juridical territoriality. The voluntarism of participating to this metaverse (and to those that will follow) and the fact that virtual reality excludes complications related to physical, effective coexistence shall be a boost for all sovereign entities to develop a new sense of shared legal approach that may at the same time fully integrate metaverse technology into different Countries, whose policy may be that of focusing on achieving or even enhancing a elevated rating in terms of the digital transformation. At these ends, it is necessary to determine what currency will be used in the metaverse in a future perspective. A conclusion may be made that, despite not seeming so impellent at this moment, it will soon be necessary to plan not only a legal policy, but to deeply understand the insufficiency of a State-based approach which will miss the opportunity of a new Grotian era.

References

1. Sitnikov MS (2024) Financial and legal development of social relations using digital currencies in metaverses. J Digit Technolo Law 2(1):200–220
2. In the sense of existing in a world where public powers have ceased their functions to privates and the only existing juridical relations is the contract. Snow Crash's situation may even be more extreme. For an analysis of the terms in which anarcocapitalism is born: Jensen, J. (2022) Repurposing Mises: Murray Rothbard and the Birth of Anarchocapitalism. Journal of the History of Ideas, 83(2), 315-332
3. Albnese C (2021) Per una critica della realtà-virtuale e dell'intelligenza-artificiale e una teoria sulla formazione linguistica dell'ossimoro 1st edn. Efesto, Roma
4. Hardy FE (2005) Krsnaism. In: Encyclopedia of Religion, vol 8. Macmillan Publishers, New York
5. Meta Website, https://about.meta.com/metaverse/
6. The definition is not mine. The idea of structuring a law explicitly done for the metaverse may be found at the following website. Metaverse Law, https://www.metaversepolicy.org/
7. Mohammed SY, Aljanabi M, Gadekallu ThR (2024) Navigating the Nexus: a systematic review of the symbiotic relationship between the metaverse and gaming. Int J Cognit Comput Eng 5:88–103
8. Bartel C (2015) Free will and moral responsibility in video games. Ethics Inf Technol 17:285–293
9. Krishnaa RR (2023) Challenges in the metaverse jurisdiction and international treaty law. Intergov Res Policy J. https://irpj.euclid.int/articles/challenges-in-the-metaverse-jurisdiction-and-international-treaty-law/. Accessed 31 Jan 2025
10. Ibidem
11. Sitnikov MS (2024) Financial and legal development
12. Moncusí Ferré A (2005) Les paradoxes de la modernitat. El Contemporani 30:75–78
13. Pellow D (1996) Setting boundaries: the anthropology of spatial and social organization, 1st edn. Bloomsbury Academic, London
14. Grotius H (1949) On the law of war and peace (De Jure Belli ac Pacis). Walter J. Black-Roslyn, New York. https://archive.org/details/lawofwarpeacedej0000hugo
15. Edwards CS (1981) Hugo Grotius, the miracle of Holland: a study in political and legal thought. Nelson Hall, Chicago
16. Gurvitch G (1927) La philosophie du droit de Hugo Grotius et la théorie moderne du droit international. Revue de Metaphysique et de Morale 34:365–391
17. EUR-Lex, https://eur-lex.europa.eu/legal-content/EN/TXT/?uri=celex%3A52023DC0442, last accessed 2025/01/31
18. DiMatteo LA, Cannarsa M, Research handbook on the metaverse and law, 1st edn. Edwar Elgan Online, Cheltenham
19. Casale-Brunet S, Mattavelli M, Chiariglione L (2023) Exploring blockchain-based metaverses: data collection and valuation of virtual lands using machine learning techniques. Digit Bus 3(2)
20. Zambardino F (2020) L'inquadramento giuridico dei token. Un'analisi coparatistica della regolamentazione italiana e sammarinese. IANUS Diritto e Finanza 22:21–35
21. Borroni A, Seghesio M (2019) Bitcoin e blockchain: un'analisi comparatistica dalla nascita alla potenziale regolamentazione. IANUS Diritto e Finanza 19:275–314
22. Kuraś P, Organiściak P, Kowal B, Strzałka D, Demidowski K (2024). Integration of payment systems in IoE and metaverse—challenges and the future of e-commerce in the virtual world. e-mentor 1(103):74–85
23. Zambardino F (2020) L'inquadramento giuridico dei token
24. World Economic Forum website: https://www.weforum.org/stories/2024/12/tokenization-blockchain-assets-finance/
25. Nakamoto S (2008) Bitcoin: a peer-to-peer electronic cash system. https://bitcoin.org/bitcoin.pdf

26. Ibidem
27. Adeniran A, Alkinoon M, Mohaisen D (2024) Understanding the utilization of cryptocurrency in the metaverse and security implications. In: Hà MH, Zhu X, Thai MT (eds) Computational data and social networks. CSoNet 2023. Lecture notes in computer science, vol 14479. Springer, Singapore
28. Ibid
29. Belk R, Humayun M, Brouard R (2022) Money, possessions, and ownership in the metaverse: NFTs, cryptocurrencies, Web3 and wild markets. J Bus Res 153:198–205
30. Gomber P, Kauffman RJ, Parker C, Weber BW (2018) J Manag Inf Syst 35(1):220–265

Seamless Experiences Across Metaverse Instances

Leonardo Chiariglione

Abstract Like it happened for digital media and the web, the metaverse is poised to occupy a highly relevant place in the life of billions of people. Its development path, however, is not clear yet, as the setting of many variables, such as technologies, business models, and standards are still open questions. This paper addresses the issue of metaverse standards taking as example the approach adopted by the MPAI Metaverse Model (MPAI-MMM) recently published by Moving Picture, Audio, and Data Coding (MPAI). MPAI has attacked the metaverse standardisation problem starting from an initial subset of functional requirements required by a significant number of application domains with the intention of gradually adding more functionalities as more applications are added. The paper describes the main steps that have led to the development of the standard starting from the analysis of the potential domains of application and the specification method. Then it explores the MMM main components: the Processes performing Actions on Items, the specification of Profiles, and the verification of the usability of the standard.

Keywords Metaverse · Applications · Standards · Open source

1 Introduction

Conceived a long time ago, named three decades ago, and implemented on a large scale two decades ago, the metaverse can be considered as the latest step to enrich the virtual space that human senses cannot perceive directly but can, if equipped with suitable "transducers".

Not unlike the space—the universe—that our senses can perceive directly and has seen a still unending history of divisions and hoardings, the virtual space has also had a similar fate. Eventually, the Morse alphabet unified the virtual world of telegraphy, but different alphabets had been used for several decades. The telephony experience was the same all over the world, but its long-distance infrastructure created many

L. Chiariglione (✉)
MPAI, Geneva, Switzerland
e-mail: leonardo@chiariglione.org

A. Borroni et al. (eds.), *Metaverses: Reshaping Law, Economy, and Society*, PoliTO Springer Series, https://doi.org/10.1007/978-3-032-08664-8_18

different virtual spaces whose number the International Telecommunication Union (ITU) had a hard time reducing. The virtual radio space was the same in the first half of the twentieth century but in the second half was divided into many different streams. It is fair to say that television was divided into multiple incompatible spaces almost by design with the Radio part of the ITU playing the role of notary public in recording the oft-incompatible television systems that member states had unilaterally decided to adopt. MPEG standards are a rare example of unification by design in the world of media and management of the virtual space created digital audio-visual media. More recently, the web played a similar role in creating a single virtual space where text, images could be enjoyed by millions—and now billions of people later with the addition of audio and video media now billions of people.

As the successor of these previous attempts at creating virtual spaces where people could seamlessly communicate and extend their lives, the metaverse faces the same problems but with a greatly augmented complexity by the vastly increased number of variables involved.

Existing technologies enable entrepreneurs to set up independently designed infrastructures where forms of enhanced virtual lives can be conducted. But the problem is—to do what? Indeed, as the real world offers so many ways to lead a life, there should probably be at least that many ways to lead a virtual life and entrepreneurs have tried a few. In the context of a fast-changing world, gaming looks like the lifestyle for which a significant user response exists in the form of more or less ambitious extensions of the many—incompatible—games that have been developed over more than half a century. In the enterprise domain, the so-called industrial metaverse where digital replicas of industrial environment are deployed for monitoring, forecasting, and other purposes seems to collect significant interest.

A predominant role, as was the case for many other virtual spaces, is played by technologies that enable satisfactory and efficient ways for humans bound to the real space to interact or even take an active part in the virtual space.

2 A Format for the Metaverse

Motivated by the enthusiastic response given by users to the launch of Second Life (June 2003) [1], the MPEG Group was the first standards committee to develop a metaverse standard—Media Context and Control (MPEG-V) [2]. This specified an architecture and associated information representation enabling interoperability between virtual worlds.

In June 2022 the Metaverse Standards Forum (MSF) was established to seek "to enable, foster and promote cooperation and coordination over metaverse-related interoperability between standards organizations, companies, and universities".

In February 2023 the IEEE started working on Standard for Metaverse, e.g. Terminology, Definitions and Taxonomy [3] and Standard for Ethically Aligned Design and Operation of Metaverse Systems [4].

In April 2023, the ITU-T established the Metaverse Focus Group (FG-MV) "to analyse the technical requirements of the metaverse to identify fundamental enabling technologies in areas from multimedia and network optimization to digital currencies, Internet of Things, digital twins, and environmental sustainability". The FG-MV has closed its activities publishing several deliverables.

The ISO/IEC System Evaluation Group—Metaverse (SEG 15) "explores the needs for standardisation, opportunities in the area of metaverse and related technologies, and proposes an initial roadmap for standardisation activities".

In January 2022, MPAI was the second standard developing organisation to organically attack the metaverse interoperability problem—defined as the ability of each of two independently developed metaverse instances (M-Instance) and clients to use data from and as intended by the other M-Instance or client. MPAI has developed and published two Technical Reports [9, 10] in January and April 2023 and a Technical Specification on metaverse architecture covering a plurality of application domains in October 2023 [11] and a second Technical Specification on metaverse technologies in October 2024 [12]. This paper will describe the MPAI approach to metaverse standardisation.

3 MPAI

MPAI—Moving Picture, Audio, and Data Coding by Artificial Intelligence—the international unaffiliated, non-profit organisation developing standards for AI-enabled data coding was established in September 2020 [5]. The founders had a long experience of the positive effects of standards and the benefits of "AI standards" for industry and consumers alike were very clear in mind, but what does it mean to develop standards for (data coding based on) AI? An answer—at least for the first phase of the MPAI work—was found in the specification of an environment where different standardised AI-based processing elements could be combined to achieve a goal that is more complex than possible with individual elements. This standard—called AI Framework (MPAI-AIF) [6]—defines AI Workflows (AIW) as structured combinations of processing elements called AI Modules (AIM) as processing elements and is at the basis of a significant share of the 15 standards that MPAI has developed so far.

A welcome consequence of a standard specifying AI Modules with standard functions and interfaces, is the promotion of an open market of interchangeable components—because they were designed for that—components from multiple sources performing well-identified functions and competing on price and performance. MPAI has designed and implemented an ecosystem composed of different players whose role is:

1. MPAI:
    a. Issues Technical, Conformance Testing, and Performance Assessment Specifications.

   b. Appoints:
      i. MPAI Store as:
         1. Conformance Tester.
         2. Implementer ID Registration Authority.
         3. Distributor of Conformance-Tested implementations.
      ii. Performance Assessors as publishers of Implementation Performance Reports.
2. Implementers develop AIW and AIM implementations.
3. MPAI Store makes available implementations:
   a. That have passed Conformance Testing.
   b. For which Performance Reports are published by Performance Assessors.
4. End Users:
   a. Download implementations.
   b. May provide experience reports.

The Governance of the MPAI Ecosystem (MPAI-GME) V2.0 standard [7] formalises the roles described above. The MPAI Store, formally an independent entity from MPAI, has been established in Scotland as a company limited by guarantee [8].

4 Preliminary Steps to a Metaverse Standard

In [11] MPAI identified 18 use cases where metaverse can represent and enhance or even be an alternative to the current way of doing business (Table 1).

Table 2 provides links to analyses of services that may play synergistic roles with future metaverse considered in [9].

Table 3 reports the areas of metaverse functionalities identified and analysed in depth in [9].

Each of the nine areas was subdivided into sub-areas and for each of them functionalities were identified and analysed. Table 4 is an example of one of the ca. 150 functionalities analysed.

Table 1 The 18 metaverse use cases analysed in [9]

Automotive	Defence	Education	Enterprise	eSports	Events
Finance	Food	Gaming	Healthcare	Hospitality	Professional training
Real estate	Remote work	Retail	Social media	Travel	Virtual spaces

Table 2 Services for metaverse

Content creation	Content creation tools	Marketplace	Crypto wallets
Cryptocurrency exchanges	Development services	Platforms	

Table 3 Areas of metaverse functionalities

Instance	Environment	Content representation
Universe perception by metaverse	Metaverse perception by universe	User
Interaction	Information search	Economy support

Table 4 The personal status in digital human functionality

Area	4. Perception of the Universe by the Metaverse
Subarea	4.2 Audio
Title	4.2.3 Personal Status in Digital Humans
Description	The audio scene description of a universe environment populated by humans is used to extract the personal status of a particular digital human who may be present in the metaverse environment or attending from another environment in the same or different metaverse or from the universe
Comment	At a meeting in a metaverse environment, a virtual secretary is tasked to draft a report using the speech uttered by the attending digital humans and their personal statuses

Table 5 includes different metaverse-enabling technologies analysed in [9]. Each entry is linked to a section in [9] where the state of the art is explored.

Finally, [9] explores three critical areas of metaverse operation: Metaverse Governance, Metaverse Regulation, and Metaverse Stakeholders.

The idea behind [9] was to identify a significant number of functionalities that are common to a sufficient number of applications, and from these to develop a standard offering the necessary functionalities. Implementations of such a standard could trigger enough interest to extend these and add more functionalities attractive to more communities.

Table 5 Technologies for the metaverse

Sensory info	Data processing	User devices	Network	Energy
Audio	Basic processing	Augmented reality	Network architecture	Storage
Visual	Computing services	Virtual reality	Network features	Consumption
Touch			Mobile networks	
Olfaction			Fixed networks	
Gustation				
Brain signals				

5 The MMM Foundations

The MPAI-AIF framework provided the conceptual basis and the context for the MPAI Metaverse Model (MPAI-MMM) standard. Indeed, the MMM assumes that implementations will be conceptually constituted by a set of agreed Processes providing the functionalities required to achieve the goal for which the metaverse instance (M-Instance) has been developed. Processes execute on a computing platform or on connected devices. The platform can be of traditional, distributed or blockchain based. MMM Processes can play a similar role to that played by AI Modules in the AI Framework and facilitate the development of MMM implementations that rely on MPAI standard components.

MPAI found it useful to identify four classes of Processes: *Devices* (i.e., processes running on a device), *Services* (metaverse-native or third-party), *Apps*, and *Users*. The last type of Process is particularly interesting because in MMM a User represents a human and acts on their behalf. An avatar—in the MPAI parlance a Persona—is nothing more than a *rendered* User, which can very well render an autonomous agent, still acting on behalf of a human. A human can deploy many Users, and each User can be rendered with a variety of Personae depending on the needs and the rules of the M-Instance.

The MMM is especially focused on Processes that enable a User—the "longa manus" of a human in the metaverse—to perform actions that are useful to a purpose. Therefore, MPAI identifies and specifies the functional requirements of a minimum but sufficiently rich set of 30 "verbs" called *Actions* that a Process should be able to perform. Of course, the Actions that a Process is actually able to perform depend on the functions that the Process has been designed for.

The action of a transitive verb is performed on an object. Therefore, MMM specifies 65 *Items* where each Item represents an entity on which a Process can perform an MMM Action. For instance, *Audio* and *Visual* are Items that humans can perceive when rendered; *Asset* is an Item that is the subject of a Transaction; *M-Location* and *U-Location* are places in the metaverse and in the universe, respectively. Particularly important are the Items called *Rights*, representing the authorisation of a Process to perform Actions on Items at Times and M-/U-Locations, and *Rules*, representing the set of metaverse governing rules.

Figure 1 graphically represents the role of the different MMM elements: human_1 uses $\text{Device}_{1.1}$ to import object_1 as $\text{Item}_{1.1}$ and deploys $\text{User}_{1.1}$ which contacts Service_1 to obtain Item_1 and is rendered as $\text{Persona}_{1.1.1}$.

The notation used is part of the MMM Use Case Description Language based on the following conventions:

1. Processes (e.g., Users) are sequentially identified by one subscript.
2. Items relevant a Process are identified by the subscript of the Process followed by a sequential number.
3. Items are followed by M, S, A, V, AV, or AVH in brackets to indicate their Synthetic, Speech, Audio, Visual, Audio-Visual, or Audio-Visual-Haptic nature, respectively.

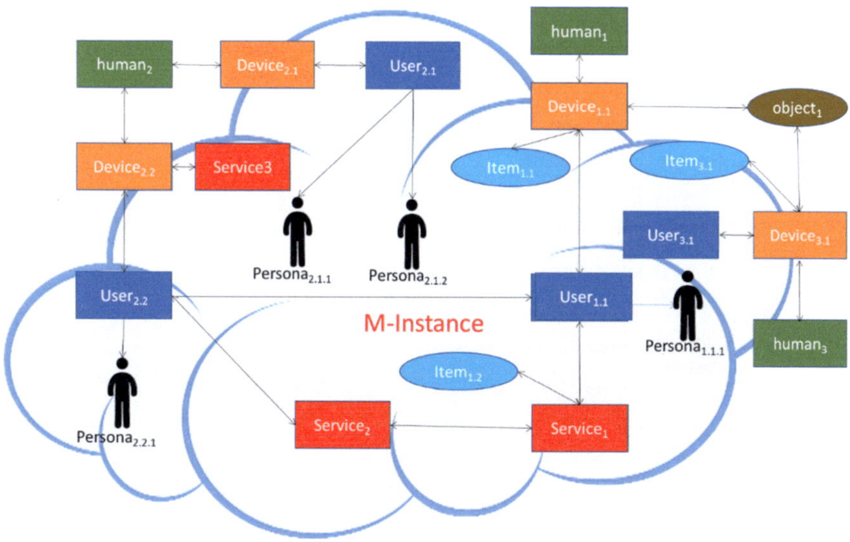

Fig. 1 The main MMM elements

4. If the Actions are performed at different M-Instances, all Processes, Items, and Locations are prefixed by a sequential capital letter indicating the M-Instance.

human$_2$ offers an example of two Users representing the human. User$_{2.1}$ can use two Personae to represent itself (Persona$_{2.1.1}$ and Persona$_{2.1.2}$) and User$_{2.2}$ only one (Persona$_{2.2.1}$).

6 The MMM Standard

Actions

Table 6 provides the names of the 30 Actions specified by [11] with the links to their online specification. The Actions are grouped for reader convenience.

General Actions

This group of Actions includes *Register* (for a human registering with an M-Instance), *Change* (to change the Rights of a Process or Item), *Hide* (to make an Item inaccessible), *Authenticate* (to verify that an Item is what it says it is), *Identify* (to convert Data and Metadata imported from the universe to an Item of the M-Instance), *Modify* (to make a copy and possibly modify elements of an Item), *Validate* (to check whether a Process has the Rights to perform the Action it intends to perform), and *Execute* (a Program).

Table 6 MMM actions

General actions	Call a service	Metaverse-to-Metaverse	Metaverse-to-Universe	Universe-to-Metaverse
Register	Author	MM-Add	MU-Actuate	UM-Animate
Change	Discover	MM-Animate	MU-Embed	UM-Capture
Hide	Inform	MM-Disable	MU-Send	UM-Embed
Authenticate	Interpret	MM-Embed	Track	UM-Send
Identify	Post	MM-Enable		
Modify	Transact	MM-Send		
Validate	Convert			
Execute	Resolve			

Call a Service

This group of Actions includes *Author* (to create Items), *Discover* (to find Items or Processes with certain characteristics), *Inform* (to know more about an Item), *Interpret* (to get help in understanding an Item), *Post* (an Item to a marketplace), *Transact* (to transfer Value to a Process in order to get Rights), *Convert* (to change an Item according to a Qualifier, see later), *Resolve* (for a Process to be able to access a Process in another M-Instance).

Metaverse-to-Metaverse

The names of the last three groups of Actions are typically prefixed by UM, MM, and MU to indicate the universe or metaverse sources and destinations. This group includes Actions that originate and have effect in the M-Instance.

MM-Add (to place an Item somewhere in the M-Instance without making it perceptible), *MM-Enable* (to make an Item perceptible), *MM-Embed* (to place an Item somewhere in the M-Instance and make it perceptible, a Composite Action of MM-Add and MM-Enable), *MM-Animate* (to use a stream of data to animate an Item), *MM-Disable* (to stop making an Item perceptible), and *MM-Send* (to send an Item to a Process).

Metaverse-to-Universe

This group includes the Actions that originate in the M-Instance and have effect in the universe.

MU-Actuate (to render an Item placed at a Device somewhere in the universe), *MU-Embed* (to render an Item in the M-Instance somewhere in the universe), *MU-Send* (to Store an Item at an address), and *Track* (a Composite Action that includes MM-Embed an Item somewhere in the M-Instance, UM-Animate it, and MU-Embed the Item somewhere in the universe).

Universe-to-Metaverse

This group includes the Actions that originate in the universe and have effect in the M-Instance: *UM-Capture* (for a Device to capture media from somewhere in the universe and convert it to Data and Metadata), *UM-Send* (for a Device to send Data and Metadata to a Process), *UM-Embed* (a Composite Action including a Device that UM-Captures media from somewhere in the universe and UM-Sends Data and Metadata to a Process and the Process that Identifies Data and Metadata as Item and MM-Embeds the Item somewhere in the M-Instance).

Items

Table 7 gives the full list of MMM Item where the links are to their specifiction online. Items are grouped for reader convenience.

General

Table 7 MMM items

General	Contract	Identifier	M-Capabilities	M-Environment
Human and user	M-Instance	Program	U-Environment	
	Account	Activity Data	Personal Profile	Personal Data
Process interaction	Message	P-Capabilities	Process Action	Request-Action
	Response-action	Rights	Rules	
Service access	Authentication	Basic Discovery	Basic Information	Basic Interpretation
	Discovery	Information	Interpretation	
Finance	Asset	Currency	Provenance	Transaction
	Value	Wallet		
Perception	Audio basic scene descriptors	Audio Object	Audio Scene Descriptors	Audio-Visual Basic Scene Descriptors
	Audio-visual event descriptors	Audio-Visual Object	Audio-Visual Scene Descriptors	3D ModelObject
	Speech basic scene descriptors	Speech Object	Speech Scene Descriptors	Visual Basic Scene Descriptors
	Visual object	Visual Scene Descriptors	Summary	Text Object
Internal state	Cognitive state	Emotion	Personal Status	Social Attitude
Space and time	Basic M-Location	Basic U-Location	M-Location	Orientation
	Point of view	Position	Space–Time	Spatial Attitude
	Time	U-Location	Universe-Metaverse Map	

This group includes: *Program* (executable code), *Contract* (Program that can be activated by a Process, e.g., another activated Contract), *M-Environment* (a subset of the M-Instance) *M-Instance* (an MMM implementation), *U-Environment* (a subset of the universe).

Human and User

This group includes Items providing information about humans and Users: *Account* (uniquely references a human Registered with an M-Instance), *Personal Profile* (Data submitted by a human when Registering with an M-Instance), *Activity Data* (the record of all the Actions made by a User), and *Personal Data* (the set of Personal Profile of a human, and Activity Data of their Users).

Process interaction

This group includes Items involving the interaction between Processes: *Message* (Items and Data sent to a Process), *P-Capabilities* (set of capabilities of a Process), *Request-Action* (a Process request to a Process), *Response-Action* (the response of a Process to a request), and *Process Action Rights*, and *Rules*, which require more details.

Process Action is the Item introduced to represent sentences like "place the replica of an object in the universe somewhere in the metaverse". This involves a combination of Actions and Items that cannot be expressed with a simple Verb-Noun combination. Therefore, Process Action enables the representation of an Action, its object, and other elements linked by complements. For instance, a human wishing to ask their User to "place the virtual table at this place in the metaverse" will say "User *MM-Embed* Table *At* M-Location". If the human wishes to render in their room the virtual table placed in the metaverse will say "User *MU-Embed* Table *At* U-Location". This is a shorthand notation for two Process Actions in sequence: "*User MM-Send* Table *To* Device" and "*Device MU-Actuate* Table *At* U-Location". A Process who needs to ask another Process to perform something should call the *Inter-Process Protocol* that uses two Items:

1. *Request-Action* representing the request of a Process to another Process—the Device in this case—to render the replica of the table in the universe.
2. *Response-Action*, and Item sent by a Process receiving a Request-Action from another Process. This is normally sent when the performance of the Action is confirmed, denied, or includes a Transaction request. In this case, the requesting Process can then decide whether it is in its best interest to make a Transaction to have its request performed by the other Process.

The Process Action "User *UM-Embed* table *At* U-Location *At* M-Location" is to be performed with an even longer sequence of Process Actions: "Device *UM-Capture* table *From* U-Location" and "Device *UM-Send* table *To* User" and "User *Identify* table *To* Table" and "User *MM-Embed* Table *At* M-Location". Note that the 3rd step ratifies the import of universe data and its conversion to a metaverse Item.

The above leads to the need to specify a scripting language able to represent all Process Actions. This is done by defining the general form of a Process Action as

Process ActsOn Item | Data&Metadata | Media where *ActsOn* represents one of the 30 Actions followed by a combination of complement each followed by one of Process, Item, Data&Metadata, or Address (M- and U-Locations are Items) as follows:

At	Process| M-/U-Location	Where item ends up being located with spatial attitude
By	Process	Used to perform action
From	Address | U-Location	URL, U-Location where source is located
Into	Item	Action leads to
Of	Process	Item refers to
To	Address | Process	Item/Process where Items ends up being placed
With	Data and metadata	Additional item required to perform Action

The MMM Scripting Language is human readable but the compact Backus Naur form is used for efficient inter-Processes communication. The way humans communicate with Processes beyond Process Action is not part of the MMM.

Rights are the obvious choice for MMM to regulate Process activity in the metaverse. Rights make it possible to create metaverses that are based on the rule of law and economy friendly. This can be seen from the various examples analysed above: Process renders a table at a U-Location or captures a table from a place in the universe, uses a Device to capture a table, or to place the Table at an M-Location. MMM represents the Rights of a Process as collections of Process Actions each having one of the following attributes: *Original* (given to the Process by the M-instance at Registration time), *Acquired* (by the Process with a Transaction), or *Granted* (to the Process by another Process). A Process has Rights on an Item—e.g., an M-Location or an object—means that the Process in the Process Action can perform the Process Actions specified in the Rights on the Item.

Rules specify what a Process *must* do, *may* do, and *may not* do. As in a society of humans, Rules applicable to an M-Instance should be publicly know (as the Roman Twelve Tables) and, because Processes should understand them, the Rules should be machine-readable. MPAI has found that an extension of Process Action enables a general form of Rule representation:

$Process_A$ Must/May/Maynot/ *ActOn Item/Process* At/By/From/Into/Of/To/With *Indirect Items/Processes* At *Time* To $Process_B$.

For instance, an M-Instance's Rule "Process *May* UM-Send Program At Time" specifies that a certain Process may import a Program that is certified if Program includes the "Certificate" element whose validity is given by Time (a duration).

Service Access

This group includes Items used in some basic Services: *Authentication* (to request authentication about an Item), *Basic Discovery/Information/Interpretation* (MMM's basic form to search for Items or Processes/request more information on Items/

request for Item interpretation), and *Discovery/Information/Interpretation* (proprietary forms of Item and Process search/request for more information on Items/request for interpretation of Items).

Finance

This group includes Items used for financial purposes: Asset (representing an Item that is the object of a Transaction), Currency (a medium of exchange enabling Transactions), Provenance (the list of all Transactions executed on an Asset), Value (the combination of Amount and Currency, and Wallet (a container of Values).

Perception

This group includes Items whose rendering makes them susceptible of a sensorial impact on humans. Currently, MMM supports the following Media: 3D Model, Audio, Speech, Visual, and Audio-Visual. Items in the Perception groups are Media Objects (i.e., Media Data and Qualifier Data, see later for Qualifier), Text Objects (Text Data and Qualifiers), Media Basic Scene Descriptors (descriptors of a Media Basic Scene made of Media Objects), Media Scene Descriptors (descriptors of a Media Scene made of Media Objects *and* Basic Scenes), and Summary (e.g., of a conversation between Digital Humans). All these Items have already been specified in other MPAI standards [13–16].

Internal State

This group includes Items that describe the internal state of a Digital Human: Cognitive State (the understanding the context), Emotion (state of mind resulting from interaction with the context), Social Attitude (positioning taken vis-à-vis the context), and Personal Status (the combination of the preceding three Items). These Items have already been specified the Multimodal Conversation (MPAI-MMC) standards [14].

Space and Time

This group includes Items about space and time: Basic M-Location/U-Location (a region of a virtual/real space that is not exposed as further subdivided), M-Location/U-Location (a region of a virtual/real space that is exposed as further subdivided, Orientation/Position, (orientation/ position of an object and its velocity/acceleration), Point of View (the combination of position and orientation without velocity and acceleration), Spatial Attitude (the combination of Position and Orientation), Time (measure of time), Space–Time (the combination of Spatial Attitude and Time).

Universe-Metaverse Map (a list of U-Locations and corresponding M-Locations and/or Items with their Spatial Attitudes). All these Items have already been specified by the Object and Scene Description (MPAI-OSD) standard [15].

7 Specification Method

MMM provides the functional requirements that specify the 30 Actions, but how are the 65 Items specified? As clear from the previous section, more than half of the MMM Items have already been specified by other MPAI standards where they are called "Data Types". The specification of all Items includes definition, functional requirements, JSON syntax, and semantics. This level of specification, however, is not sufficient for some Items produced by a Process if we expect that they will be understood by another Process. For instance, there are many formats for media, files, and streaming. Therefore, if MMM does not specify that, say, glTF and not Universal Scene Descriptions is used to represent avatars and graphic content, how can the animation stream generated by a Process be understood by a receiving Process and used to MM-Animates a 3D Model? To answer this question, MMM relies on a separate MPAI standard called Data Types, Formats, and Attributes (MPAI-TFA) [17]. MPAI-TFA specifies Qualifiers, a data type that provides additional technical information beyond the data of a specific data type. Qualifiers are organised in three parts: Sub-Types (e.g., colour space information), Format (e.g., compression and transport), and Attributes (e.g., semantic information on the data). Jointly with the data, Qualifiers provide information informing receiving clients or M-Instances enabling them to handle a specific data type if it has the necessary information or to rely on a conversion service.

MPAI has developed Qualifiers for several data types. These are used by MMM in addition to some which were developed for other MPAI standards. For instance, the Contract Qualifier conveys information on the Contract type (block chain/not block chain) and format (non-executable/executable, in the latter case, if non block chain, the programming language and, if block chain, which language).

The *Inter-M-Instance Protocol* enabling a Process$_A$ in M-Instance$_A$ to request Process$_B$ in M-Instance$_B$ to perform an Action on, say, an Item$_A$—offers a good example of how Qualifiers are used. Process$_A$ first requests M-Instance$_A$'s *Resolution Service* to establish a communication with M-Instance$_B$'s *Resolution Service*. Figure 2 shows how, in case the two MMM implementations do not use the same Qualifier for the same data type, Process$_B$ may request a Conversion Service to perform the necessary conversion of the data received. Once this is done, Process$_B$ can perfom the Action requested and respond back through the same route to reach Process$_A$. In its turn, Process$_A$ may need to request the help of a Conversion Service to understand the data received.

8 Profiles

The MMM standard has already addressed the problem that MPEG addressed some thirty-five ago when the many application domains of the future MPEG-2 standard had made so many requests that it was not economically viable to specify a

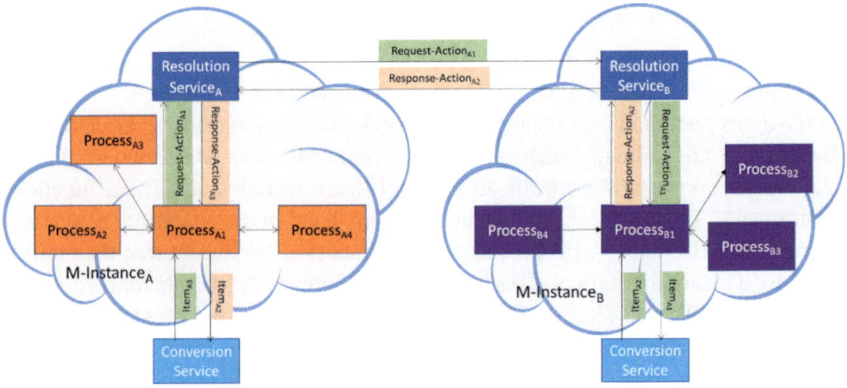

Fig. 2 Inter-Process/M-Instance communication

single codec supporting all the functionalities. The problem was solved by using the notion of Profile that defines sets of tools that must be included in an implementation to enable interoperability with another implementation of the same Profile. The application of the Profile notion to the metaverse has led to the identification of 4 profiles:

1. Baseline Profile supports applications where such as lecture, meeting, and hang-out.
2. Finance Profile supports trading activities without using perceptible Items.
3. Management Profile includes the Baseline and Finance Profile functionalities enhanced by security functionalities.
4. High Profile supports all functionalities.

As was done for MPEG Profiles, in MMM, too, Levels are used to signal the level of the user experience provided by a Profile.

Figure 3 depicts the current features or MMM:

1. Four Profiles: Baseline, Management, Finance, and High.
2. High Profile including Management Profile that includes the Baseline and Finance Profiles.
3. The Baseline, Management, and High Profiles have Audio, Audio-Visual, and Audio-Visual-Haptic Levels.
4. The Finance Profile does not have Levels.

The first three Profiles use a subset of the Actions and Items specified by MMM. While the identified four Profiles serve well the needs conveyed by the identified Functionalities, the consideration of more Functionalities in the future may lead to an increased number of Profiles and potentially Levels.

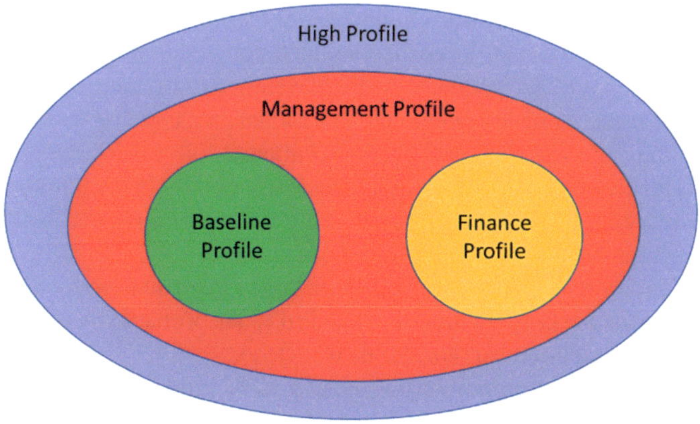

Fig. 3 The profiles of MPAI-MMM—architecture

9 Verifying the MMM Standard

The MMM standard specifies 30 Actions and 65 Items, but are they sufficient for at least an initial set of applications? To respond to this question, MPAI has developed a set of use cases and used the MMM Scripting Language to verify that the considered use cases can indeed be implemented using the MMM standard.

Verification Use Cases

These are the use cases considered:

1. Virtual Lecture: a school manager builds a facility with meeting rooms where they pay teachers to hold lectures with haptic objects and pupils pay to attend lectures.
2. Virtual Meeting: a meeting manager builds a meeting room where they deploy a Virtual Meeting Secretary producing a summary of the conversation that is displayed in the room for people to comment. A participant pays to get a translation of another participant's utterances.
3. Hybrid working: a company adopts a hybrid in-presence and remote working style. The metaverse is where the two types of workers can work together.
4. eSports Tournament: a site manager develops a game landscape and rents it to a game manager who deploys autonomous characters, places virtual cameras and microphones in the landscape and displays onto a dome screen / streams online the captured Audio-Visual from the game.
5. Virtual performance: an impresario acquires Rights to a parcel, authors an Auditorium, embeds Auditorium on Parcel. A participant buys a ticket and utters a private speech to another participant. The impresario collects participants' preferences, interprets participants' mood, generates mood-based special effects (FX).

6. AR Tourist Guide: AR Tourist Guide Service Provider commissions $human_1$ the development of a virtual landscape suitable for a tourist application, $human_2$ the development of scenes, $human_3$ the development of autonomous agents for the different places of the landscape, and $human_4$ an app that enables customer running the app to interact with Digital Humans.
7. Virtual Dance: a dance teacher places a Persona animated by an autonomous agent as assistant and dances with students using a Persona animated by themselves and a Persona animated by an autonomous agent.
8. Virtual Car Showroom: uses a virtual showroom attendant to welcome customers and takes over the conversation when a deal looks probable. A customer drives a haptic 3D model of a car, and the deal is closed.
9. Drive a Connected Autonomous Vehicle: a CAV implementing the Connected Autonomous Vehicle (MPAI-CAV) standard [18] exchanges its perception of the environment using the MMM scene representation. A passenger on CAV_A can join a colleague in CAV_B with their Persona and have a meeting with the colleague's CAV_B.

The Virtual Dance Use Case

MMM Use Cases are described with the MMM Use Case Description Language and consist of three parts: Description (in natural language), Declarations, and Operation.

Description

1. Dance teacher places their virtual secretary Persona animated by an autonomous agent in the dance school.
2. Student #1:
   - Shows up at school.
   - Greets the secretary.
3. Virtual secretary reciprocates greetings.
4. Dance teacher:
   - Places a haptic Persona of theirs in the dance school.
   - Dances with Student #1.
5. Student #2:
   - Is at home.
   - Shows up at school.
6. Teacher:
   - Places their haptic Persona close to student #2.
   - Places (replaces) another haptic Persona of theirs close to student #1.
   - Animates the new haptic Persona with autonomous agent dancing with student #1.
   - Dances with student #2.

Declarations

Declare	User1	// Dance teacher //
	AVHPersona1.1	// Dancing persona#1 //
	MLoc1.1	// Place#1 (Teacher's Office) //
	AVHPersona1.2	// School Secretary //
	MLoc1.2	// Place#2 (Dancing School //
	AVH Persona1.3	// Dancing persona#2 //
	MLoc1.3	// Place#3 (dancing area) //
	SObject1	// Speech Object#2 (Greetings) //
	MLoc1.4	// Place#4 (dancing area) //
	User2	// Dance student #1 //
	AVHPersona2.1	// Student's Persona/
	MLoc2.1	// Student#1's home //
	MLoc2.2	// Place#5 in dancing area //
	User3	// Dance Student #2 //
	AVHPersona3.1	// Student's Persona //
	MLoc3.1	// Dance Student#2's home //
	MLoc3.2	// Place#6 in dancing area //

Operation

Note: MLoc = M-Location, SA = Spatial Attitude

Process_A	Action	Item	Secondary item or process
User1	Tracks	AVHPersona1.1	At MLoc1.1With SA
	MM-Embeds	AVHPersona1.1	At MLoc1.2 With SA
	MM-Embeds	AVHPersona1.2	At MLoc1.3 With SA
	MM-Animates	AVHPersona1.2	
User2	Tracks	AVHPersona2.1	At User2's MLoc2.1With SA
	MM-Embeds	AVHPersona2.1	At MLoc2.2 With SA
	MM-Disables	AVHPersona2.1	From MLoc2.1
User1	MM-Embeds	AVHPersona1.3	At MLoc1.3 With SA
User3	Tracks	AVHPersona3.1	At MLoc3.1 With SA
	MM-Embeds	AVHPersona3.1	At MLoc3.2 With SA
	MM-Disables	AVHPersona3.1	From MLoc3.1
User1	Tracks	AVHPersona1.1	At MLoc1.4 With SA
	MM-Disables	AVHPersona1.1	From MLoc1.3
	Tracks	AVHPersona1.3	At MLoc1.3 With SA

10 Conclusions

Metaverse is a grand idea whose implementation will certainly contribute to the well-being of humankind. The degree of this contribution will largely depend on the degree users will be able to seamlessly experience different metaverse instances. A standard is the obvious answer to this need, but metaverse standardisation is not easy to achieve. One reason is given by the fact that many business players dream of establishing their own standard in their business domain. Another is the fact that many domains share functional requirements that risk being differently responded to because different business players may differently succeed in making their solutions prevail.

MPAI has attacked the metaverse standardisation problem by considering an initial subset of functional requirements still satisfying a significant number of application domains. The MPAI-MMM metaverse standard is based on a general model of a platform populated by a society of interacting processes—some infrastructural, some autonomous, and some human-driven—performing Actions on Items. The platform capabilities increase by increasing the number of standard Actions and Items. Profiles can be used to define the particular subsets of metaverse capabilities required by different various domains and applications inside domains.

MPAI is developing an open-source MMM implementation that will soon be released. The goal is to facilitate the development of metaverse instances for different applications, to validate the specification using real cases, and to receive inputs for future enhanced versions of the MMM standard.

References

1. Second Life; https://secondlife.com/
2. MPEG; Media Context and Control (MPEG-V); https://mpeg.chiariglione.org/standards/mpeg-v
3. IEEE; Standard for Metaverse: Terminology, Definitions, and Taxonomy; P2048; https://standards.ieee.org/ieee/2048/11169/
4. IEEE; Standard for Ethically Aligned Design and Operation of Metaverse Systems; P7016; https://standards.ieee.org/ieee/7016/11178/
5. MPAI; Statutes; https://mpai.community/about/statutes/
6. MPAI; Artificial Intelligence Framework (MPAI-AIF) V2.1; https://mpai.community/standards/mpai-aif/
7. MPAI; Governance of the MPAI Ecosystem (MPAI-GME) V2.0; https://mpai.community/standards/mpai-gme/
8. MPAI Store; https://mpai.store
9. MPAI; Technical Report: MPAI Metaverse Model (MPAI-MMM)—Functionalities (MMM-FNC) V1.0; January 2023; https://mpai.community/standards/mpai-mmm/fnc/v1-0/
10. MPAI; Technical Report: MPAI Metaverse Model (MPAI-MMM)—Functionality Profiles (MMM-FPR) V1.0; April 2023; https://mpai.community/standards/mpai-mmm/fpr/v1-0/
11. MPAI; MPAI Metaverse Model (MPAI-MMM)—Architecture (MMM-ARC) V1.2; https://mpai.community/standards/mpai-mmm/arc/v1-2/

12. MPAI; MPAI Metaverse Model (MPAI-MMM)—Technologies (MMM-TEC) V1.0; https://mpai.community/standards/mpai-mmm/tec/v1-0/
13. MPAI; Context-based Audio Enhancement (MPAI-CAE)—Use Cases (CAE-USC) V2.3; https://mpai.community/standards/mpai-cae/usc/v2-3/
14. MPAI; Multimodal Conversation (MPAI-MMC) V2.3; https://mpai.community/standards/mpai-mmc/v2-3/
15. MPAI; Object and Scene Description (MPAI-OSD) V1–2; https://mpai.community/standards/mpai-osd/v1-2/
16. MPAI; Portable Avatar Format (MPAI-PAF) V1.3; https://mpai.community/standards/mpai-paf/v1-3/
17. MPAI; Data Types, Formats, and Attributes (MPAI-TFA) V1.2; https://mpai.community/standards/mpai-tfa/v1-2/
18. MPAI; Connected Autonomous Vehicle (MPAI-CAV)—Architecture (CAV-ARC) V1.0 https://mpai.community/standards/mpai-cav/arc/v1-0/

MIX
Papier aus verantwortungsvollen Quellen
Paper from responsible sources
FSC® C105338

If you have any concerns about our products,
you can contact us on
ProductSafety@springernature.com

In case Publisher is established outside the EU,
the EU authorized representative is:
Springer Nature Customer Service Center GmbH
Europaplatz 3, 69115 Heidelberg, Germany

Printed by Libri Plureos GmbH
in Hamburg, Germany